MODIFYING BITTERNESS

HOW TO ORDER THIS BOOK

M–5PM Eastern Time

terCard

PERMISSION TO PHOTOCOPY–POLICY STATEMENT

Mechanism, Ingredients, and Applications

Fred,
With my compliments
Glenn

Edited by

GLENN ROY, Ph.D.

Pepsi-Cola Co.
Valhalla, New York

LANCASTER · BASEL

Modifying Bitterness

a **TECHNOMIC**® publication

Published in the Western Hemisphere by
Technomic Publishing Company, Inc.
851 New Holland Avenue, Box 3535
Lancaster, Pennsylvania 17604 U.S.A.

Distributed in the Rest of the World by

Printed in the United States of America

10 9 8 7 6 5 4 3 2 1

Main entry under title:
Modifying Bitterness: Mechanism, Ingredients, and Applications

A Technomic Publishing Company book
Bibliography: p.
Includes index p. 325

Library of Congress Catalog Card No. 97-60221
ISBN No. 1-56676-491-2

To my dad,
Twenty-two years gone but not forgotten.
And to my mom,
With effervescent inspiration, ever present.

Table of Contents

Preface

THE oral sensation of bitter taste is often unpleasant to the human palate and, therefore, formulations of foods, beverages, and oral pharmaceuticals attempt to alleviate or ameliorate bitter taste perception. The Calorie Control Council (CCC) reported that food research into bitter taste modulation has reached a new level of importance to the flavor industry by using knowledge learned from the pharmaceutical industry and rational flavor design (*CCC Focus*, July 31, 1995, p. 5).

Only recently has bitterness control become of commercial importance to a food or pharmaceutical formulation chemist. Over the years, an increasing interest in more palatable food and beverage products with low fat and low sugar content has arisen, thus creating a market need for the control of bitterness perception. With this information, a detailed account of bitterness modification was continually compiled from the literature.

This comprehensive review of the most recent literature on modifying bitterness is organized primarily by ingredients or processing approaches affecting the bitter taste reduction or inhibition. The index includes the specific substances being taste-modified. An in-depth and thoroughly referenced review of mechanisms, ingredients and applications of bitter taste reduction or inhibition is the mainstay of the book. A literature clinic provides an understanding of the mechanism by which bitter taste is perceived and its similarity to sweet taste perception. The utility and advances of taste sensors are briefly reviewed to point out their unique and potential ability to screen for bitter-tasting substances and formulations without recruiting a willing and human taste panel. The inspiring works of Belitz and Rouseff premise the botanical understanding and importance of bitter taste source, reason and amelioration. Since the published works of Belitz and Rouseff, additional bitter chemical principles and sources in bitter food products are summarized for their relevance to the topic of future end product palatability. Some discussion will also address bitterness in salt substitutes (KCl) and the taste control of sweeteners with a bitter aftertaste (e.g., Sweet N' Low or saccharin, Sunett or acesulfam-K).

Section I (Chapters 1–5) was substantially and significantly updated with permission from two previously published scientific journal articles: Roy, G., The applications and future implications of bitterness reduction and inhibition in food products. *CRC Critical Reviews in Food Science and Nutrition,* 29(2):59–71, Boca Raton, Florida, 1990, and Roy, G., Bitterness: reduction and inhibition. *Trends in Food Science & Technology,* 3(4):85–91, Elsevier Science Ltd, The Boulevard, Langford Lane, Kidlington 0X5 1GB, United Kingdom.

Some portion of this book (Section II) has been provided by authors of the proceedings from the 1995 Institute of Food Technologists' Annual Meeting & Expo in Anaheim, California, Technical Symposium Session No. 58, entitled *Modifying Bitterness: Mechanism, Ingredients, Applications.* Several distinguished scientists offer their perspective on various topics in bitter taste sensitivity and control. Dr. D. Eric Walters of The Chicago Medical School offers "The Interaction of Sweet and Bitter Substances" as a discussion on how evolution may have governed bitterness perception and its relationship with sweetness perception. "The Factors Affecting Bitter Taste Perception" is comprehensively discussed by Dr. John Thorngate of Moscow University, Idaho. Of intrinsic importance to our ever-increasing aging population is the discussion by Dr. Claire Murphy and Jill Razani on "Bitterness Across the Life Span." Intense sensory research with the aged has proliferated for over five years. For example, Monell Chemical Senses Center in Philadelphia, Pennsylvania, has devoted considerable research effort toward understanding bitter taste sensitivity. Contributing authors to this book, P. A. S. Breslin and G. K. Beauchamp, clearly and statistically demonstrate the "Suppression of Bitterness by Sodium: Implications for Flavor Enhancement." On a practical level, they suggest it may be extremely difficult or impossible to develop a non-sodium salt substitute that does not create food off-flavors. But, a glimpse into "Development of a Low-Sodium Salt: A Model for Bitterness Inhibition" is provided by Dr. Robert Kurtz and William Fuller of BioResearch, Inc. The market demand and preparation of low fat cheeses revealed considerable bitterness in the taste profiles, due to peptides previously masked by fat. New enzymic technology came to the rescue. "The Use of Exopeptidases in Bitter Taste Modification" by Graham Bruce and Denise Pawlett of Imperial Biotechnology, Ltd., describes the control of hydrolysis in the preparation of cheeses (control of bitter peptide development). Dr. Kurihara and Katsuragi announce a "Specific Inhibitor of Bitter Taste." Their world-renowned, revolutionary, and natural ingredient-pair approach is an elegant discovery to the solution of bitter taste masking in oral pharmaceuticals and foods. Domestic food use approval would be expected promptly for this "supermolecule."

The book continues with a natural progression of applications and techniques in oral pharmaceuticals with many formulae used to control bitter taste (Section III). Oral pharmaceuticals are the most successful marketplace for the possible use of a specific inhibitor for bitter taste. Chapter 13 was substantially and significantly updated with permission from a previously published article: Roy, G., Taste masking in oral pharmaceuticals. *Pharmaceutical Technology,* 18(4):84–99, April 1994, and 18(5):36, May 1994, Advanstar Publications, Eugene, Oregon.

In closing, the editor offers a cautionary and futuristic perspective of what could be done to add another staple to the table, i.e., a bitterness inhibitor to complement a salt and pepper shaker. However, when using bitter taste reduction and inhibition, the editor offers a fair warning to those practicing the molecular design and synthetic chemistry for bitterness inhibition and reduction.

Acknowledgement

THIS book has been in the "works" for ten years and typed within three states (NY, IL, PA, and finally in NY again). This book's topic arose by inspiring food-industry research that prompted literature reviews. In particular, General Foods Corp. in Tarrytown, NY (1980–1986), and The NutraSweet Co. in Mt. Prospect, IL (1986–1990), were the foundation of interest in bitterness control in foods and beverages. Oral pharmaceutical application arose from market applied research. Over the years there have been many exceptional scientists with whom I have had the distinguished pleasure of working in food science and research. Most notably, Dr. Paul R. Zanno (currently at Campbell Soup Co., Camden, NJ), Dr. Grant E. DuBois (currently at Coca-Cola in Atlanta, GA), and Dr. Michael S. Kellogg (currently at Monsanto Co., Chesterfield, MO) were especially inspiring in their encouragement to "bootleg-research" the topic of this book. I extend my special thanks to Dr. D. Eric Walters (currently at The Chicago Medical School), Dr. George Muller (currently at Celgene, Warren, NJ), and J. Chris Culberson (currently at Merck, Rahway, NJ) for their skills in the collaborative effort of molecular design, synthesis, and testing of substances for bitterness reduction and inhibition. I am grateful to Eric Walters and Bill Fuller (Bioresearch, Inc.) for chairing the Institute of Food Technologists' Sensory Evaluation Division's Symposium No. 58 which I organized in June 1995. To all the above, I am indebted to their enthusiasm, interest, and friendship during our careers.

I also sincerely thank the contributing symposium authors for their pioneering and keen insight, interest, and travel to California at their own expense. Their collective contributions appear in the Proceedings portion (Section II) of this book. Finally, I thank The Institute of Food Technologists' Sensory Evaluation Division (Dallas Hoover, Univ. of DE and Barbara Klein, University of Illinois, Urbana, IL) for sponsoring the symposium on this topic. I have greatly enjoyed editing this book.

Introduction

BITTERNESS is, of course, only one aspect of a total flavor profile. The use of the term "bitters" in a generic bartender sense is applied to flavoring essences. Some similarity is found with the description "bitter" applied as a flavor attribute to many types of products. In our current world market, several classes of everyday food and oral drug items are wrought with bitter taste in addition to the other taste modalities of sweet, sour, salt, and umami (MSG). In most cases, the bitter attributes are consumer- or producer-masked by sugar, other ingredients, or process.

Numerous food and beverage products, and oral pharmaceuticals have pleasant as well as unpleasant bitter-tasting components in their taste profile. In some cases, the bitter taste modality is an undesirable trait. Bitter characteristics found in some food systems have been removed or diminished by various ingredients and/or processes, but no universally applicable bitter inhibitor has ever been commercialized. Each application requires a tailored technique of bitterness reduction and inhibition in the formulation of foods, beverages, and oral pharmaceuticals.

The past several years have been marked by articles, patents and books on bitterness inhibition and reduction in foods, beverages and oral pharmaceuticals. The seminal contributions of structure activity relationships by Belitz et al. (1979, 1985) led the way to a comprehensive text on bitterness. The late Hans-Dieter Belitz pioneered research in studying and reviewing bitterness source and reason. He contributed greatly to advancing the chemistry and concept of value-added palatability in products for the consumer marketplace. Hans-Dieter Belitz' passing was announced in German. The paradox was that, in all of this outstanding career, his technical papers were published in English. Also important to the science is "Bitterness in Foods and Beverages," including citrus bitterness inhibition and reduction, as edited by Rouseff (1990).

Certainly, an ideal solution to bitterness inhibition or reduction would be the development of a universal inhibitor that does not affect the other tastes. Comprehensive reviews of efforts to control bitter taste in foods and beverages have been published (Roy, 1990, 1992) and expanded to include oral pharmaceuticals (Roy, 1994). Since then, there have been several developments in the research for bitterness inhibitors that deserved titles as chapters. Most noteworthy is the report in the Journal *Nature* by Kao Corporation and Hokkaido University of mixtures of naturally occurring phosphatidic acid and β-lactoglobulin that offer bitterness inhibition without affecting other taste modalities.

Bitterness is generally viewed as an undesirable attribute of foods and beverages, yet segments of the population regularly ingest items with a prominent bitter taste, e.g., caffeine, quinine, beer, liqueurs, and unusual vegetables. The influence of taste sensitivity, exposure, selected personality traits (i.e., neophobia, variety-seeking, sensation-seeking), and pharmacological reactivity to alcohol and caffeine consumption, two widely consumed bitter substances, was assessed in 20 healthy adults, 10 males and 10 females. Self-reported alcohol use was positively correlated with measured ethanol taste-detection threshold and pharmacological reactivity (self-reported behavioral effects). The latter accounted for 23% of the variance in alcohol intake. Caffeine intake was associated with personality traits. Sensation-seeking status and self-reported reactivity to caffeine accounted for 46% of the variance in caffeine intake. Pleasantness ratings for novel bitter and sour foods were unaffected by 10 exposures whereas increased ratings were given to sweet and salty items. Variation in the influence of these factors between individuals and across products may explain individual differences in the acceptability and use of foods and beverages containing alcohol, caffeine and other bitter compounds (Mattes, 1994). One study, among many, has measured thesholds for a number of bitter compounds (Tanimura, 1993). Additional related information on inhibition, cross-adaptation, structure-activity relationships, and gustatory transduction is available (van der Heijden, 1993).

Bitter substances have been used medicinally for generations as is attested to by many old herbal and official pharmacopeias. There are several products in which the bitter attribute is both characteristic and desirable (e.g., vermouth, aperitifs, tonic water, and other "mixers"), and others in which some element of bitterness is necessary to achieve a desirable flavor balance (e.g., seasonings, sake, bitter lemon). For example, flavor- and bitterness-rich sake was manufactured with a yeast mutant producing β-phenethylalcohol and its esters as flavors and tyrosol as a bitter substance (Arikawa et al., 1993).

The indications of a receptor-mediated phenomenon for sweetness and bitterness led to research on sweet compounds that led to knowledge of sweetness inhibitors and has also led to specific bitterness inhibitors. Development of a universally applicable cocktail of food and pharmaceutical debittering ingredients could provide more palatable bitter food products. Those products originally masked with sugar or other calorie-laden substances could then be formulated with the bitter inhibitor as truly low-calorie products. Ordinarily, the consumer or producer uses some of the items with a chosen dressing (e.g., vinegar and oil, salt, sauces) and in effect masks the true overall flavor and bitterness of the food item. A novel, branded bitterness-reducing ingredient would allow more versatile use of otherwise disliked or unpalatable foodstuffs. Food items discussed in Chapter 4 would be ideally suited for a "table-top" bitter inhibitor whose activity would be perceived by mastication and conceivably produce non-bitter and desirable food flavors. Several opportunities exist for commercial application of such a novel table-top debittering agent in categories where known bitterness may be detrimental to the mainstream and widespread use and acceptability of the ingestible products.

As we read in these following chapters, be mindful that the technique of the ingredient and process approaches in modifying bitterness simply prevent contact or "binding" of the bitter on the taste buds, markedly improve the taste of the bitter, or just plain overwhelm our palate with cooling, flavor, or sweetness.

In the future, a better knowledge of the partition of odor/flavor compounds in the food matrix may allow us to master them by technology (Etievant, 1995). A review with 105 references discusses the interaction between flavor compounds and major food components as well as the influence of food composition and structure on release of flavor and perception (Bakker, 1995). A recent American Chemical Society Symposium Series addresses the interactions of volatile flavoring substances and food constituents (McGorrin and Leland, 1996). The growing market for improved palatability without bitter taste prompts the review of bitter taste considerations. Numerous details are presented in an organized review (Glendinning, 1994).

- rejection response
- bitterness and toxicity
- costs and benefits of the bitter rejection response
- interspecific variation in bitter threshold and diet composition
- comparative tolerance to poisons

REFERENCES

Arikawa, Y., Baba, S. and Oguri, I. 1993. Manufacture of alcoholic beverages with yeast mutants. In JP 05 49, 465 to Nagano Prefecture; *Chem. Abstr.* 118: 253818w.

Bakker, J. 1995. Flavor interactions with the food matrix and their effects on perception. *Food Sci. Technol. 66,* 411–39.

Belitz, H.-D., Chen, W., Jugel, H., Treleano, R., Weiser, H., Gasteiger, J., and Marsili, M. 1979. Sweet and bitter compounds: structure and taste relationship, in *Food Taste Chemistry, Vol. 115,* Boudreau, J. C., Ed., American Chemical Society Symposium Series, *115,* 193–131.

Belitz, H.-D. and Weiser, H. 1985. Bitter compounds: occurrence and structure-activity relationships. *Food Rev. Intl., 1*(2), 271–354.

Etievant, P. X. 1995. Knowledge of the identity of molecules involved in food taste. *C.R. Acad. Agric. Fr. 81*(5), 17–24 in French.

Glendinning, J. I. 1994. Is the bitter rejection response always adaptive? In *Physiol. Behav. 56*(6), 1217–27.

Mattes, R. 1994. Influence on acceptance of bitter foods and beverages. *Physiol Behav. 56*(6):1229–1236.

McGorrin, R. J. and Leland, J. V. Editors. 1996. Flavor-food interactions, ACS Symposium Series, *633,* Amer. Chem. Soc. Washington, D.C.

Rouseff, R. L. 1990. *Developments in Food Science—25: Bitterness in Foods and Beverages,* Elsevier.

Roy, G. 1990. The applications . . . of bitterness reduction and inhibition in food products. *Crit. Rev. Food Sci. Nutr. 29*(2), 59–71.

Roy, G. 1992. Bitterness: reduction and inhibition. *Trends in Food Science and Nutrition, 3*(4), 85–91.

Roy, G. 1994. Taste masking in oral pharmaceuticals. *Pharmaceutical Technology, 18*(4), 84–99, *18*(5), 36.

Tanimura, S. and Mattes, R. D. 1993. Relationships between bitter taste sensitivity and consumption of bitter substances. *J. Sensory Studies 8,* 31–42.

van der Heijden. 1993. Sweet and bitter tastes. Flavor Sci. Discuss. *Flavor Res. Workshop.* 1990 (Publ. 1993), 67–115, edited by Acree, T. E., Teranishi, R., Amer. Chem. Soc.: Washington, D.C.

SECTION I

MECHANISM, INGREDIENTS, APPLICATIONS

CHAPTER 1

Recent Overview of the Mechanism of Bitter Taste

GLENN ROY[1]

TRANSDUCTION AND PERCEPTION

ALTHOUGH many mechanisms (Kumazawa et al., 1988; Koyama and Kurihara, 1972) have been advanced for bitter taste, measurement of surface-pressure increases on forming in vitro complexes with lipid monolayers of the taste receptor membrane suggests the lipid layers may be involved in the reception of bitter stimuli. Receptors are likely imbedded within a lipid membrane of the taste bud (Figure 1). However, the data do not exclude the possibilities that specificity of the recognition for bitter may lie subsequent to the initial binding of quinine at a receptor, or that specificity may be a property of the relative orientation of the stimulus molecule at the receptor membrane (Brand et al., 1976; Kuribara, 1973). What is inescapable from abundant literature is that a pattern of contributions, arising from one or multiple sites distributed over an epithelial area having a characteristic topology, results in sweet or bitter taste. Thus, bitter and sweet modalities for a number of compound classes are related by topographical overlap of their interaction patterns. But a report also shows no significant correlations were found among stimuli, consistent with the existence of multiple receptors, or a receptor mechanism of low specificity, i.e., a nonspecific membrane mechanism as proposed by Teeter and Brand (1987).

A review of 71 references discussing bitter-taste signal transduction considers extracellular targets for bitter compounds such as ion channels and receptors and intracellular targets for membrane-permeable compounds. The latter group includes G proteins, the family of phospholipases and in-

[1]Pepsi-Cola Co., 100 E. Stevens Ave., Valhalla, NY 10595, U.S.A.

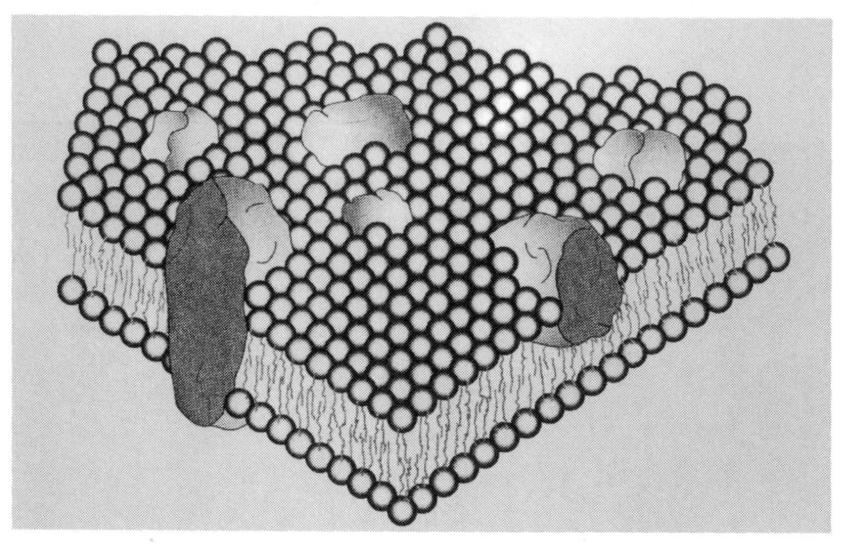

Figure 1 Lipid bilayer membrane matrix possibly involved in bitter taste perception. The "blobs" are the receptor proteins imbedded in the membrane. (Adapted from Singer, S. J. and Nicholson, G. L. *Science,* 1972, *175,* 720. Reproduced with permission of the publisher.)

tracellular Ca^{+2} stores. Many of these mechanisms converge to a single second messenger, inositol triphosphate and probably involve release of Ca^{+2} (Spielman et al., 1992). It is probable that a diversity of mechanisms is involved in the transduction of bitter taste but a G protein-linked receptor model has been proposed (Figure 2). One of the mechanisms uses the second messengers, inositol 1,4,5-triphosphate (IP3) and diacylglycerol

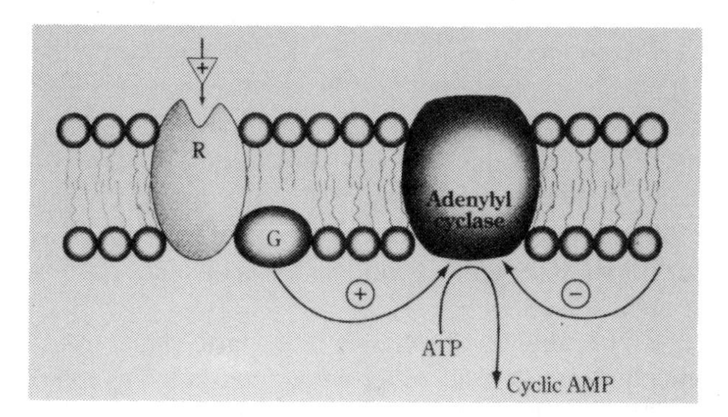

Figure 2 A G protein-linked receptor model of taste receptor cells. The recognition process consists of a series of reactions in which chemical compounds react and create new compounds. (Adapted from Rawls, R. L. *Chem. Eng. News,* 1987, *65*(51), 28. Reproduced with permission of the publishers.)

(DAG). In the absence of an isolated human taste receptor, animal studies have provided most understanding of taste mechanism.

RATS AND MICE

Animal studies of innervated species provided the mainstay of current knowledge in implied human bitter taste mechanisms. Recently though, more attention to human sensory studies in bitter taste has been a knowledge builder and will be discussed in following text.

Cloning of G protein and phosphodiesterase from rat taste tissue was accomplished using degenerate primers corresponding to conserved regions of G protein α subunits. The polymerase chain reaction produced eight distinct c-DNAs including α-gustducin, which is only found in taste tissue. The study of these proteins suggested that gustducin and transducin regulate phosphodiesterase activity in taste cells and that this may promote bitter transduction and inhibit sweet transduction (McLauglin et al., 1994). The distribution of evoked expression of the proto-oncogene c-fos was immunohistochemically examined in the parabrachial nucleus of water-deprived rats after free ingestion of palatable liquids and after intraoral infusion of adversive taste solutions including bitter substances (Yamamoto et al., 1994).

The effects of low doses of d-amphetamine (0.25–0.5 mg/kg, IP) on taste reactions elicited by quinine solutions in a 5–10 min. taste reactivity test were assessed in a series of three experiments. Amphetamine consistently suppressed aversive reactions elicited by quinine solutions. The results suggest that amphetamine, like morphine, attenuates the aversiveness of the taste of quinine solution (Parker and Leeb, 1994). Morphine pretreatment attenuates aversive taste reactions elicited by quinine solutions in rats. Neither tolerance nor sensitization developed to morphine-induced attenuation of quinine aversiveness; morphine-suppressed quinine elicited aversive reactions on each trial. In addition, when tested in the absence of morphine, rats displayed a reduced aversion to quinine, suggesting that quinine became conditionally less aversive following previous pairings with morphine (Clarke and Parker, 1995).

The effect of nicotine pretreatment on the palatability of flavored solutions was assessed using the taste reactivity test. In experiment 1, low doses of nicotine (0.2–0.4 mg/kg, s.c.) suppressed the aversive taste of quinine and quinine-sucrose mixture and enhanced the hedonic taste properties of sucrose (0.4 mg/kg, s.c.) in rats that were nicotine naive. In experiment 2, rats were chronically pre-exposed to nicotine or saline over a period of 21 pretreatment days. Tolerance developed to the ability of nicotine to enhance the palatability of sucrose. Furthermore, rats that were chronically preexposed to nicotine displayed enhanced hedonic evaluation of sucrose 24 h after nicotine was withdrawn. These results confirm human self-reports that withdrawal from nicotine dependency enhances the palatability of sweet-tasting foods (Parker and Doucet, 1995).

Von Ebner's gland protein (VEGP) is a secretory protein that is abundantly expressed in the small von Ebner's salivary glands on the tongue in rats but not mice. It is presumed VEGP is a component of the perireceptor environment around the taste papillae that might function as transporter of hydrophobic molecules; for example, bitter substances. A new approach is described to investigate the physiological role of VEGP by expression of the cloned rat VEGP gene in transgenic mice. Taste papillae of mice do not contain VEGP. The founder transgenic mouse and offspring can carry the gene as shown by PCR analysis and saliva of the transgenic mice contained high levels of VEGP. In two bottle preference tests, transgenic and nontransgenic siblings show significantly different capabilities to taste the bitter compound denatonium benzoate at 10 μm. The reduced sensitivity of transgenic mice to denatonium benzoate points to a clearance function or "repulsion of the substance VEGP" and the specificity of other taste compounds remains to be studied (Koch et al., 1994).

Posterior tongue receptors are buried within the papillary trenches that are filled with von Ebner gland secretions, thereby providing a barrier to stimulation in anesthetized preparations. Rat and human lingual 18-kDa von Ebner gland proteins as carriers for lipophilic bitter substances have been reviewed (Schmale et al., 1993).

Salivary secretions modulate taste perception. Taste buds in the circumvallate and foliate papillae are bathed in secretions of unique lingual salivary glands, von Ebner's glands (VEG). Johns Hopkins University has identified a rat cDNA encoding a novel protein of 1290 amino acids, Ebnerin, that is specifically expressed in VEG and released onto the tongue surface along the apical region of taste buds in the clefts of circumvallate papillae. Ebnerin possesses a putative single transmembrane domain at the C-terminus with 17 amino acids in the cytoplasmic area. The extracellular region of Ebnerin contains a number of repeated domains with homology to the scavenger receptor cysteine-rich domain and to a repeated domain of bone morphogenetic protein-I and other related proteins. Western blot analysis reveals that Ebnerin exists in both particulate and soluble forms in VEG and is present in secretions from VEG. In situ hybridizations and immunohistochemistry demonstrate the Ebnerin is located in secretory duct epithelial cells of VEG and is released onto the tongue surface along the apical region of taste buds in the clefts of circumvallate papillae. The unique structure and localization of Ebnerin suggest that it may function as a binding protein in saliva for regulation of taste sensation (Li and Snyder, 1995).

A persistent problem with attempts to examine bitter taste mechanisms has been the lack of adequate behavioral methodology providing data that parallel those obtained from physiological investigations. Florida State University has developed a brief, non-surgical contact procedure to assess the ability of rats to detect the presence of weak bitter compounds dissolved in a strong sucrose solution. Male Fischer 344 rats were trained to drink immediately to multiple 10s presentations of acetaminophen (2, 8, 32, 128 mM), chlorpheniramine maleate (1, 3, 9, 27 mM), L-tryptophan (13.5, 27, 54, 108 mM), pseudoephedrine hydrochloride (1, 4, 16, 64 mM) and quinine hydrochloride (0.008, 0.04, 0.2, 1.0 mM) dissolved in 0.8 M sucrose. The number of licks to sucrose and water were also measured as baseline, while a microcomputer controlled stimulus presentations and measured the animal's licks of each bitter-sweet solution during each 10s presentation. The response to the bitter + sucrose mixture were significantly decreased at most concentrations with increasing levels of the bitter component. This was true for all five bitter-tasting compounds, but over different concentration ranges relatively unique to each compound. The rats' mixture-detection threshold for the bitter compounds was estimated to be within the range where mixture licking diverged significantly from sucrose licking. The higher threshold obtained with a two-component mixture presentation compared to two-bottle behavioral and neural thresholds may be explained by mixture suppression. Mixture suppression is also shown psychophysically in humans when the taste intensities of the components in a mixture are suppressed, although their qualities may be identified.

This study is the first to characterize the sensory effects of acetaminophen, psuedoephedrine and chlorpheniramine maleate; all taste bitter to humans. The results demonstrate rats' acute ability to discriminate by taste, not only the presence but the concentration of a dilute bitter compound dissolved in a strong sucrose solution (Contreras et al., 1995). Princeton University researchers have investigated the presence of multiple classes of bitter receptors in rats, but evidence supports a single receptor site containing perhaps one or more protein surfaces (Scott and Farley, 1988).

Partial membrane preparations from circumvallate and foliate taste regions of mice tongues responded to the addition of known bitter taste stimuli by increasing the amount of inositol phosphates produced after 30 sec. incubation. Addition of both the bitter stimulus, sucrose octaacetate and G protein stimulant, GTPγS, led to an enhanced production of inositol phosphates compared with either alone. Pretreatment of the tissue samples with pertussis toxin eliminated all response to sucrose octaacetate plus GTPγS, whereas pretreatment with cholera toxin was without effect. Wester blots of solubilized tissue from circumvallate and foliate regions probed with antibodies to the α-subunit of several types of G proteins revealed bands reactive to antibodies against Gao, with no apparent activity to antibodies against Gai3. Given the results from the immunoblots and those of the toxin experiments, it is proposed that the transduction of bitter taste of sucrose octaacetate in mice involves a receptor-mediated activation of a Gi-type protein, which activates a phospholipase C to produce the signal to a second messenger, IP3 and DAG (Spielman et al., 1994).

The above experiment would also have been interesting in the presence of denatonium benzoate as stimulus. Inositol phosphate concentrations in mouse circumvallate papilla and lingual epithelium increased markedly in response to mouse lingual stimulations with 20 mM quinine hydrochloride and 20 mM denatonium benzoate. The increase was not affected by Pronase treatment (Nakashima and Ninomiya, 1994). Monell Chemical Senses (Philadelphia, PA) and their Japanese counterparts [Suntory (Beverages) Ltd.] support the fugitive increase in inositol triphosphate generation observed at 7.5 ms after the administration of sucrose octaacetate into mouse taste cells. The addition of GTP induced the increase in the amount of IP3 generation, whereas the addition of GDPβS was ineffective (Nagai et al., 1994).

GERBILS

Environmental pollutants were shown to alter taste responses in the gerbil. Nine insecticides (organophosphorus, carbamate and pyrethroid types) and two herbicides (paraquat and glyphosphate) was used to study the effect on integrated chorda tympani recordings from gerbils to a range of tastants before and after a 4-minute application of one of the 11 pollutants. The taste stimuli were: sodium chloride (100 mM), calcium chloride (300 mM), magnesium chloride (100 mM), HCl (10 mM), potassium chloride (500 mM), monosodium glutamate (MSG, 50 mM), sucrose (100 mM), fructose (300 mM), sodium saccharin (10 mM), quinine HCl (30 mM) and urea (2 M). All of the environmental pollutants tested had some suppression effect on all of the taste stimuli except for urea (Schiffman et al., 1995).

The adenylate cyclase system is implicated in taste transduction (cf. Figure 2). Application of certain modulators of the adenylate cyclase system to the gerbil tongue alters taste responses. Integrated chorda tympani recordings were made in gerbils to bitter, sweet, salty, sour and glutamate tastants before and after a 4-min. application of four types of modulators of the adenylate cyclase system. The four types of modulators tested were:

(1) Sodium fluoride (NaF), a compound that promotes dissociation of GTP binding protein

(2) Forskolin, a powerful stimulant of adenylate cyclase
(3) 8-Bromoadenosine 3′:5′-cyclic monophosphate sodium salt (8BrcAMP) and N6, 2′-O-dibutyryladenosine 3′:5′-cyclic monophosphate sodium salt (DBcAMP), two membrane permeable forms of cAMP
(4) d-(5-Isoquinolinylsulfonyl)-2-methylpiperazine dihydrochloride (H-7), and N-(2-(methylamino)ethyl)-5-isoquinolinesulfonamide) dihydrochloride (H-8), which are protein kinase inhibitors

The taste compounds tested were NaCl (30 mM), monosodium glutamate–MSG (50 mM), sucrose (30 mM), HCl (5 mM and 10 mM), KCl (300 mM), quinine hydrochloride (30 mM), $MgCl_2$ (30 mM), erythromycin (0.7 mM and 1 mM) and urea (2 M). The results clearly indicated NaF at 20 mM significantly inhibited responses to bitter compounds up to 35% and enhanced the response to sucrose by 30%. NaCl (20 mM), used as a control for NaF, inhibited most responses up to 78% with no enhancement of sucrose as seen with NaF. 8BrcAMP (1.16 mM) reduced the responses to bitter-tasting quinine HCl, $MgCl_2$ and erythromycin but not to urea. It also blocked the responses of KCl and HCl, which have bitter components. There was slight enhancement of the sucrose response. It had no significant effect on NaCl or MSG. A similar trend was found for five stimuli. These data indicate that modulation of the adenylate cyclase system can affect the amplitude or neural responses of some bitter and sweet taste responses, at least in the gerbil (Schiffman et al., 1994).

Integrated chorda tympani (CT) recordings of gerbils were made to salty, sour, sweet, bitter and glutamate tastants before and after a 4-minute application of modulators of lipid-derived second messenger systems. The modulators included two membrane-permeable analogs of DAG; 1-oleoyl-2-acetyl glycerol (OAG); thapsigargin, which releases Ca^{+2} from intracellular stores; ionomycin, a calcium ionophore; lanthanum chloride, an inorganic calcium channel blocker; nifedipine, a dihydropyridine calcium channel blocker (DiC8); quinacrine diHCl, a phospholipase A2 antagonist; mellitin, a phospholipase A2 agonist; and indomethacin, which decreases the release of prostaglandins by inhibiting the enzyme cyclooxygenase.

The main findings were that OAG (125 μM) and DiC8 (100 μM) blocked the responses of several bitter compounds while enhancing the taste response of several sweeteners. Lanthanum chloride blocked all responses, which may be due to the fact that it blocks tight junctions. Quinacrine (1 mM) suppressed several bitter responses while enhancing the response to several sweeteners. The enhancement of sweet taste responses by DAG analogs suggests that there is cross-talk between the adenylate cyclase system and one or more pathways involving lipid-derived second messengers in taste cells (Schiffman et al., 1995).

BOVINE

Early experiments with bovine circumvallate papillae (bcp) as a model system for the gustatory receptor membrane suggest bitter taste is induced by penetration of brucine and caffeine *into* the nonpolar region of the lipid layer. Similarly, the lipid fraction from bcp, which contains many taste buds, adsorbed more quinine than quinidine when compared with the fraction from tongue epithelium without taste buds (Faull and Halpern, 1969).

RABBIT

Guanylyl cyclase activity was cytochemically demonstrated in rabbit foliate taste buds. The enzymic activity was localized in the apical portion (microvilli and neck) of taste bud cells. Especially strong activity was observed on the microvillous membrane of type I (dark) cells and often on a blunt process of type III cells. The microvilli of type II (light) cells showed weak enzymic activity. Considering that the apical portion of taste cells is a likely site of interaction between taste stimuli and the cells, the results support the idea that cGMP is involved in taste transduction (Asanuma and Nomura, 1995).

FROG

Research has also investigated responses of stimuli with frog taste cells. The dorsal epithelium of the frog (*Rana catesbeiana*) tongue responded to bitter-tasting compounds as a transepithelial decrease in tissue resistance response across the tissue. The changes in response and resistance were much smaller than those produced by salt or acid. Although the mechanism of the response could not be demonstrated, the similarity of the response to the taste nerve activity suggests that the response is related to bitter sensation (Okamoto et al., 1995).

Salty and sour tastes are directly transduced by apical channels, whereas sweet and bitter tastes utilize cyclic nucleotide second messengers. The Roche Institute of Molecular Biology has shown that rod transducin is present in mammalian taste receptor cells, where it is activated by a bitter receptor and in turn activates a phosphodiesterase. The researchers introduced peptides derived from transducin's phosphodiesterase-interaction region into frog taste cells, which caused an inward whole-cell current in a subset of cells. They found that the peptides effects are reversibly suppressed by isobutyl methylxanthine (IBMX) and forskolin, indicative of a transducin-activated phosphodiesterase. Cyclic nucleotides suppress whole-cell current, indicating that cyclic nucleotides may regulate taste cell conductance. IBMX modifies taste cell responses to two taste stimuli, implicating phosphodiesterase in taste transduction. Submicromolar cyclic nucleotides directly suppress the conductance of inside-out patches derived from the taste cell plasma membrane, independently of protein phosphorylation. The channels are unusual in that they are suppressed rather than activated by cyclic nucleotides. The authors propose that transducin, via phosphodiesterase, decreases cyclic nucleotide levels to activate the cyclic-nucleotide-suppressible conductance, leading to Ca^{+2} influx and taste cell depolarization (Kolesnikov and Margolskee, 1995).

After frog taste cells were adapted to 1 mM quinine-HCl (Q-HCl) for 10s, modification of receptor potentials in the taste cells induced by salt, acid, sugar and bitter stimuli was studied with microelectrodes. The phasic component of receptor potentials induced by 0.1 M NaCl, KCl, NH_4Cl and $MgCl_2$ was enhanced following adaptation to Q-HCl. The rate of rise of receptor potentials in response to the salts was increased after Q-HCl adaption. The amplitude and the rate of rise of receptor potentials induced by 1 mM acetic acid were larger after Q-HCl adaption that after water adaption. The amplitude of the phasic component and the rate of rise of receptor potentials for 0.5 M sucrose after Q-HCl were the same as those after water. The amplitude of tonic receptor potentials for 1 mM Q-H_2SO_4, brucine and picric acid after Q-HCl adaptation was the same as after 1 mM NaCl adaptation. Correlation coefficients between taste cell responses induced by 1 mM Q-HCl and 1 mM Q-H_2SO_4 were very high, but those between 1 mM Q-HCl and 1 mM brucine responses and between 1 mM Q-HCl and 1 mM picric acid were low. This indicates that Q-HCl and Q-H_2SO_4 bind to the same receptor site, but brucine and picric acid bind to different receptor sites to which Q-HCl does not bind (Sato and Sugimoto, 1995).

The resting potential, salt signal transduction in frog taste cells, acid signal transduction, bitter signal transduction, sweet signal transduction, water transduction and ion channels in frogs taste cell membranes, has been reviewed (Sato et al., 1995; Sato, 1988). The molecular mechanisms of generation of the receptor potential in the bullfrog cell for bitter stimuli indicate that a Na^+/Cl^- cotransport occurs through the basolateral membrane of QHCl-sensitive taste cells. It is concluded that the receptor potential for QHCl stimulation is produced by an active secretion of intracellularly accumulated Cl^- through Cl^- pumps of the apical receptive membrane (Sato et al., 1994). The receptor potential in response to quinine HCl was measured by the patch clamp method in isolated taste cells of the bullfrog. The magnitude of the receptor potential was increased by excluding extracellular Ca^{+2} concentration on the taste cell membrane, suggesting a new taste transduction mechanism (Tsunenari et al., 1994).

Also investigated were the characteristics of 80 pS K^+, 40 pS K^+, Ca^{+2} and cAMP-activated channels by the Tip-dip method (Kumazawa et al., 1994).

The initial events in chemosensory transduction that involve a transient increase or decrease of second messengers such as inositol 1,4,5-triphosphate, total phosphate, cAMP and cGMP may be measured by commercially available kits or modifications of methods described in the literature based on RIA (Spielman et al., 1995). The role of ion channels in taste excitability, ion channels in signal transduction involving ionic stimuli and non-ionic stimuli (e.g., saltiness and sweetness, respectively), and the effects of amino acids and bitter substances is discussed (Avenet, 1992). Intracellular free Ca^{+2} concentration in the single taste receptor cells in mice increased by the addition of denatonium or sweet tastes (Uchida et al., 1994). A model of Ca^{+2} ion channel was used to explain the active potential changes of taste receptor cells. The model involved the basolateral membrane, apical membrane, membrane potential, calcium channel and calcium ion concentration (Sasaki and Kambara, 1995).

PRIMATES AND HUMANS

It is believed chimpanzee chorda tympani taste fibers fall into groups that conform with the human taste qualities. The relationship between bitter and sweet taste stimuli was studied and eight fibers were classified as bitter according to their responses to 31 stimuli including quinine, denatonium benzoate and caffeine. The results indicate a clear dichotomy between the sweet and bitter fibers; i.e., the sweet fibers never responsed to the bitter compounds. However, in addition to their responses to the above compounds, some of the bitter fibers were stimulated by other compounds, including sodium chloride-amiloride mixture, potassium chloride and xylitol, perhaps by some bitter-tasting component of the salts. The results suggest that the bitter and sweet tastes are conveyed in specific and separate groups of nerve fibers in the chimpanzee. Because of the closeness between chimpanzee and human, this finding has implications for the question of taste coding in humans and the concept of taste qualities (Hellekant and Ninomiya, 1994).

Sucrose octaacetate (SOA) is the defining stimulus for one of the best characterized examples of genetic control of taste sensitivity in a mammalian system. When compared to the bimodal or trimodal taste genetics of phenylthiocarbamide (PTC) and 6-*n*-propylthiouracil (PROP), SOA studies have been used in few human studies. However, PTC and PROP testing is accomplished primarily by the filter paper method. Standard sip and spit methods with SOA recently have demonstrated the SOA tresholds of 4 μM for humans to be unimodally distributed over a fairly small range, similar to that reported for brucine, strychnine and quinine sulfate. The filter paper test with PTC and PROP indicated PTC nontasters correlate with SOA tasters. A tentative conclusion is that SOA and PTC involve two distinctly different bitter transduction mechanisms (Boughter and Whitney, 1993). SOA bitterness arises from the specificity for the number or placement of the acetates. Partial hydrolysis gives substances that exhibit taster and demitaster genetic preferences SOA (Capeless et al., 1994).

Mixture suppression is a common explanation for the perceived reduction in a taste modality within a mixture of tastants. A review discusses the mixtures of quinine hydrochloride and sodium chloride as eliciting heterogeneous taste percepts. Each such percept consists of a bitter and a salt sensation. Using such functional measurement in combination with a two-stimulus procedure, it was found that the salt suppresses the bitterness and that the bitterness has almost no suppressive effect on the saltiness. In addition, it was shown that the total intensity of the mixture of perceptions is almost identical to the sum of the intensities of the bitterness and saltiness sensations found with the percept. As was found in early experiments with mixtures of other tastants, central sensory integration within a heterogeneous percept seems to be a fairly simple additive process (Frijters and Schifferstein, 1994).

Single third-order neurons in the hamster parabrachial nucleus (PbN) were studied for anterior tongue stimulation with binary mixtures of heterogeneous taste stimuli, including sucrose + QHCl, sucrose + citric acid and NaCl + citric acid. More recently, NaCl + QHCl and NaCl + sucrose mixtures elicited action potentials of single PbN neurons when recorded extracellularly. All stimuli were tested alone and in mixture at four concentrations: NaCl and sucrose at 0.001, 0.01, 0.1 and 1 M; QHCl at 0.00032, 0.0032, 0.032 and 0.1 M. The NaCl + QHCl and NaCl + sucrose mixtures were formed by pairing the four concentrations of each simulus with the strongest concentration of the other stimulus. The response frequencies evoked by the mixtures did not differ markedly from those evoked by the other stronger single component. The response frequencies of hamster PbN neurons to heterogeneous taste mixtures were either equivalent to the response of the strong single component or exhibited a small mixture suppression. There was little evidence of neuronal-affected mixture enhancement (Vogt and Smith, 1994).

Chlorhexidine gluconate at a dose used to control bacteria in the mouth has a reversible effect on taste perception. Taste-intensity ratings and taste-quality identification for a concentration series of sucrose, sodium chloride, citric acid and quinine hydrochloride were obtained from 15 healthy humans. The participants rinsed with 0.12% chlorhexidine for three minutes twice a day. Each individual was tested three times: before the four-day rinse period, 30 minutes after the final rinse and four days after the rinse period. Chlorhexidine rinses reduced the perception intensity of sodium chloride and quinine hydrochloride, but not sucrose or citric acid. No effects on taste perception were detected four days after the rinse period. Further, the identification of sodium chloride as salty was seriously impaired by chlorhexidine, but the *identification* of quinine hydrochloride as bitter was not affected. Specific sites of action of chlorhexidine on the taste epithelium are not known, but its effects on salty taste may be related to its strong positive charge and its effect on bitter taste may be related to its amphiphilicity. Chlorhexidine has promise as a probe of taste transduction, as well as for the management of salty/bitter dysgeusias in humans (Helms et al., 1995).

Amphiphilic substances may stimulate cellular events through direct activation of G proteins. Experiments indicate that several amphiphilic sweeteners and the bitter stimulant, quinine, activate transducin and Gi/Go proteins. Concentrations of taste substances required to activate G proteins in vitro correlated with those used to elicit taste. These data support the hypothesis that amphiphilic taste substances may elicit taste through direct activation of G proteins (Naim et al., 1994).

Recent evidence suggests similar G protein messenger mechanisms may be involved in human taste perception and vision. The rod and cone transducins are specific G proteins originally thought to be present only in photoreceptor cells of the vertebrate retina. Transducins convert light stimulation of photoreceptor opsins into activation of cyclic GMP phosphodiesterase. The authors report that rod transducin is also present in vertebrate taste cells, where it specifically activates a phosphodiesterase isolated from taste tissue. Furthermore, the bitter compound denatonium in the presence of taste cell membrane activates transducin, but not Gi.

Evidence mounts for a peptide that competitively blocks taste-cell membrane activation of transducin, arguing for the involvement of a seven-transmembrane-helix G protein-coupled receptor. These results suggest that rod transducin communicates signals of bitter taste by coupling taste receptor(s) to taste cell phosphodiesterase. Phosphodiesterase-mediated degradation of cyclic nucleotides may lead to taste cell depolarization through the cyclic nucleotide suppressive conductance (Ruiz-Avila et al., 1995). Prior hypotheses on the mechanism of taste perception proposed a receptor protein fashioned from an α-helix (Suami and Hough; 1991, 1992, 1993). The hypothesis of proteinaceous character was also presented by Lee (1987) and was proposed long ago by Fisher (1971).

Denatonium benzoate, the most bitter substance known to man, will be mentioned frequently. It causes a rise in the intracellular calcium concentration from internal stores in taste cells. The transduction of bitter taste may occur via a receptor-second messenger mechanism leading to neurotransmitter release and may not involve depolarization-mediated calcium entry (ion channels) (Akabas et al., 1988).

A transducin-like G protein, gustducin, has been previously identified and cloned from rat taste cells. Neither receptors nor effectors that interact with α-gustducin in taste are known. However, α-gustducin has an 80% structural similarity to the visual G protein, α-transducin. When α-gustducin was reconstituted in a visual system, identical interactions occurred with the receptor bovine rhodopsin, and the effector bovine retinal cyclic GMP-phosphodiesterase and with bovine brain and retinal G protein $\beta\gamma$-heterodimers; receptor-catalyzed GDP-GTP exchange and the intrinsic GTPase activity of α-gustducin and α-transducin. Therefore, the functional equivalence of α-gustducin and α-transducin suggests that taste buds are likely to contain receptor and effector proteins that share many properties with their retinal equivalent (Hoon et al., 1995).

Some hereditary traits, perhaps in receptor elements, may provide certain individuals with transduction sequences that select sensitivity to certain bitter substances (see Chapter 5). Also, with some evidence to suggest age differences in bitter taste perception exist, more research will be done (see Chapter 8). This is key, since an ever-growing elderly population will be dependent upon good tasting products. The sensitivity of bitter substances in aging populations has been studied (Yokomukai et al., 1993, 1994). The taste worlds of humans vary because of selected taste blindness to phenylthiocarbamide (PTC) and its chemical relative, 6-*n*-propylthiouracil (PROP). Modern statistical analyses of early PTC results show that a higher frequency of women tasted PTC at threshold classification. In the laboratory, scaling of PROP bitterness led to the identification of a subset of tasters (supertasters) who rate PROP as intensely bitter. Supertasters also perceive stronger tastes from a varity of bitter and sweet substances, and perceive more burn from oral irritants (alcohol and capsacin). The density of taste receptors on the anterior tongue (fungiform papillae, taste buds) correlates significantly with perceived bitterness of PROP and supports the supertaster concept. Psychophysical data from studies in the lab also show a sex effect; women are supertasters more frequently than men. The anatomical data also support the sex difference; women have more fungiform papillae and more taste buds. Further investigations of PTC/PROP tasting and food behaviors should include scaling to identify supertasters and separate sex effects (Bartoshuk et al., 1994).

There is considerable evidence to suggest the existence of multiple bitter transduction sequences. This suggests individual differences in response to various bitter compounds and may reflect differences in the relative availability of specific transduction sequences. When 52 subjects were asked to judge the intensity of the bitterness of quinine sulfate and urea separately, a statistically significant portion (1/3) of the test population rated quinine sulfate more bitter than urea, while another 1/3 rated urea more bitter. More molecules of urea were necessary to have its effect. Hence, molecular concentration may indicate a separate mechanism of bitter taste perception. Urea is suggested to denature a taste protein triggering perception, while quinine just has an affinity for the protein surface of the receptor. Previous research and this research indicated the results did not correlate specifically with sweet/bitter selective individuals known as PROP tasters. PROP tasters are said to be sweet dislikers whereas PROP nontasters were almost always sweet likers (Looy and Weingarten, 1992).

So the ability to taste low concentrations of propylthiouracil (PROP) and related bitter compounds is heritable. Current analysis determines whether the distribution of PROP taste thresholds is consistent with an additive or dominant mode of Mendelian transmission. To that end, the lowest concentration of PROP detectable was determined for 1015 subjects and models of bi- and tri-modal distribution of PROP taste thresholds were tested. The model with the greatest likelihood had three distributions and followed an additive model of PROP taste sensitivity if the variances associated with the distributions were assumed to be equal. However, if the taste thresholds were transformed to remove skewness, or if the variances were unequal, then three- or two-distribution models were equally likely. Resolution of the mode of inheritance for bitter taste perception awaits additional family studies and the characterization of the molecular basis of taste perception for these bitter compounds (Reed et al., 1995).

Cross-culturally for Japanese and Australians, there does not appear to be selection sensitivity towards ranges of concentrations of the taste modalities sweet, salt, sour, umami and bitter. However, only caffeine was tested for bitterness differences (Prescott et al., 1992).

REFERENCES

Akabas, M. H., Dodd, J. and Al-Awqati, Q. 1988. A bitter substance induces a rise in intracellular calcium in a subpopulation of rat tates cells. *Science, 242,* 1047.

Asanuma, N. and Nomura, H. 1995. Cytochemical localization of guanylyl cyclase activity in rabbit taste bud cells. *Chem. Senses, 20*(2), 231–7.

Avenet, P. 1992. Ion channels and taste. *Pennington Cent. Nutr. Ser., 2* (Science of Food Regulation), 374–89.

Bartoshuk, L. M., Duffy, B. V. and Miller, I. J. 1994. PTC/PROP tasting: anatomy, psychophysics, and sex effects. *Physiol. Behav., 56*(6), 1165–71.

Boughter, J. D., Jr. and Whitney, G. 1993. Human taste thresholds for sucrose octaacetate. *Chem. Senses, 18*(4), 445–8.

Brand, J. G., Zeeberg, B. R. and Cargan, R. H. 1976. Biochemical studies of taste sensation. V. Binding of quinine to bovine taste papillae and taste bud cells. *Int. J. Neurosci., 7,* 37.

Capeless, C. G., Boughter, J. D. and Whitney, G. 1994. Hydrolysis of sucrose octaacetate; qualitative difference in taster and demistaster avoidance phenotypes. *Chem. Senses, 19*(6), 595–607.

Clarke, S. N. D. A. and Parker, L. A. 1995. Morphine-induced modification of quinine palatability: effects of multiple morphine-quinine trials. *Pharmacol., Biochem. Behav., 51*(2/3), 505–8.

Contreras, R. J., Carson, C. A. and Pierce, C. E. 1995. A novel psychophysical procedure for bitter taste assessment in rats. *Chem. Senses, 20,* 305–12.

Faull, J. R. and Halpern, B. P. 1969. Removal of alkaloids from aqueous solution by lipids from taste bud bearing epithelium calves, *Feb. Proc., 28,* 275.

Fisher, F. 1971. *Gustation and Olfaction;* C. Ohloff and A. Thomas, Eds.; Academic Press: New York, p. 198.

Frijters, J. E. R. and Schifferstein, H. N. J. 1994. Perceptual interactions in mixtures containing bitter tasting substances. *Physiol. Behav., 56*(6), 1243–9.

Hellekant, G. and Ninomiya, Y. 1994. Bitter taste in single chorda tympani taste fibers from chimpanzee. *Physiol. Behav, 56*(6), 1185–8.

Helms, J. A., Della-Fera, M. A., Mott, A. E. and Frank, M. E. 1995. Effects of chlorhexidine on human taste perception. *Arch. Oral Biol., 40*(10), 913–20.

Hoon, M. A., Northup, J. K. Margolskee, R. F. and Ryba, N. J. P. 1995. Functional expression of the taste specific G-protein, α-gustducin. *Biochem. J., 309*(2), 629–36.

Kolesnikov, S. S. and Margolskee, R. F. 1995. A cyclic-nucleotide-suppresible conductance activated by transducin in taste cells. *Nature (London), 376*(6535), 85–8.

Kock, K., Moreley, S. D., Mullins, J. J. and Schmale, H. 1994. Denatonium bitter tasting among transgenic mice expressing rat von Ebner's gland protein. *Physiol. Behav., 56*(6), 1173–7.

Koyama, N. and Kurihara, K. 1972. Mechanism of bitter taste perception: interaction of bitter compounds with monolayer of lipids from bovine circumvallate papillae. *Biochim. Biophys. Acta, 288,* 22.

Kumazawa, T., Nomura, T. and Kurihara, K. 1988. Liposomes as model for taste cells: receptor sites for bitter substances including N−C=S substances and mechanism of membrane potential changes. *Biochemistry, 27,* 1239.

Kumazawa, T., Miyamoto, T., Okada, Y. and Sato, T. 1994. Ion channels in bullfrog taste cells. *Nippon Aji to Nioi Gakkaishi, 1*(3), 196–9.

Kuribara, Y. 1973. Effect of taste stimuli on the extraction of lipids from bovine taste papillae. *Biochim. Biophys. Acta, 306,* 478.

Lee, C. K. 1987. *Adv. Carbohydr. Chem. Biochem., 45,* 199.

Li, X.-J. and Snyder, S. H. 1995. Molecular cloning of Ebnerin, a von Ebner's gland protein associated with taste buds. *J. Biol. Chem., 270*(30), 17674–9.

Looy, H. and Weingarten, H. P. 1992. Facial expression and genetic sensitivity to 6-*n*-propylthiouracil predict hedonic response to sweet. *Physiol. Behav., 52*(10), 75–82.

McLaughlin, S. K., McKinnon, P. J., Spickofsky, N., Danho, W. and Margolskee, R. F. 1994. Molecular cloning of G proteins and phosphodiesterases from rat taste cells. *Physiol. Behav., 56*(6), 1157–64.

Nagai, H., Spielman, A. I., Dasso, M., Huque, T. and Brand, J. B. 1994. Real time measurement of the second messenger in bitter taste transduction. *Nippon Aji to Nioi Gakkaishi, 1*(13), 220–3, in Japanese.

Naim, M., Seifert, T., Nuernberg, B., Gruenbaum, L. and Schultz, G. 1994. Some taste substances are direct activators of G-proteins. *Biochem. J., 297*(3), 451–4.

Nakashima, K. and Ninomiya, Y. 1994. The IP3 level of mouse circumvallate papilla in response to various taste stimulations. *Nippon Aji to Nioi Gakkaishi, 1*(3), 216–19, in Japanese.

Okamoto, F., Kajiya, H., Okabe, K. and Soeda, H. 1995. Transepithelial responses across the tongue epithelium to bitter-tasting compounds in frogs. *Fukuoka Shika Daigaku Gakkai Zasshi, 22*(1), 9–14, in English.

Parker, L. and Leeb, K. 1994. Amphetamine-induced modification of quinine palatability: analysis by the taste reactivity test. *Pharmacol., Biochem. Behav., 47*(3), 413–20.

Parker, L. A. and Doucet, K. 1995. The effects of nicotine withdrawal on taste reactivity. *Pharmacol., Biochem. Behav., 52*(1), 125–9.

Prescott, J., Laing, D., Bell, G., Yoshida, M., Gillmore, R., Allen, S., Yamazaki, K. and Ishii, R. 1992. Hedonic responses to taste solutions: a cross-cultural study of Japanese and Australians. *Chemical Senses, 17*(6), 801–9.

Reed, D. M., Bartoshuk, L. M., Duffy, V., Marino, S. and Price, R. A. 1995. Propylthiourea tasting: determination of underlying threshold distribution using maximum likelihood. *Chem. Senses, 20,* 529–33.

Ruiz-Avila, L., McLaughlin, S. K., Wildman, D., McKinnon, P. J., Robichon, A., Spickofsky, N. and Margolskee, R. F. 1995. Coupling of bitter receptor to phosphodiesterase through transducin in taste receptor cells. *Nature (London), 376*(6535), 80–5.

Sasaki, N. and Kambara, T. 1995. Generating system of active potential in taste receptor cells. II. Effect of Ca ion channel. *Nippon Shika Daigaku Kiyo, Ippan Kyoikukei, 24,* 37–54.

Sato, T. 1988. In the *International Conference on Chemosensory Transduction: receptor events and transduction in taste and olfaction,* Monell Chemical Senses Center, Philadelphia, PA, March 23–25.

Sato, T., Okada, Y. and Miyamoto, T. 1994. Receptor potential of the frog taste cell in response to bitter stimuli. *Physiol. Behav., 56*(6), 1133–9.

Sato, T., Okada, Y. and Miyamoto, T. 1995. Molecular mechanisms of gustatory transduction in frog taste cells. *Prog. Neurobiol. (Oxford), 46*(2/3), 239–87.

Sato, T. and Sugimoto, K. 1995. Quinine-HCl-induced modification of receptor potentials for taste stimuli in frog taste cells. *Zool. Sci., 12*(1), 45–52.

Schiffman, S. S., Gatlin, L. A., Suggs, M. S., Heiman, S. A., Stagner, W. C. and Erickson, R. P. 1994. Modulators of the adenylate cyclase system can alter electrophysiological taste responses in gerbil. *Pharmacol., Biochem. Behav., 48*(4), 983–90.

Schiffman, S. S., Suggs, M. S., Losee, M. L., Gatlin, L. A., Stagner, W. C. and Bell, R. M. 1995. Effect of lipid-derived second messengers on electrophysiological taste responses in the gerbil. *Pharmacol., Biochem. Behav., 52*(1), 49–58.

Schmale, H., Ahlers, C., Blaeker, M., Kock, K. and Spielman, A. I. 1993. *Ciba Found. Symp. 179 (Molecular Basis of Smell and Taste Transducton),* 167–85.

Scott, P. E. and Farley, I., 1988. Cross-adaption of bitter compounds: single receptor site, Abstract #189 of the *10th Annual Meeting Assoc. Chemoreception Sci.,* Sarasota, FL, April 27–May 1.

Spielman, A. I., Nagai, H., Sunavala, G., Dasso, M., Hugue, T. and Brand, J. G. 1995. Second messenger assays. *Exp. Cell Biol. Taste Olfaction,* 203–10.

Spielman, A. I., Huque, T., Whitney, G. and Brand, J. G. The diversity of bitter taste transduction mechanisms. *Soc. Gen. Physiol. Ser. 47* (Sensory Transduction), 307–24.

Spielman, A. I., Huque, T., Nagai, H., Whitney, G. and Brand, J. G. 1994. Generation of inositol phosphates in bitter taste transduction. *Physiol. Behav., 56*(6), 1149–55.

Suami, T. and Hough, L. 1991. Molecular mechanism of sweet taste. 1. Sweet and non-sweet tasting amino acids. *J. Carbohydr. Chem., 10*(5), 851–860.

Suami, T. and Hough, L. 1992. Molecular mechanism of sweet taste. 2. Glucopyranose, fructopyranose and sucrose. *J. Carbohydr. Chem., 11*(8), 953–67.

Suami, T. and Hough, L. 1993. Molecular mechanism of sweet taste. 3. Aspartame and its non-sweet isomers. *Food Chem., 46*(3), 235–8.

Tanimura, S., Shibuya, T. and Ishibashi, T. 1994. Neural responses of the glosopharyngeal nerve to several bitter stimuli in mice. *Comp. Biochem. Physiol., A: Physiol., 108A* (2–3), 189–94.

Teeter, J. H. and Brand, J. G. 1987. Peripheral mechanisms of gustation: physiology and biochemistry. In *Neurobiology of Taste and Smell;* T. E. Finger and W. L. Silver, Eds., John Wiley & Sons, NY: pp. 299–330.

Tsunenari, T., Hayashi, Y., Orita, M. and Mori, T. 1994. Receptor potential to quinine HCl in isolated taste cells of bullfrog *Rana catesbeiana. Nippon Aji to Nioi Gakkaishi, 1* (3), 192–5.

Uchida, Y., Fujiyama, R., Miyamoto, T., Okada, Y. and Sato, Y. 1994. Molecular mechanism of the change from sweet taste stimulation to gustatory receptor potential. *Nippon Aji to Nioi Gakkaishi, 1* (3), 208–11.

Vogt, M. B. and Smith, D. V. 1994. Responses of single hamster parabrachial neurons to binary taste mixture of NaCl with sucrose or QHCl. *J. Neurophysiol., 71* (4), 1373–80.

Yamamoto, T., Shimura, T., Sakai, N. and Ozaki, N. 1994. Representation of hedonics and quality of taste stimuli in the parabrachial nucleus of the rat. *Physiol. Behav., 56*(6), 1197–202.

Yokomukai, Y., Cowart, B. J. and Beauchamp, G. K. 1993. Individual differences in sensitivity to bitter-tasting substances. *Chemical Senses, 18*(6), 669–81.

Yokomukai, Y., Cowart, B. J. and Beauchamp, G. K. 1994. Gustatory responses to bitter substances in the aged. *Nippon Aji to Nioi Gakkaishi, 1* (3), 328–31.

CHAPTER 2

General Correlation between Models of Sweetness and Bitterness Perception

GLENN ROY[1]

THEORETICAL STUDIES

IT is widely accepted that certain chemically charged and space-filling features determine whether a substance tastes sweet, bitter or tasteless. From a theoretical standpoint, some bitterness may also be accounted for in models as spatial disorientations that do not provoke sweet taste in chemically identical but configurationally different molecules. Finally steric hindrance may cause distortion of the required dispersion binding of hydrophobic groups.

Several models have been proposed to describe the mechanism of perception of sweetness and bitterness; one of these also claims to explain potency effects. All assume the existence in the sweet or bitter molecule of particular arrangements of certain types of chemical groups. Over many years of tasting substances from organic synthesis, Hokkaido University has concluded that small differences in functional group organization are responsible for changing a sweet taste to a bitter taste and vice versa. They describe taste behavior in mixed sweet and bitter solutions.

The current accepted theory suggests that a bitter compound and a sweet compound bind independently at specific receptors. This is referred to as "independent." The independent receptor mechanism would be expected to yield taste data in which the intensities of bitter and sweet would be unaffected by mixing the two tastes, regardless of the concentration. The data demonstrate that a bitter compound and a sweet compound bind at the same receptor in a "competitive" manner. In the competitive receptor mechanism, one would expect both flavors to become altered, i.e., one

[1]Pepsi-Cola Co., 100 E. Stevens Ave., Valhalla, NY 10595, U.S.A.

stronger and other weaker, as component concentrations vary; the latter would occur because of competition of the substances for the same site. Extensive mixed sensory evaluations of varied concentrations of a sweet and bitter substance (D-Phe and L-Phe; sucrose and L-Phe) clearly indicate that sweetness and bitterness act in a competitive manner and should be considered as competing for the binding sites at the same receptor (Nakamura and Okai, 1993a, 1993b).

Some studies on the structure-bitterness relationship suggest that a lecithin-like lipid located on the cell membrane may be part of a receptor for bitterness, but helical peptides or proteinaceous receptors are rational to explain the taste of D- and L-amino acids (Lee, 1987; Fischer, 1971) (Figure 1).

Goodman at UCSD has conducted taste research and finds L-aspartyl-D α-methylphenylalanine methyl ester and L-aspartyl-D-alanyl-2,2,5,5-tetramethylcyclopentanyl ester elicit bitter and sweet tastes, respectively. The C-terminal residues of the two analogs adopt distinctly different conformations in the solid state by X-ray diffraction analysis. The bitter substance has an extended structure, while the sweetener adopts the typical L-shaped structure shown by other sweeteners. Solution NMR and computer simulation agree well with the authors' model for sweet and bitter tastes. Thus, several unique features signify the structural conformation and other requirements of dipeptides and dipeptide mimetics for sweet and bitter taste (Yamazaki et al., 1994; Kent, 1995) (Figure 2).

A further interesting experiment involves a hydrophobic substance possessing a sour taste. That substance was used to pretreat or load the receptor with sour taste stimulus. Then when either 2 mM caffeine or 0.05 mM

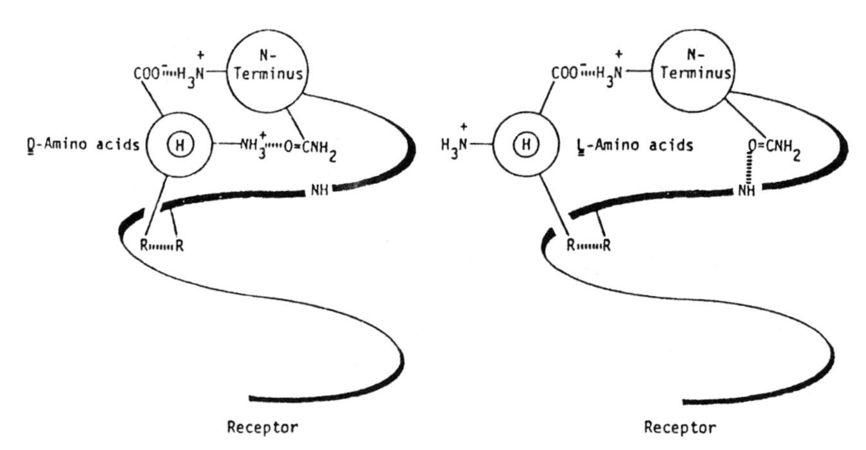

Figure 1 A hypothetical helical receptor protein that recognizes D or L amino acids to account for taste responses.

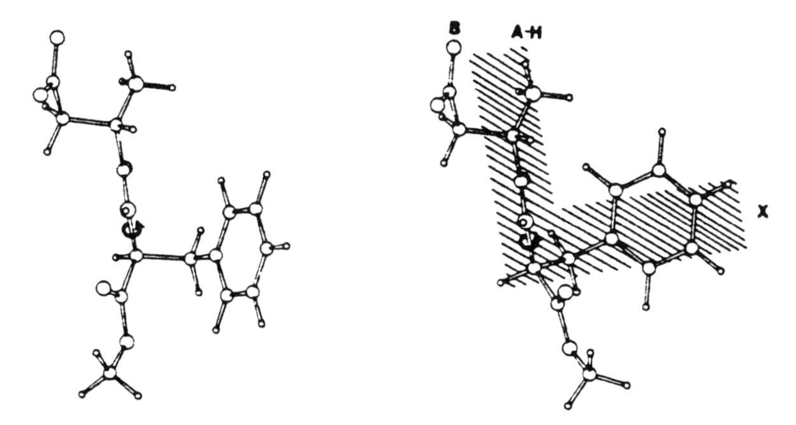

Figure 2 Stick models of the X-ray crystal structure of an isolated aspartame molecule illustrating a coplanar L-shape consistent with the Goodman model of sweet taste. The traditional AH-B and X regions are identified.

strychnine were given after the sour stimulus, the weak bitterness of caffeine was eliminated and the strong bitterness of strychnine was weakened. These data are suggestive in support of the competitive receptor hypothesis. This same research has led to efforts to develop inverted aspartame-type sweeteners without bitterness (Takahashi et al., 1994).

The chemical signatures of sweetness perception were derived using the Shallenberger-Acree-Kier nomenclature; these are AH (an electrophile), B (a nucleophile) and X (a hydrophobic group). Many taste researchers use the basic tenet of the Shallenberger-Acree-Kier model to explain the bitterness phenomenon (Tamura et al., 1990a, 1990b; Ishibashi et al., 1987). A summary of the nomenclature of taste models is given in Figure 3.

In a model, the AH (electrophilic) species binds to the receptor site via an amino or hydrophobic group. Concomitant binding of a hydrophobic region (X) in the compound potentiates a bitter taste and is the minimum requirement for bitterness (Shinoda et al., 1987). For example, trehalose X-ray crystal data show an absence of a sweet AH-B glucophore, which would explain the absence of sweetness. Additionally, evidence indicating the importance of hydrophobic character to bitterness is presented (Lee et al., 1995). Only these two groups (AH and X) are required in this simple model for the bitter receptor, which defines an optimum AH-B distance of 4.1 Å. In this unified model of sweet/bitter taste, a third receptor site (B′) must be free in order to produce bitterness. If a sulfonic acid group such as in taurine blocks it, no bitterness is perceived. The authors further claim that high overall hydrophobicity of the molecule is the determinant of the potency of bitterness, and that there is an optimum molecular di-

Model	Requirements for sweetness	Requirements for bitterness
Temussi	AH (electrophile) B (nucleophile) X (hydrophobic group)	B } (orientation AH } reversed) X
Belitz	p+ (electrophile) p− (nucleophile) a (apolar group) in a region where (x, $z > 0$)	p+ p− or no p− a in a region where (x, $z < 0$), if p− group present
Okai[a]	AH (electrophile) B (nucleophile) X (hydrophobic group)	AH no B X

[a] The AH group is also referred to as SU. The X group is also referred to as BU

Figure 3 Comparison of nomenclature systems used in three models of the molecular requirements for sweetness and bitterness.

ameter (15 Å) to fit the receptor complex to produce bitterness (Tamura et al., 1990b). A simplified sweet/bitter taste receptor model is illustrated in Figure 4.

Belitz (Belitz and Weiser, 1985; Belitz et al., 1979) used Cartesian (x, y, z) coordinates to indicate the relative positions of the electrophilic (p^+), nucleophilic (p^-) and apolar hydrophobic groups. This model predicts whether a compound will taste sweet or bitter, as seen in Figure 5. The p^- and p^+ groups of the compound are assumed to lie on the ($-x$) and $-y$) axes, respectively. If the hydrophobic group of the compound lies in a re-

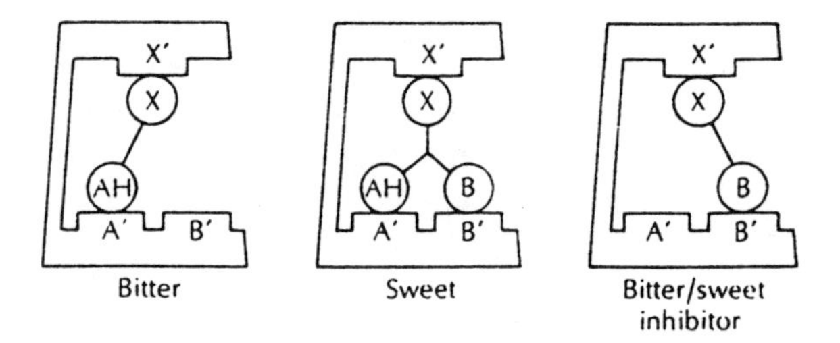

Figure 4 The Okai unified bitter/sweet taste receptor model. The electrophilic group AH binds to the receptor A′ by an amino or hydrophobic group. In order for a compound to be perceived as bitter, the AH group must bind to A′; a second hydrophobic group X must bind to a second site X′, and a third site B′ must be left open. An inhibitor fills only B′ and X′.

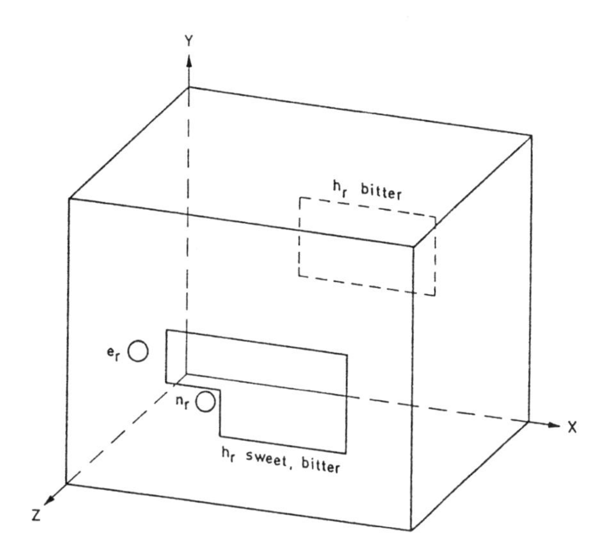

Figure 5 The Belitz model bipolar system with the $-x$ or $-y$ regions creating bitter responses.

gion where the x or z coordinate is negative, the compound will produce a bitter, rather than a sweet, response. Compounds that lack the p^- group are assumed to be bitter, regardless of the location of the hydrophobic group. For many classes of bitter substances, the distance between the p^- and p^+ groups is 2.5–8 Å. Thirty iminylsulfamates ($R-C{=}NHSO_3\text{-}Na^+$) have been synthesized and evaluated for structure/taste relationships. Fifteen were predominantly bitter, five were bitter with a sweet aftertaste, three were sour with a sweet aftertaste, and the remainder a mixture of all taste modalities. The length (x), height (y), and width (z) of the R substituents were determined for the tastants, and no reliable structure/taste relationship was found (Spillane and Walsh, 1996). A model by Temussi is said to account for sweetness or bitterness when a molecule can be reversed within the AH-B unit to "fit" better (Figure 6).

Belitz et al. (1988), Belitz and Weiser (1985), Beets (1978) and Ishibashi and co-workers (1987, 1988a, 1988b) have pointed out that a monopolar electrophile with a hydrophobic substituent is a sufficient condition for bitterness. The following example demonstrates the criticality of hydrophobic location for this taste effect. The qualitative similarity between the sweetener S (Nofre and Tinti, 1989) and denatonium benzoate B (the most bitter compound known to man) leads one to believe, at least in this case, that a single receptor site triggers bitter taste responses dependent upon the hydrophobic substitution. Superimposition of the two structures (B + S, Figure 7) supposes bitter taste-favorable hydrophobicity where "half moons" are indicated.

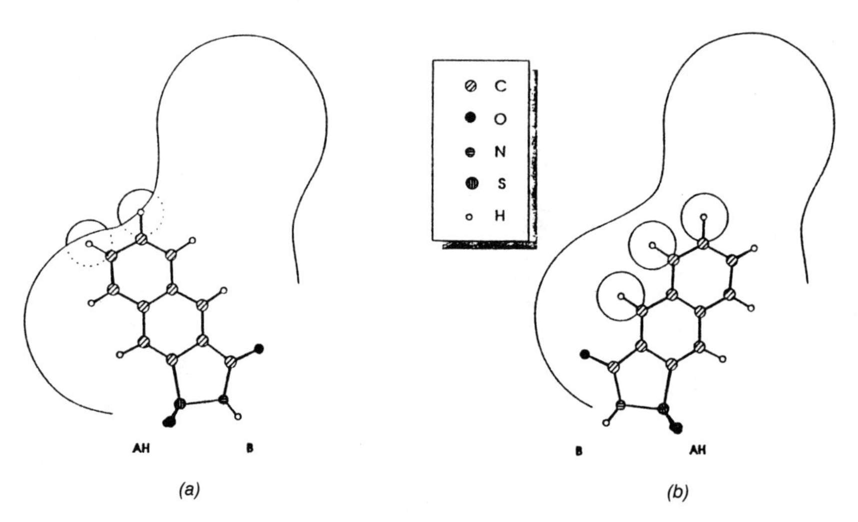

Figure 6 The Temussi model comparing the fit of benzo-saccharin in the sweet (a) model and bitter (b) model. Note the reversed binding role of AH-B.

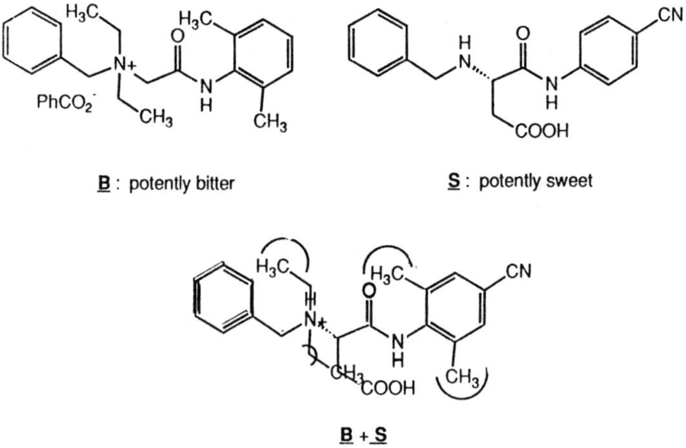

B : potently bitter

S : potently sweet

B + **S**

Figure 7 A potently bitter and potently sweet chemical pair with assumed similarity in structural orientation. Overlap of the structures shows where hydrophobicity accounts for bitterness.

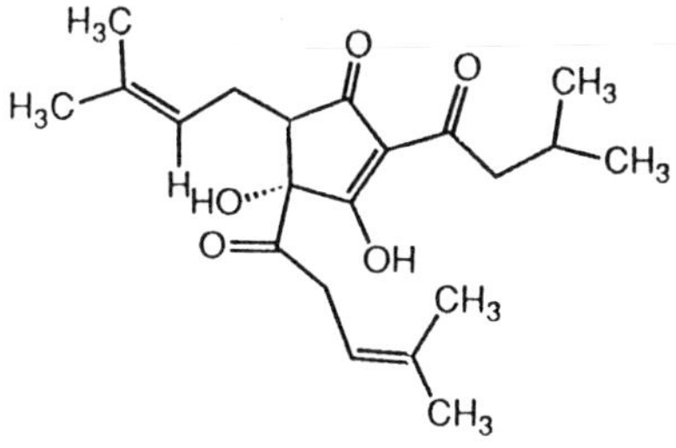

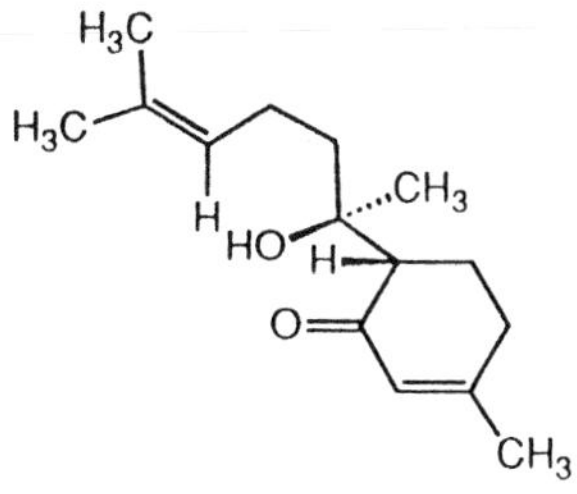

I : isohumulone H : hernandulcin

Figure 8 A natural bitter from beer and a natural sweetener from a plant showing similarity of structure between bitterness and sweetness.

Belitz has described "allowed for sweetness or forbidden for sweetness" (bitter) zones (Belitz et al., 1988). These can be likened also to isohumulone I, the bitter principle of beer that relates somewhat closely (Figure 8) to hernandulcin H (Kinghorn and Soejarto, 1989), which is 1000 times sweeter than sugar. The opposite stereochemistry of the hydroxyl group in H gives a bitter material.

COMMON SIMILARITIES, EXPERIMENTAL DATA

Sweetness and bitterness inhibitors derived from aminomethane sulfonic acid have been synthesized and evaluated (Roy et al., 1990, 1991). Based on molecular modeling of known sweeteners and sweetness inhibitors, a structure was proposed that fits the requirements for sweetness inhibition. The NutraSweet Co. patented a number of sweetness inhibitors based on the suosan structure and comprised of tasteless aryl urea sulfonic acids. These structures were designed by Muller and Culberson at The NutraSweet Company as idealizing the B and X filling of the receptor (cf. Figure 4). Since the sulfonic acid groups bind to the B′ receptor site, and the compounds have no AH group per se, tasteless aryl urea sulfonic acids were correctly predicted to be bitterness as well as sweetness inhibitors. In organoleptic studies with sweeteners (sucrose, aspartame, saccharin, 6 chloro-D-tryptophan) and bitter substances (caffeine, quinine, denatonium benzoate and naringin), certain aryl ureas of aminomethane sulfonic acid exhibited sweetness and bitterness reduction. These studies provide evidence of a competitive relationship between bitterness perception and sweetness perception.

The observed sweetness potencies of the suosan sweetener series and calculated Mulliken population analysis of the electric fields near the carbonyl oxygen of the COOH group are strongly correlated (Santhosh and Mishra, 1994). The competitive inhibition observed between bitterness perception and sweetness perception suggests that populations of nearly identical receptor surfaces or pathways are present in the sweet-detecting and bitter-detecting regions of the tongue, which are grouped in two distinct areas. Sweetness is detected on the tip of the tongue, whereas bitterness is detected at the back of the tongue (Figure 9).

A rational proposal was advanced that an optimal bitterness inhibitor could be designed if the requirements for high-potency sweetness were fulfilled without an AH group. The NutraSweet Co. synthesized a large structure-activity relationship (SAR) in search of sweeteners and bitter inhibitors. The University Claude Bernard in Lyons, France, reported the rational design and discovery of sucrononic acid, the sweetest known substance, which is 200,000 times as sweet as sucrose (Tinti and Nofre, 1991a, 1991b). Two hydrophobic regions were required for optimal high-potency sweetness: one hydrophobic region is strictly aromatic (Xaryl), the other is more aliphatic in nature (Xaliphatic). Therefore, hypothetically, a potent and perhaps more universal bitterness inhibitor could be realized by examining compounds that:

- do not have an AH group (now termed the AH1 group) that binds to the A′ receptor site
- possess a nucleophilic B group
- have both Xaryl and Xaliphatic hydrophobic groups

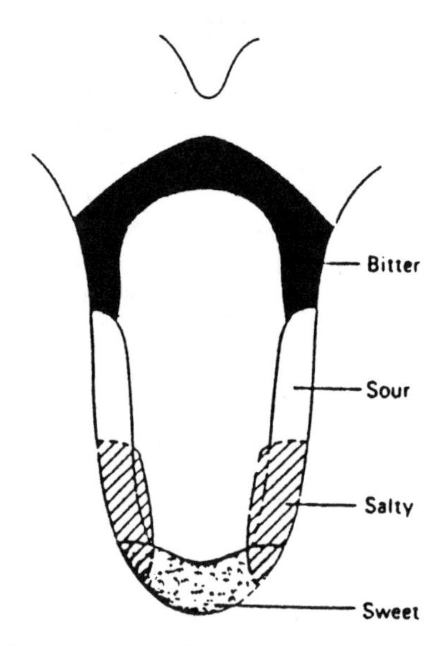

Figure 9 Regional perception of taste on the human tongue.

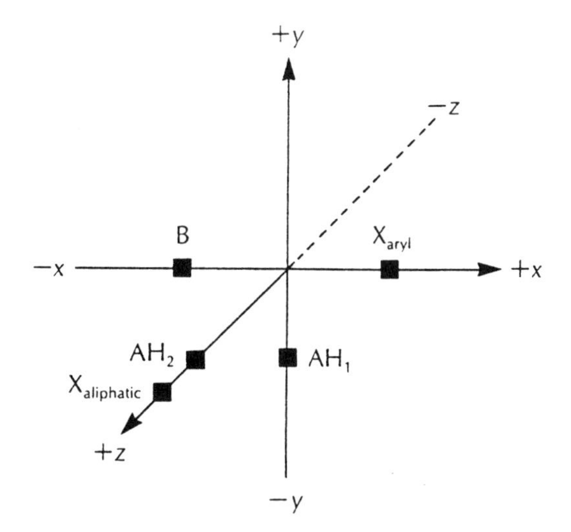

Figure 10 A complement to Figure 5 model of bitter inhibitors where an additional hydrophobic group at +x or +z in a molecule may potentiate and universalize inhibition.

- possibly possess a secondary electrophilic group, termed AH2 (e.g., the amide bond of a dipeptide) as a means of "tethering" the B group to the two hydrophobic regions
- are tasteless (Figure 10)

Unfortunately, The NutraSweet Co. research was terminated.

Ideally, a consortium of shared libraries of compound structures could be used to test models and expedite the discovery of useful bitterness inhibitors. Tasteless peptides of taurine or aminomethane sulfonic acids that have several hydrophobic regions are a logical starting point. A tasteless sulfonaphthoimidazole that adheres well to the above criteria has been reported (Castiglione-Morelli et al., 1990), and analogous functional group series should be evaluated as potential bitterness inhibitors.

Bioresearch Inc. (Fuller and Kurtz, 1991, 1993) (San Diego, CA) has patented sweetness inhibitors for use in bitterness inhibition (see Chapter 10). Their premise is that substantially tasteless substances whose structure closely mimics that of sweeteners do in fact possess sweet and bitter inhibition properties.

SENSORY MEASUREMENTS

Time vs. intensity sensory measurements are very important for the determination of aftertaste in various food and beverage products (see also

Chapter 7). Efforts began to research sensory profiling of bitterness in the 1980s (Pilkova and Pokorny, 1992; Leach and Noble, 1986). Certainly, now many industrial and academic facilities privately excel in the science of sensory measurement with a language all its own.

CASE STUDIES

Time-intensity (TI) measurements and multivariate statistics were applied in the study of 23 sweet and/or bitter stimuli with 25 volunteers. Time-intensity is a method that measures and records the dynamic relationship between the onset (appearance time, AT), intensity and duration (extinction time, ET) of the perception of a sensory attribute (Figure 11). The method has been used widely in the search for intense sweeteners to replace sucrose in various products. However, few studies have considered the time-intensity properties of bitter compounds whose AT and ET are generally longer than sweeteners. The results of principal-component (PCA) and cluster (CA) analyses of sweet/bitter stimuli vs. subject matrices for maximum intensity (I_{max}), time-to-maximum intensity (T_{max}), total duration (T_{dur} or T_{fin}) and area under the curve (area) (Figure 12) suggest that there are at least two receptor mechanisms each for sweet taste (one for sugars and other small compounds and another for large sweeteners) and bitter taste (one for PTC/PROP and one for other bitter compounds).

Summary of PCA Results

	Sweet Stimuli	Bitter Stimuli
Time-to-maximum intensity (T_{max}) probably proportionally related to lipophilic content	Shorter	Longer
Total duration (T_{dur})	Shorter	Longer
Area under the curve (area)	Smaller	Larger

There was no trend in the range of intensities, and the order of stimuli shows no relationship between intensity and taste quality for maximum intensity (I_{max}). However, significant interindividual differences in sensitivity to the sweet and/or bitter stimuli were found. These may be attributed to physiological differences in receptor populations, saliva flow and composition and the way the TI equipment was operated. Overall, it is concluded by these authors that sweet and bitter stimuli do not share common receptor mechanisms. Furthermore, the uncommon structural homo-

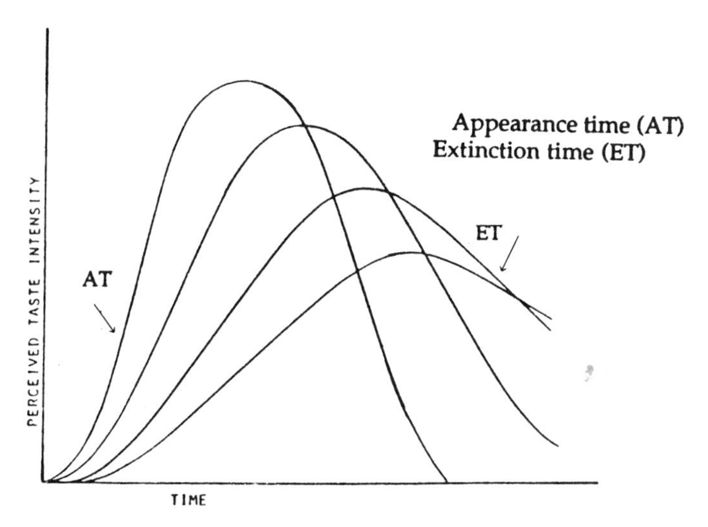

Figure 11 A hypothetical perceived intensity and time curve.

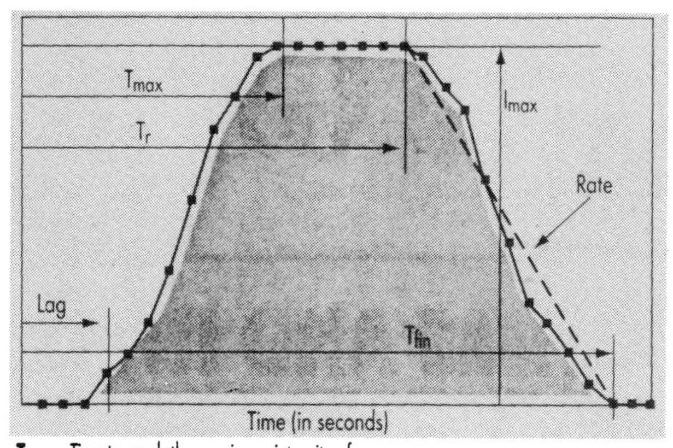

T_{max}: Time to reach the maximum intensity of response
T_r: Time at which the curve starts to decline from maximum intensity
Lag: Onset time of response
Tfin: Total duration time of response
I_{max}: Maximum intensity of response
Rate: Rate of release (maximum intensity/time to reach maximum intensity)

Source: Leatherhead Food Research Assoc.

■ The perception of a sensory attribute over time can be defined by various parameters.

Figure 12 The parameters of a time-intensity curve.

geneity for the many bitter-tasting compounds in nature indicates that unique receptors may not exist for bitterness (Guinard et al., 1995).

Monell Chemical Senses Center (Philadelphia, PA) and Yokomukai have proposed the existence of three separate, but perhaps overlapping, bitter transduction sequences particularly sensitive to (1) PTC and related compounds, (2) quinine sulfate, quinine hydrochloride, caffeine and sucrose octaacetate and (3) urea and magnesium sulfate, respectively (Yokomukai et al., 1993). Protein molecules that make up the receptor membrane may also play a role in discrimination of taste compounds entering the lipid layer.

Dr. Susan Schiffman, a psychophysiologist, has been instrumental in sensory studies to quantify bitter perception and investigate avenues to bitter reduction and inhibition. For example, a study has quantified the degree of reduction in perceived bitterness by sweeteners at both threshold and suprathreshold concentrations of bitter compounds. Detection and recognition thresholds were determined for six bitter compounds, including caffeine, denatonium benzoate, magnesium chloride, quinine hydrochloride, sucrose octaacetate, and urea in the absence and presence of several suprathreshold concentrations of five sweeteners. The sweeteners were sucrose, aspartame, sodium saccharin, mannitol and sorbitol. Polycose was also tested with the sweeteners.

The degree to which bitter thresholds were affected by the addition of sweeteners was found to be dependent on the chemical classification of the sweeteners and their concentrations. In general, the natural sweeteners, sucrose, mannitol and sorbitol, were more effective than the noncaloric sweeteners, aspartame and and sodium saccharin, in elevating the detection and recognition thresholds of the bitter compounds. A sweetness intensity approximating that of 6% sucrose (0.175 M sucrose) or greater was required to elevate thresholds. For elderly subjects, sweeteners did not significantly elevate thresholds for denatonium benzoate and sucrose octacetate. The degree to which sorbitol and sucrose can decrease the perceived bitterness intensity of suprathreshold concentrations of the six bitter compounds was also determined. The concentrations of sweeteners and bitter compounds were selected to be of moderate to high subjective intensity. The levels of sweeteners used in the mixtures were sucrose (none, 0.946 M and 2.13 M) and sorbitol (none, 2.1 M, and 3.68 M). Both sweeteners significantly reduced the bitterness ratings of almost every concentration of the six bitter compounds. The greatest reductions in bitterness were 87% for 0.192 mM denatonium benzoate mixed with 2.13 M sucrose and 84.7% for 1.8 M urea mixed with 3.68 M sorbitol (Schiffman et al., 1994a, b).

Trained tasters ($n = 16$) have determined that high-potency sweeteners tended to become more bitter with increasing concentration, while low-potency sweeteners tended to become less bitter with increasing concentration. No bitter data were observed for alitame, aspartame, glycine, maltitol, monoammonium glycyrrhizinate, neosugar (fructo-oligosaccharides), palatinit (isomalt), sodium cyclamate, sucralose and thaumatin (Schiffman et al., 1995a).

Threshold and suprathreshold sensitivities to 13 bitter compounds were determined for 16 young adults and 18 elderly persons. Half of the subjects in each age group were tasters of the bitter compound phenylthiocarbamide (PTC) and half were nontasters. Both detection and recognition thresholds, determined by a forced-choice ascending detection method, were elevated in older subjects. There were no significant differences in threshold values between tasters and nontasters of PTC. A strong relationship was found between bitter threshold values and the logarithm of the octanol/water partition coefficient for both young and elderly subjects. For young subjects, suprathreshold bitterness ratings were more intense for tasters of PTC than nontasters. Significant losses in suprathreshold sensitivity to bitter tastants with age were also found. However, unlike threshold sensitivity, no relationship was found between suprathreshold bitter taste intensity and lipophilicity (Schiffman et al., 1994a, b).

Taste changes as a result of aging are discussed in Chapter 8. Threshold sensitivity to and perceived intensity of two bitter compounds, quinine sulfate and urea, were assessed in 52 young and 60 elderly adults. Consistent with previous literature, age-related declines in sensitivity to the bitterness of quinine were observed at both threshold and suprathreshold levels. In contrast, the same young and elderly subjects showed comparable sensitivity to the bitterness of urea.

These results provide further support for the existence of multiple bitter taste transduction sequences in humans, and indicate that they may be differentially affected by aging (Cowart et al., 1994). The aged, taste-deficient population is growing and literature of the topic is as well. Although bitter inhibition or reduction is important, we must not compromise discovery in the area of bitter taste potentiation as that can enhance the sensory appeal of a variety of cheeses, coffee, tea and beer. More evidence suggests the olfactory acuity declines first, followed by gustation of bitterness, primarily. CRC Press devoted an entire issue of *Critical Reviews in Food Science and Nutrition* to the profound effect of sensory perception in the elderly. Also, Elsevier featured *Chemosensory Changes in Aging* (Tepper and Genillard-Stoerr, 1991). Further Dr. Claire Murphy contributes to the topic in this book with "Bitterness Perception across the Life Span" in Chapter 8.

Masking experiments have been carried out with monosaccharides, disaccharides and polyols in various concentrations with quinine (Mogensen and Adler-Nissen, 1988). Quinine equivalent values (QEVs) decrease with masking component in agreement with Birch and Lee (1976).

Inbred and congenic strains of mice exhibited several patterns of relative specificity to bitter tastants in 48-h, two-bottle preference tests. With segregation analyses of descendents of crosses between contrasting strains, these patterns suggested at least three genetic loci influencing bitter perception. The extensively characterized sucrose octaacetate (SOA) locus underlies one pattern. Variation at this locus had pleiotropic effects on avoidance of other acetylated sugars, plus such structurally dissimilar bitter tastants as brucine, denatonium benzoate, and quinine sulfate. Unlike SOA, however, sensitivity to quinine sulfate was polygenetically determined, and produced a second characteristic pattern. At least one, possibly several, additional unlinked loci contributed to quinine differences. Phenylthiocarbamide (PTC)-aversion differences exemplified a third pattern. Segregation consistent with monogenic control of PTC aversion did not covary with SOA or quinine sulfate avoidance. Variants of the three major patterns may be useful for analysis of specific mechanisms. While both showed the SOA pattern, strychnine differences were markedly smaller than brucine (dimethoxystrychnine) differences. Likewise, a hop extract containing primarily iso-alpha acids (e.g., isohumulone) produced an SOA-like pattern, while an extract with nonisomerized alpha acids (e.g., humulone) did not (Whitney and Harder, 1994). Beer products will also be discussed in Chapter 4.

Psychophysical experiments were conducted to determine whether isohumulones share a common receptor mechanism with other bitter compounds, and whether parotid saliva flow affects perception of their bitterness. Findings from a study of interindividual differences in sensitivity to 23 sweet and/or bitter compounds among 25 subjects using the time-intensity (TI) method suggest that isohumulone and tetrahydroisohumulone may share a common receptor mechanism with other bitter compounds, except those with the thiourea moiety. Isohumulone and tetrahydroisohumulone displayed a unique dome-shaped TI profile. The bitterness of the two compounds took longer to develop, but it lasted as long as for other bitter stimuli. In a study of the relationship between the perception of bitterness in beer and parotid saliva flow in 20 young adults, no significant difference was found among the mean saliva flows triggered by 0, 15, and 30 mg/mL of isohumulones added to beer, and no significant correlation was found between saliva flow and maximum intensity or total duration of bitterness (Guinard et al., 1994).

REFERENCES

Asanuma, N. and Nomura, H. 1995. Cytochemical localization of guanylyl cyclase activity in rabbit taste bud cells. *Chem. Senses, 20*(2), 231–7.

Avenet, P. 1992. Ion channels and taste. *Pennington Cent. Nutr. Ser., 2* (Science of Food Regulation), 374-89.

Beets, M. G. J. 1978. Structures and modalities in gustation, in *Structure Activity Relationships in Human Chemoreception, Part D,* Applied Science Publishers, London, 295.

Belitz, H.-D., Rohse, H., Stempfl, W., Wieser, H., Gasteiger, J. and Hiller, C. 1988. Schematic sweet and bitter receptors—a computer approach, *Proc. 5th Int. Flavor Conf.* July 1-3, 1987, in Frontiers of Flavor, Charalambous, G., Ed., Elsevier, Amsterdam, 49.

Belitz, H.-D., Chen, W., Jugel, H., Treleano, R., Weiser, H., Gasteiger, J. and Marsili, M. 1979. Sweet and bitter compounds: structure and taste relationship, in *Food Taste Chemistry, Vol. 115,* Boudreau, J. C., Ed., *American Chemical Society Symposium Series, 115,* 93-131.

Belitz, H.-D. and Weiser, H. 1985. Bitter compounds: occurrence and structure-activity relationships. *Food Rev. Intl., 1*(2), 271-354.

Birch, G. G., Cowell, N. D. and Young, R. H. 1972. *J. Sci. Food Agric., 23,* 1207.

Birch, G. G. and Lee, C. K. 1976. Structure functions and taste in the sugar series: the structural basis of bitterness in sugar analogues. *J. Food Sci., 41*(6), 1403.

Boughter, J. D., Jr. and Whitney, G. 1993. Human taste thresholds for sucrose octaacetate. *Chem. Senses, 18*(4), 445-8.

Castiglione-Morelli, M. A., Lelj, F., Naider, F., Tallon, M., Tancredi, T. and Temussi, P. A. 1990. *J. Med. Chem., 33,* 514-520.

Clydesdale, F. M., Ed. 1993. *Crit. Rev. in Food Science and Nutrition, 33*(1), CRC Press Inc., Boca Raton, FL, 1-101.

Cowart, B. J., Yokomukai, Y. and Beauchamp, G. K. 1994. Bitter taste in aging: compound-specific decline in sensitivity, in *Physiol. Behav., 56*(6), 1237-41.

Fisher, F. 1971. *Gustation and Olfaction;* C. Ohloff and A. Thomas, Eds.; Academic Press, New York, 198.

Frijters, J. E. R. and Schifferstein, H. N. J. 1994. Perceptual interactions in mixtures containing bitter tasting substances, *Physiol. Behav., 56*(6), 1243-9.

Fuller, W. D. and Kurtz, R. J. 1991. Ingestibles containing substantially tasteless sweetness inhibitors as bitter taste reducers, in World Patent Cooperation Treaty WO 91/18523 to Bioresearch Inc.

Fuller, W. D. and Kurtz, R. J. 1993. World Patent Cooperation Treaty WO 93/10677 to Bioresearch Inc.

Goldstein, H. and Ting, P. 1994. Post kettle bittering compounds: analysis, taste, foam, and light stability. *Monogr.-Eur. Brew. Conv., 22,* 141-64.

Guinard, J.-X., Hong, D. Y., Zoumas-Morse, C., Budwig, C. and Russell, G. F. 1994. Chemoreception and perception of the bitterness of isohumulones, in *Physiol. Behav., 56*(6), 1257-63.

Guinard, J.-X., Hong, D. Y. and Budwig, C. 1995. Time-intensity properties of sweet and bitter stimuli: implications for sweet and bitter taste chemoreception. *J. Sensory Studies, 10,* 45-71.

Hellekant, G. and Ninomiya, Y. 1994. Bitter taste in single chorda tympani taste fibers from chimpanzee, in *Physiol. Behav., 56*(6), 1185-8.

Hermans-Lokkerbol, A. C. J. and Verpoorte, R. 1994. Preparative separation and isolation of three bitter acids from hop, *Humulus lupulus* L., by centrifugal partition chromatography. *J. Chromatogr., A, 664*(1), 45-53.

Hoon, M. A., Northup, J. K., Margolskee, R. F. and Ryba, N. J. P. 1995. Functional expression of the taste specific G-protein, α-gustducin. *Biochem. J., 309*(2), 629–36.

Hughes, P. S. and Simpson, W. J. 1994. Sensory impact of hop-derived compounds. *Monogr.-Eur. Brew. Conv., 22,* 128–40.

Ishibashi, N., Arita, Y., Kanehisa, H., Kouge, K., Okai, H. and Fukui, S. 1987. *Agric. Biol. Chem., 51,* 2389.

Ishibashi, N., Ono, I., Kato, K., Shigenaga, T., Shinoda, I., Okai, H. and Fukui, S. 1988a. Role of the hydrophobic amino acid residue in the bitterness of peptides. *Agric Biol. Chem., 52*(1), 91.

Ishibashi, N., Kouge, K., Shinoda, I., Kanehisa, H. and Okai, H. 1988b. A mechanism for bitter taste sensibility in peptides. *Agric Biol. Chem., 52*(3), 819.

Kent, D. R. 1995. A stereoisomeric approach to the molecular basis of taste: peptides and peptidomimetics (sweet, bitter). *Dissertation Abstr. Int., B, 56*(9), 4887. No. DA9601758.

Kinghorn, A. D. and Soejarto, D. D. 1989. Hernandulcin, intensely sweet compounds of natural origin. *Med. Res. Reviews, 9*(1), 91.

Kock, K., Moreley, S. D., Mullins, J. J. and Schmale, H. 1994. Denatonium bitter tasting among transgenic mice expressing rat von Ebner's gland protein, in *Physiol. Behav., 56*(6), 1173–7.

Leach, E. J. and Noble, A. C. 1986. Comparison of bitterness of caffeine and quinine by a time-intensity procedure. *Chem. Senses, 11*(3), 339–345.

Lee, C. K. 1987. *Adv. Carbohydr. Chem. Biochem., 45,* 199.

Lee, C. K., Koh, K. L. and Xu, Y. 1994. Structure of trehalose derivatives. Part 6. Structure and taste of 3,6:3′,6′-dianhydro-α,α-trehalose monohydrate. *Food Chem., 52*(2), 153–6.

Li, X.-H. and Snyder, S. H. 1995. Molecular cloning of Ebnerin, a von Ebner's gland protein associated with taste buds. *J. Biol. Chem., 270*(30), 17674–9.

Looy, H. and Weingarten, H. P. 1992. Facial expression and genetic sensitivity to 6-*n*-propylthiouracil predict hedonic response to sweet. *Physiol. Behav., 52*(10), 75–82.

Mogensen, L. and Adler-Nissen, J. 1988. Evaluating bitterness masking principles by taste panel studies, *Proc. 5th Int. Flavor Conf.,* July 1–3, 1987, in Frontiers of Flavor, Charalambous, G., Ed., Elsevier, Amsterdam, 79.

Moriwaki, M. 1993. Alcoholic beverages and bitter substances (Saneigen FFI K.K.). *Foods Food Ingredient J., 155,* 64–75, in Japanese; *Chem. Abstr.,* 121: 155776y.

Nagai, H., Spielman, A. I., Dasso, M., Huque, T. and Brand, J. B. 1994. Real time measurement of the second messenger in bitter taste transduction. *Nippon Aji to Nioi Gakkaishi, 1*(3), 220–3, in Japanese.

Naim, M., Seifert, R., Nuernberg, B., Gruenbaum, L. and Schultz, G. 1994. Some taste substances are direct activators of G-proteins. *Biochem. J., 297*(3), 451–4.

Nakamura, K. and Okai, H. 1993a. Molecular recognition of sweet compounds and bitter compounds. *Pept. Chem., 31st,* 329–32.

Nakamura, K. and Okai, H. 1993b. Do we recognize sweetness and bitterness at the same receptor? *ACS Symp. Ser., 528* (Food Flavor and Safety), 28–35.

Nakashima, K. and Ninomiya, Y. 1994. The IP3 level of mouse circumvallate papilla in response to various taste stimulations. *Nippon Aji to Nioi Gakkaishi, 1*(3), 216–19, in Japanese.

Nofre, C. and Tinti, J.-M. 1989. S: *N*-alkylderivatives of L-aspartic acid or L-glutamic acid α-monoamides, European Patent Application EP 0338946, University of Claude Bernard, October 25.

Pilkova, L. and Pokorny, J. 1992. *Die Nährung, 36*(3), 309–310.

Roy, G., Culberson, C., Muller, G. and Nagarajan, S. 1990. *N*-(Sulfomethyl)-*N'*-aryl ureas. Patent Cooperation Treaty WO 90/11,695.

Roy, G., Culberson, C., Muller, G. and Nagarajan, S. 1991. *N*-(Sulfomethyl)-*N'*-aryl ureas as inhibitor of bitter and sweet taste. In U.S. Pat. No. 4,994,490 to The NutraSweet Co.

Ruiz-Avila, L., McLaughlin, S. K., Wildman, D., McKinnon, P. J., Robichon, A., Spickofsky, N. and Margolskee, R. F. 1995. Coupling of bitter receptor to phosphodiesterase through transducin in taste receptor cells. *Nature (London), 376*(6535), 80–5.

Santhosh, C. and Mishra, P. C. 1994. Electrostatic potential and electrical field mapping of some sweeteners of the suosan series: a search for the structure-activity relationship. *J. Quantum Chem., 51*(5), 335–41.

Schiffman, S. S., Gatlin, L. A., Sattely-Miller, E. A., Graham, B. G., Heiman, S. A., Stagner, W. C. and Erickson, R. P. 1994a. The effect of sweeteners on bitter taste in young and elderly subjects. *Brian Res. Bull., 35*(3), 189–204.

Schiffman, S. S., Gatlin, L. A., Frey, A. E., Heiman, S. A., Stagner, W. C. and Cooper, D. C. 1994b. Taste perception of bitter compounds in young and elderly persons: relation to lipophilicity of bitter compounds. *Neurobiol. Aging, 15*(6), 743–50.

Schiffman, S. S., Gatlin, L. A., Suggs, M. S., Heiman, S. A., Stagner, W. C. and Erickson, R. P. 1994c. Modulators of the adenylate cyclase system can alter electrophysiological taste responses in gerbil. *Pharmacol., Biochem. Behav., 48*(4), 983–90.

Schiffman, S. S., Booth, B. J., Losee, M. L., Pecore, S. D. and Warwick, Z. S. 1995a. Bitterness of sweeteners as a function of concentration. *Brain Res. Bull., 36*(5), 505–13.

Schiffman, S. S., Suggs, M. S., Losee, M. L., Gatlin, L. A., Stagner, W. C. and Bell, R. M. 1995b. Effect of lipid-derived second messengers on electrophysiological taste responses in the gerbil. *Pharmacol., Biochem. Behav., 52*(1), 49–58.

Schmale, H., Ahlers, C., Blaeker, M., Kock, K. and Spielman, A. I. 1993. *Ciba Found. Symp.,* 179 (Molecular basis of smell and taste transduction), 167–85.

Shiba, T. 1990. Chemical structures and functions of bitterness. *Koryo 166,* 73–82, in Japanese; *Chem. Abstr.,* 114: 4921w).

Shinoda, I., Nosho, Y., Kouge, K., Ishibashi, N., Okai, H., Tatsumi, K. and Kikuchi, E. 1987. Variation in bitterness potency when introducing Gly-Gly residue into bitter peptides. *Agric. Biol. Chem., 51*(8), 2103.

Spielman, A. I., Huque, T., Whitney, G. and Brand, J. G. 1992. The diversity of bitter taste transduction mechanisms. *Soc. Gen. Physiol. Ser. 47* (Sensory Transduction), 307–24.

Spielman, A. I., Huque, T., Nagai, H., Whitney, G. and Brand, J. G. 1994. Generation of inositol phosphates in bitter taste transduction. *Physiol. Behav., 56*(6), 1149–55.

Spielman, A. I., Nagai, H., Sunavala, G., Dasso, M., Hugue, T. and Brand, J. G. 1995. Second messenger assays. *Exp. Cell Biol. Taste Olfaction,* 203–10.

Spillane, W. J. and Walsh, M. R. 1996. Synthesis and taste analysis of iminylsulfamates (hydrazonylsulfonates) of *N*-benzaldehyde, *N*-acetophenone, *N*-benzophenone and *N*-isobutyrylphenone. *Food Chem.*, *55*(3), 265–9.

Tagashira, M., Watanabe, M. and Uemitsu, N. 1995. Antioxidative activity of hop bitter acids and their analogs. *Biosci.*, *Biotech.*, *Biochem.*, *59*(4), 740–2.

Takahashi, M., Nakamura, K. and Okai, H. 1995. Can we obtain strong sweetness according to sweet peptide model? *Pept. Chem.*, *32nd*, 57–60.

Tarnura, M., Miyoshi, T., Mori, N., Kinomura, K., Kawaguchi, M., Ishibashi, N. and Okai, H. 1990a. *Agric. Biol. Chem.*, *54*, 1401–1409.

Tarnura, M., Mori, N., Miyoshi, T., Koyama, S., Kohri, H. and Okai, H. 1990b. *Agric. Biol. Chem.*, *54*, 41–51.

Tepper, B. J. and Genillard-Stoerr, A. 1991. Chemosensory changes in aging. Trends, in *Food Sci. and Technol.*, *2*(10), 244–6.

Tinti, J.-M. and Nofre, C. 1991a. In *Sweeteners: Discovery, Molecular Design and Chemoreception* (Walters, D. E., Orthoefer, F. T. and DuBois, G. E., eds), 88–99, American Chemical Society.

Tinti, J-M. and Nofre, C. 1991b. In *Sweeteners: Discovery, Molecular Design and Chemoreception* (Walters, D. E., Orthoefer, F. T. and DuBois, G. E., eds), 206–213, American Chemical Society.

Tomlinson, J. B., Ormrod, I. H. L. and Sharp, F. R. 1995. A novel method for bitterness determination in beer using a delayed fluorescence technique. *J. Inst. Brew.*, *101*(2), 113–18.

Wackerbauer, K. and Balzer, U. 1992. Hop bitter principles in beer. Part. 3. Influence of non-isohumulone bitter principles on beer quality. *Brauwelt*, *132*(16/17), 734–7, in German.

Whitney, G. and Harder, D. B. 1994. Genetics of bitter perception in mice. *Physiol. Behav.*, *56*(6), 1141–7.

Yamamoto, T., Shimura, T., Sakai, N. and Ozaki, N. 1994. Representation of hedonics and quality of taste stimuli in the parabrachial nucleus of the rat. *Physiol. Behav.*, *56*(6), 1197–202.

Yamazaki, T., Benedetti, E., Kent, D. and Goodman, M. 1994. Conformational prerequisites for the sweet taste of dipeptides and dipeptide mimics. *Angew. Chem.*, *106*(14), 1502–17, in German; *Chem. Abstr.* 121: 177948h.

Yasukawa, K., Takeuchi, M. and Takido, M. 1995. Humulone, a bitter in the hop, inhibits tumor promotion by 12-*O*-tetradecanoylphorol-13-acetate in two-stage carcinogenesis in mouse skin. *Oncology*, *52*(2), 156–8.

Yokomukai, Y., Cowart, B. J. and Beauchamp, G. K. 1993. Individual differences in sensitivity to bitter-tasting substances. *Chemical Senses*, *18*(6), 669–81.

CHAPTER 3

The Evolution of in vitro Taste Sensors

GLENN ROY[1]

UNFORTUNATELY, the large volume of sensor assay literature is in Japanese with few or no English translations to date. Only recently have significant remote nonhuman sensing efforts been directed towards quantifying bitter perception. The knowledge gained may ultimately provide assay avenues to bitter reduction and inhibition.

The utility and advances of taste sensors are reviewed to point out their unique potential ability to screen for bitter-tasting substances without recruiting a willing and human taste panel. Some modern assays are being developed to alert a processor of formulations to the bitter taste of certain substances. For example, applications have been investigated for beer, coffee, carbonated soft drinks, tomatoes and amino acids.

A taste sensor consists of a base plate, a taste-detection membrane, and an electrode equipped with a gas-escaping hole and a gas-impermeable membrane to prevent the blockade of electrical circuit caused by the formation of a gas layer. A nonmetal (graphite) electrode is used to increase the compatability with organic sensing membranes. The sensor is said to be useful in determining the taste of beer and carbonated drinks (Santo et al., 1993). Lipid membranes are useful for transforming information about taste substances into electrical signals. A lipid monolayer membrane modified by absorbing to a polymer membrane responds to such electrolytic taste substances as HCl (sourness), or NaCl (saltiness) with large response magnitudes, and such nonelectrolytic substances as caffeine (bitterness) or sugar (sweetness) with sensitivities. The researchers have published their findings in English (Hayashi et al., 1995).

[1]Pepsi-Cola Co., 100 E. Stevens Ave., Valhalla, NY 10595, U.S.A.

An organic membrane that electronically detects the nature of tastants, especially sweetness, is capable of distinguishing equiconcentrations of tastants by measuring electrode potential of the acidic taste of tartaric acid, HCl, NaCl and quinine HCl (Santo et al., 1995). The tastes of amino acids were classified into several groups by the principal component analysis of the output response patterns from a multichannel taste sensor with lipid membranes (Toko et al., 1994b). The changes in electrical potential of biological membranes and liposomes in response to the addition of various bitter substances, amino acids and umami taste substances are reviewed (Kumazawa, 1994).

A review with 24 references discusses differences between taste and olfaction and chemical sensing, self-excited oscillation of potential in a water-oil-water system, effect of taste substances on the hysteresis loop of the kinetic π-A curve (π: surface pressure; A: surface area) of a monomolecular membrane, forced oscillation sensors and applications of carryover phenomena characteristic of nonlinear oscillators for taste sensing (Kumazawa and Yosikawa, 1995).

In order to clarify by what mechanism the lipid bilayer membrane changes its potential under the stimulation of bitter substances, a microscopic model for the effects of the substances on the membrane is presented and studied theoretically. It is assumed that the substances are adsorbed on the membrane and change the partition coefficients of ions between the membrane and the stimulation solution, the dipole orientation in the polar head and the diffusion constants of ions in the membrane. Based on a comparison of the calculated results with the experimental ones, it is shown that the response arises, mainly from a change in the partition coefficients. Protons play an essential role in the membrane potential variation due to the change in their partition coefficients. The present model reproduces the following observed unique properties in the response of lipid bilayers to bitter substances, which cannot be accounted for by the usual channel model for the membrane potential: (1) the response of the membrane potential appears even under the condition that there is no ion gradient across the membrane, (2) the response remains even when the salt in the stimulating solution is replaced with a salt made of an impermeable cation and (3) the direction of the polarization of the potential is not reversed, even when the ion gradient across the bilayer is reversed (Naito et al., 1993).

The expressions of the membrane potential and membrane electrical resistance for lipid membranes were derived according to the theoretical framework of the charged membrane-aqueous electrolyte system. Experimental results on the interaction between taste substances and a lipid membrane were analyzed. The physiochemical parameters determining the reception of a taste substance in the lipid membrane system (i.e., the partition coefficient of a taste substance between the membrane and aqueous phases, the association constant between a taste substance and lipid molecules, and the mobility of a taste substance) were obtained by fitting the theory to the experimental data. The agreement was satisfactory except for high concentration ranges of sweet and bitter substances. The contributions of the phase boundary potential and internal discussion potential to the transmembrane potential were theoretically evaluated for each taste substance (Nomura and Toko, 1992).

Oscillations of electric potential across a liquid membrane consisting of picric acid in nitrobenzene between two aqueous layers were studied. When fully described, the oscillations were found to be characteristic of the structural class of a tastant present in the liquid membrane. Smaller variations were observed in the pattern of oscillations and were apparently related to variations in the taste qualities within that class (Shaw and Coddington, 1995).

Benzalkonium chloride plus alcoholic nitrobenzene containing picric acid/sucrose solutions were studied for their electric oscillation as a liquid membrane. Sweet, sour, bitter and salty substances change the waveform and frequency of the electric oscillation. Qualitative and quantitative determination of berberine hydrochloride was made. The mechanism of the response and modification of the electrode for the method were discussed (Jin et al., 1994a). The discrimination of tastes by a multichannel taste sensor is reviewed with applications to beer, coffee, sake and tomatoes (Toko, 1994a, 1994b, 1995a, 1995b, 1995c).

The characteristic effect of bitter quinine hydrochloride on the electrical potential across a cholesterol lipid membrane is linear in the concentration range of 75–1250 mmol/L. Hence, a quantitative determination of the bitter substance is established. The mechanism of the sensor is discussed (Jin et al., 1994b).

Optical transmittance of a monoolein PVC membrane to five basic taste substances such as sodium chloride, hydrochloric acid, saccharose, L-monosodium glutamate and quinine hydrochloride was examined. The optical transmittance of the membrane was changed remarkably to the taste solutions, and the taste detection threshold of the membrane was improved by adjusting the mixed quantity of the monoolein, cholesterol and plasticizer (Misawa et al., 1995).

A surface photovoltage (SPV) technique has been applied to construct a taste sensor by combining modified Langmuir-Blodgett (LB) methods to immobilize taste-sensitive membranes. The contactless approach of the SPV method provides a simple sensing system with considerable patterning flexibility. Several kinds of artificial lipid membranes were monolithically integrated on a semiconductor surface as taste-sensitive materials, and the surface potential change caused by the reactions with taste substances was detected by scanning a light beam along the semiconductor surface. The uniformly oriented lipid membranes exhibited different responses to five taste substances with high sensitivity and fast response rate. In a preliminary experiment, commercial drinks were identified (Kanai et al., 1994).

An optical-fiber chemical sensor comprises an immobilized liposome membrane with protein and/or biotin-avidin or biotin-streptavidin pair. The biosensor is said to mimic taste- or smell-sensing and may have use in food and cosmetics. In the example, phosphatidylserine and/or phosphatidylcholine liposome encapsulated with bis-(1,3-dibutylbarbiturate)-pentamethine oxonol and alamethicin, biotin and streptavidin was prepared and immobilized on a quartz optical fiber (Umibe et al., 1995).

A new differential measurement method for a light-addressable potentiometric (LAP) sensor has been developed and applied to fabricate an integrated taste sensor with artificial lipid membranes as the ion-sensitive material. The differential measurement procedure is based on a very sensitive and highly stabilized response due to the noise-compensation effect. Sensitivity enhancement is further achieved by cancelling the base component of the differential response current. These techniques improve the sensitivity by at least two orders of magnitude compared to a conventional LAP system. The sensor shows highly sensitive responses to various taste substances, which makes it possible to identify a sweet taste through pattern-recognition routines. Miniaturization of the LAP system is also attained by using a small metal pseudoreference electrode instead of a glass electrode (Sasaki et al., 1995).

Reports have examined dynamic responses of chaotic self-contained oscillations of Millipore membranes infiltrated with dioleyl phosphate (DOPH) and dioleoyl phosphatidylethanolamine, and DOPH and cholesterol to taste substances. The changes in correlation dimensions of the oscillations of these membranes in response to taste substances were different from those observed in a membrane infiltrated with only DOPH. Apparently, more information for taste sensing can be obtained from the dynamic responses of those membranes in addition to the membrane infiltrated with only DOPH (Saito, 1992).

An artificial excitable membrane composed of a porous filter and dioleoyl phosphate exhibits a self-sustained oscillation of the membrane potential in the presence of a salt-concentration difference, pressure difference and d.c. electrode current across the membrane. When an a.c. current was superimposed on the d.c. current, chaotic (irregular) oscillation appeared. The effect of taste substances on the chaotic state was investigated. Characterization of the dynamic behavior of the membrane potential using chaotic attractors was shown to be an effective method of measuring taste (Saida et al., 1992).

The lipid/polymer membranes for transforming information of taste substances into electrical signals are interpreted by computer. The sensor responds quantitatively to different tastants and interprets data when synergistic and suppression effects are present. The sensitivity, reproducibility and durability are said to be superior to those of human taste (Toko, 1995a, 1995b).

A flow-injection analyzer for determining the bitterness of wort and beer was developed. A degassed beer sample or a diluted, filtered wort sample is injected into a carrier stream of water containing a surfactant, immediately followed by simultaneous addition of 6N HCl and isooctane. After mixing in a Teflon tube, the organic phase is separated with a Teflon membrane, and the absorbance at 275 nm is determined. Using this automated method, it is possible to determine the bitterness of wort or beer within a range of 0–50 bitterness units (BU). The coefficient of variation was 0.5% for beer and 0.7% for wort. The correlation coefficient between this method and the ASBC method was 1.000. Values obtained with this method and the ASBC method generally vary by <0.5 BU (Sakuma et al., 1993).

A multichannel taste sensor using a molecular membrane of an amphipathic substance or a bitter substance is provided. The intensity of the basic tastes can be determined by calculation of the membrane potential. Each channel of the sensor is prepared using different lipid membranes to reflect the responsiveness of the human sense of taste to sweetness, bitterness, saltiness and acidity. The determination of the tastes of sucrose, quinine, hydrochloric acid and sodium chloride was demonstrated and the simultaneous equations for determination are given in review. The durable and sensitive sensor for bitterness of a substance such as quinine is constructed, consisting of a base plate carbon electrode and a hydrophobic layer coated on the test solution-contacting surface of the electrode via amido or other chemical linkage. A *n*-octadecyl mercaptan hydrophobic monolayer has detection based on changes in the membrane potential (Santo et al., 1992a, 1995b, 1995c). A bitter-tasting alkaloid sensor is a glass plate coated with stearic acid or plasticized polyvinyl chloride and then with a monolayer of amphoteric molecules such as caprylic acid or other sensing material with hydrophilic groups on the surface. The sensor is dipped into a 1 mM KCl solution for four minutes, the membrane potential is measured, the test substance such as quinine is added to the solution and the membrane potential is again measured for determination of the bitter taste (Santo et al., 1992b, 1992c). A copolymer of vinyl chloride and acrylamide 2-methylpropanesulfonic acid sodium salt (or acryloxyphenylsulfonic acid sodium salt) was formed, added with ELVALOY 742 (plasticizer) and used as a biosensor membrane to detect the bitter taste of quinine HCl. The hydrophobic membrane responds to amphoteric ions (bitter taste) but not hydrophilic ions (acidic and basic tastes) (Shirahama and Toko, 1995).

Tastes and their characteristics are reviewed along with a measurement of taste, especially umami. The review includes basic devices for taste measurement using artificial lipid membranes, multichannel taste sensors, food and beverage recognition and flavor identification and taste detection using excitable lipid membranes (Toko, 1993a, 1993b). A similar lipid membrane-type apparatus for detection of organic chemicals and substances associated with flavors and tastes consists of multiple concave, lipid membrane sensors that identify changes in membrane potential (Matsuno and Mikuriya, 1992).

Astringent substances and pungent substances were studied using a multichannel taste sensor with lipid membranes. The electric potential pattern constructed of eight outputs from the membranes yields information on taste quality and intensity. Pungent substances, such as capsaicin, piperine and allyl isothiocyanate, had no effect on the membrane potentials of the lipid membranes. On the other hand, astringent substances such as tannic acid, catechin, gallic acid and chlorogenic acid changed the potentials remarkably. A principal component analysis of the patterns in electrical potential changes caused by the taste substances revealed that astringency is located between bitterness and sourness (Iiyama et al., 1994a, 1994b, 1995).

Capsaicin pretreatment (1–1000 ppm) daily in humans followed by taste stimuli including three concentrations of NaCl, sucrose, citric acid, quinine and 6-*n*-propylthiouracil resulted in desensitization effects on taste. Following 100 ppm capsaicin desensitization, the magnitude estimates of the two bitter tastes, in particular, and citric acid showed significant decrements. Following 10 ppm capsaicin desensitization or an ethanol control procedure, there were no such effects. Recovery was complete in 1–3 days. It seems possible that the taste decrements are due to effects on both the taste and tactile components of taste, though there is a stronger case for effects on the tactile component (Karrer and Bartoshuk, 1995).

Similar results to artificial lipid membranes and other biological systems have been observed in electrical responses from the roots of adzuki bean. Solutions of five (sour, salty, bitter, sweet, umami) basic taste substances were injected into the aqueous solution around the root of the adzuki bean. The root responded to the substances in different ways by the depolarization of the membrane potential at the root surface. Little change of electric potential was noted by nonelectrolytic taste substances (Ezaki et al., 1993).

Multichannel taste sensors use lipid membranes as a transducer of taste substances and a computer as a data analyzer. The transducer plays the role of transforming taste information generated by chemical substances to electrical potential change. Iiyama has also studied the temperature dependence of the membranes. The responses of lipid membranes were proportional to the logarithm of the concentration at temperatures, indicating the validity of the taste sensors at high temperature. Some evidence suggests lipid membranes are preservable more than half a year at room temperature (Iiyama et al., 1994b).

Multichannel lipid membranes as transducers for the determination of complex taste sensations are described. The feasibility of developing artificial taste compounds using a taste sensor was developed using the discrimination of 40 kinds of Japanese beer (Toko, 1993c).

A test element for detecting bitter flavor of substances such as quinine consists of a base plate and a membrane structure containing lipids and fluorescent substances (e.g., 3,3′-dioctadecyl-2,2′-oxacarbocyanin perchlorate) on the base plate. A test substance is applied to the membrane structure, the test substance adsorbed on the membrane is irradiated with an excitation light, and the fluorescence from the excited fluorescent substance is measured for bitter flavor detection (Minami and Takazawa, 1992a).

The authors also report the sulfa version of a fluorescence-based sensor for bitter taste as a base plate of transparent quartz, a layer of fluorescent substances such as 3,3′-dioctadecyl-2,2′-thiacyanine and a lipid thin layer. A test solution is placed on the surface and measured at 495 nm with excitation at 410 nm (Minami and Takazawa, 1992b).

The taste intensity of the components responsible for the undesirable taste of soybean seeds was measured by electrophysiological methods. Bitterness and astringency in soybeans were caused by soybean glycosides, such as saponins and isoflavones, and the soybean saponin A group contributes most strongly to the undesirable taste. The three electrophysiochemical methods were: membrane potential change across neuroblastoma cells, response to the chorda tympani nerve of the rat and response of the glossopharyngeal nerve of the frog. Only the latter was induced by soybean saponins, forcing the authors to conclude the mechanism of the undesirable tastes caused by soybean saponins was likely to be different from that of the human basic taste mechanisms (sweet, salty, sour and bitter) (Okubo et al., 1992).

Specific adsorption of bitter substances on lipid bilayer-coated piezoelectric crystals has correlated well between partition coefficients of the bitter material (between the aqueous phase and the synthetic lipid bilayer film on the crystal) and the bitter taste threshold concentration in man (Okahata et al., 1987; Okahata and En-na, 1987). The work reviewed is a sensing system of bitter and odorous compounds by using a synthetic lipid-coated quartz-crystal microbalance based on their molecular structure-dependent adsorption on the lipid membrane (Okahata, 1991).

Such simple membrane systems are useful as models of chemoreceptors in biological membranes.

REFERENCES

Ezaki, S., Iiyama, S. and Toko, K. 1993. Electrical response of a root of the higher plant to taste substances. *Kinki Daigaku Kyushu Kogakubu Kenkyu Hokoku Rikogaku-hen, 22,* 31–6, in Japanese.

Hayashi, K., Toko, K., Yamanaka, M., Yoshihara, H., Yamafuji, K., Ikezaki, H., Toukubo, R. and Sato, K. 1995. Electric characteristics of lipid-modified monolayer membranes for taste sensors. *Sens. Actuators, B23*(1), 55–61.

Iiyama, S., Toko, K., Matsuno, T. and Yamafuji, K. 1994a. Responses of lipid membranes of taste sensor to astringent and pungent substances. *Chemical Senses, 19*(1), 87–96.

Iiyama, S., Ezaki, S., Toko, K. and Yamafuji, K. 1994b. Temperature-stability and preservative stability of a lipid membrane electrode in a taste sensor. *Kinki Daigaku Kyushu Kogakubu Kenkyu Hokoku Rikogaku-hen, 23,* 1–6, in Japanese.

Iiyama, S., Ezaki, S., Toko, K., Matsuno, T. and Yamafuji, K. 1995. Study of astringency and pungency with multichannel taste sensor made of lipid membranes. *Sens. Actuators, B24*(1–3), 75–9.

Jin, L., Mao, Y. and Liu, T. 1994a. Study of taste electrochemical sensor response of berberine hydrochloride to mimic biomembrane. *Fenxi Kexue Xuebao, 10*(2), 16–20, in Chinese.

Jin, L., Sun, W., Sun, X. and Fan, Y. 1994b. Study on taste electrochemical sensor. *Fenxi Huaxue,* 22(1), 64–6, in Chinese; *Chem. Abstr.* 120: 293359v.

Kanai, Y., Shimizu, M., Uchida, H., Nakahara, H., Zhou, C. G., Maekawa, H. and Katsube, T. 1994. Integrated taste sensor using surface photovoltage technique. *Sens. Actuators, B, 20*(2–3), 175–9.

Karrer, T. and Bartoshuk, L. 1995. Effects of capsaicin desensitization on taste in humans. *Physiol. Behav., 57*(3), 421–9.

Kumazawa, N. and Yosikawa, K. 1995. Chemical sensors using nonlinear responses. Construction of model systems for taste and smell. *Tanpakushitsu Kakusan Koso, 40*(4), 427–39.

Kumazawa, T. 1994. Initial reception of taste. Receptor and change in membrane potential. *Denki Kagaku oyobi Kogyo Butsuri Kagaku, 62*(3), 196–201, in Japanese; *Chem. Abstr.* 121: 76254y.

Matsuno, G. and Mikuriya, K. 1992. Lipid membrane-type sensor for detection of chemicals. In JP 04,215,052 to Yokogawa Electric Corp.; *Chem. Abstr.,* 118: 168012j.

Minami, K. and Takazawa, Y. 1992a. Test elements for detection of substance with bitter flavor. In JP 04,215,042 to Sekisui Chemical Co., Ltd., *Chem. Abstr.,* 117: 248128w.

Minami, K. and Takazawa, Y. 1992b. Fluorescence-based sensor for bitter taste. In JP 04,340,444 to Sekisui Chemical Co., Ltd., *Chem. Abstr.,* 118: 164703n.

Misawa, K., Yakubo, M., Arisawa, J. and Matsumoto, G. 1995. Fundamental study of fiber-optic chemical sensor. *Hokkaido Kogyo Daigaku Kenkyu Kiyo, 23,* 289–96, in Japanese.

Naito, M., Sasaki, N. and Kambara, T. 1993. Mechanism of the electric response of lipid bilayers to bitter substances. *Biophys. J.*, *65*(3), 1219–30.

Nomura, K. and Toko, K. 1992. A theoretical consideration of interactions between lipid membranes and taste substances. *Sens. Mater.*, *4*(2), 89–99.

Okahata, Y., Ebato, H. and Tagushi, K. 1987. Specific adsorption of bitter substances on lipid bilayer-coated piezoelectric crystals. *J. Chem. Soc. Chem. Commun.*, *1987*, 1363.

Okahata, Y. and En-na, G.-I. 1987. Electric responses of bilayer-immobilized films as models of a chemoreceptive membrane. *J. Chem. Soc. Chem. Commun.*, *1987*, 1365.

Okahata, Y. 1991. Molecular recognition on synthetic lipid membranes. *Maku*, *16*(1), 26–33, in Japanese; *Chem. Abstr.* 114: 242909t.

Okubo, K., Iijima, M. and Kobayashi, Y. 1992. Components responsible for the undesirable taste of soybean seeds. *Biosci., Biotechnol., Biochem.*, *56*(1), 99–103.

Saida, Y., Matsuno, T., Toko, K. and Yamafuji, K. 1992. Taste detection using chaos in excitable lipid membrane. *Sens. Mater.*, *4*(3), 135–44.

Saito, M. 1992. Dynamic responses at artificial membranes composed of a mixture of lipids to taste substances. *Sens. Mater.*, *4*(2), 73–9.

Sakuma, S., Kikuchi, C., Kowaka, M. and Mawatari, M. 1993. A flow-injection analyzer for determining the bitterness of wort and beer. *J. Am. Soc. Brew. Chem.*, *51*(2), 51–3.

Santo, K., Toko, K., Hayashi, K., Ikezaki, H., Higashikubo, R. and Sato K. 1992a. Multichannel taste sensors. In JP 04,297,863 to Anritsu Corp.; *Chem. Abstr.*, 118: 58467d.

Santo, K., Toko, K., Hayashi, K., Ikezaki, H., Higashikubo, R. and Sato, K. 1992b. Construction of a sensor for bitterness and other flavors. In JP 04,324,351 to Anritsu Corp.; *Chem. Abstr.*, 118: 164702m.

Santo, K., Toko, K., Hayashi, K., Ikezaki, H., Higashikubo, R. and Sato, K. 1992c. Biosensor for bitter taste. In JP 04,238,263 to Anritsu Corp.; *Chem. Abstr.*, 118: 3391u.

Santo, K., Toko, K., Hayashi, K., Ikezaki, H., Higashikubo, R. and Sato, K. 1995. Organic membranes for taste sensor. In JP 07 05,147 to Anritsu Corp.; *Chem. Abstr.*, 122: 212569e.

Santo, K., Toko, K., Hayashi, K., Ikezaki, H., Higashikubo, R., Sato, K., Minatoguchi, T., Kobayashi, Y. and Takasugi, K. 1993. Taste sensor. In JP 05 34,311 to Anritsu Corp.; *Chem. Abstr.*, 119: 4432r.

Sasaki, Y., Kanai, Y., Uchida, H. and Katsube, T. 1995. Highly sensitive taste sensor with a new differential LAPS method. *Sens. Actuators, B*, *25*(1–3), 819–22.

Shaw, P. and Coddington, J. M. 1995. Possible prediction of taste quality using a liquid membrane. *Biophys. Chem.*, *55*(3), 209–13.

Shirahama, K. and Toko, K. 1995. Hydrophobic polymer membrane for taste sensor. In JP 07 83,873 to Anritsu Corp.

Toko, K. 1993a. Taste sensors. *Kagaku to Seibutsu*, *31*(6), 352–3, in Japanese; *Chem. Abstr.*, 119: 93800z.

Toko, K. 1993b. Taste environment created by a taste sensor. *Shokuhin Kogyo*, *36*(4), 66–75, in Japanese; *Chem. Abstr.*, 119: 201891r.

Toko, K., Kikkawa, Y. and Yamafuji, K. 1993c. Quantification of taste using a multichannel taste sensor with lipid membranes. *Kyushu Daigaku Chuo Bunseki Senta Hokoku* (Publ. 1994), *11*, 17–22, in Japanese; *Chem. Abstr.*, 121: 33410d.

Toko, K. 1994a. Taste sensors. *Baiosaiensu to Indasutori, 52*(2), 101–5, in Japanese; *Chem. Abstr.*, 121: 132417t.

Toko, K. 1994b. Multichannel taste sensor using electric potential changes in liquid membranes. *Biosens. Bioelectron.*, *9*(4–5), 359–64.

Toko, K. 1995a. Taste sensor. *Nippon Kagaku Kaishi* (5), 334–42, in Japanese.

Toko, K. 1995b. Taste sensor. *Charact. Food,* 337–401. Edited by Gaonkar, A. G., Elsevier.

Toko, K. 1995c. Food sensors. *Kagaku Kogaku, 59*(11), 801–2, in Japanese.

Umibe, K., Myamoto, H., Saito, M. and Kato, M. 1995. Optical fiber immobilized liposome encapsulated with dye sensitive to membrane electric potential for biosensor. In JP 07 35,691 to Oki Electric Ind., Co. Ltd.

CHAPTER 4

Newer Chemical Identification of Bitter Principles and Their Sources

GLENN ROY[1]

BITTERNESS is a taste modality well known to man since the earliest days of recorded civilization. Survival historically appeared to be dependent on the realization that ingesting a bitter taste warned of possible poisoning. Evolution and adaptation of a sensitivity and negative hedonic response to bitter taste evolved, we assume, mainly through inadvertently foraging alkaloids found in many plant species. We will read that foraging monkeys "tell" a different story, and some bitter citrus products have health benefits.

Numerous food and beverage products possess bitter substances from nature. Several bitter palatables are masked by the consumer-at-home or producer-at-plant with sugar or numerous other cariogenic, caloric monosaccharides: tea (powdered), coffee (instant, decaffeinated, processed), cocoa (powdered), nuts (packaged, loose, bulk), chocolate (chocolate substitutes), beverages (beer, soft drinks, fruit and vegetable juices, ciders, wines, spirits) and bakery products. Recently, sugar alcohols have gained popularity despite laxation at high ingestion levels. Other bitter palatables presented at restaurants may be "unconscientiously" masked by the consumer or producer with low-fat, low-calorie (celluloses and gums) "dressings": avocado, brussel sprouts, bitter horseradish, bitter almonds, clover, white wine, olives, buckbean, rosemary, watercress, cardoon, vermouth, bamboo shoots, wort, vanilla, cereals, rapeseed flour, lettuce, rhubarb, dried peas, peanuts, cucumbers, balsam pear, gentian, centaury, kahweol, garden sage, corn salad, endive, chicory, artichoke leaves, dandelion, benedict, yam, saffron, guarirobe, palmheart, rapini, squeezed

[1]Pepsi-Cola Co., 100 E. Stevens Ave, Valhalla, NY 10595, U.S.A.

poppy, soybeans, sarsaparilla, cascarilla and whey products. The recent global support for a "5-a-day good health in vegetables plan" should be educating consumers of the importance of the natural phytochemicals in our produce. My favorites are rapini and endive. More palatable bitter taste-masking products and ingredients are here and to come.

With the recent death of Hans-Dieter Belitz, general scientific writings have diminished in the literature of new bitter principles found in nature. Dr. Belitz was the pioneer in preparing such comprehensive natural product reviews, followed by Dr. Russell Rouseff of the Florida Department of Citrus. Since the Belitz and Rouseff review of bitter principles, additional research has identified newer principles which impart a bitter taste when masticated. Some have proven health benefits. New product development should attempt to mask taste rather than remove healthful natural products. Yet, should we ingest borderline toxins that are toxic to cancerous cells in our body? Is the ingestion of a toxic substance abusable? The paradigm of what is toxic and healthful requires continuing FDA examination of phytochemicals. Certain health benefits have been clinically proven in citrus products.

HEALTH BENEFITS

The USDA has mounting evidence to show that the bitter principles of juices may be good for you as they behave in a controlled fashion to detoxify your body. The American Institute of Cancer Research reports similar findings. Citrus limonoids such as nonbitter obacunone and the bitter glucoside of nomilin induce synthesis of glutathione-S-transferase (a detoxifying enzyme) in the liver and small intestine of mice, as well as inhibit the benzo(a)pyrene-induced tumors of the forestomach in mice. Citrus fruit and juice, as a source of these natural products, may provide chemopreventive benefits against carcinogenesis when consumed on a regular basis (Lam et al., 1989). Numerous chemicals from plants (phytochemicals) are being found to have an inhibitory or slowing effect on breast cancer initiation (Figure 1).

The effects of grapefruit juice and naringenin on the activity of the human cytochrome P450 isoform CYP1A2 were evaluated using caffeine as a probe substrate. In vitro naringenin was a potent competitive inhibitor of caffeine 3-demethylation by human liver microsomes (Ki = 7–29 μm). In vivo grapefruit juice (1.2 L per day containing 0.5 g/L naringin, the glycone form of naringenin) decreased the oral clearance of caffeine by 23% (95% Cl: 7–30%) and prolonged its half-life by 31% (95% Cl: 2–44%) (n = 12). The authors conclude that grapefruit juice and naringenin inhibit CYP1A2 activity in man. However, the small effect of caffeine clear-

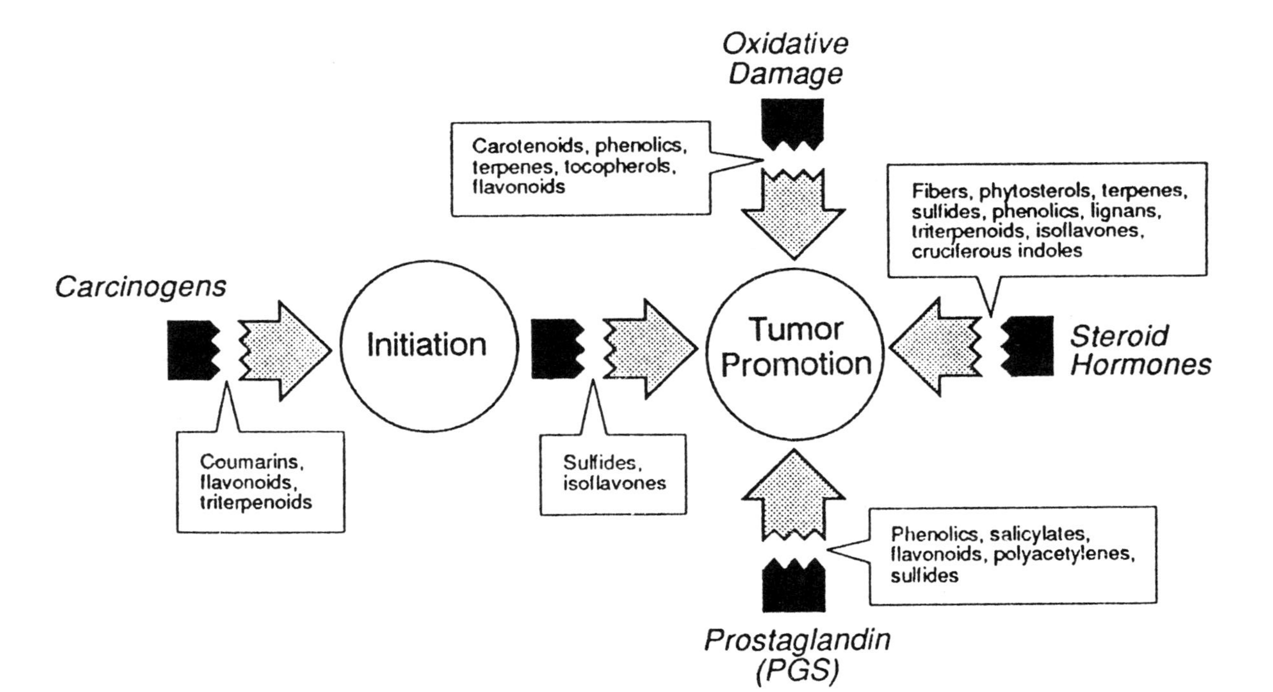

Figure 1 The dietary phytochemicals that can slow or halt metabolic pathways associated with breast cancer.

ance in vivo suggests that in general the ingestion of grapefruit juice should not cause clinically significant inhibition of the metabolism of other drugs that are substrates of CYP1A2 (Fuhr et al., 1993).

The increase in concentration of many drugs when administered concomitantly with grapefruit juice has been attributed to inhibition of cytochrome P450 enzymes by naringenin, aglycone of the grapefruit naringin. However, formation of naringenin after ingestion of grapefruit juice had not been proven. In a study, urine was collected from healthy adults for 24 h after administration of 20 mL grapefruit juice (621 μM naringin) per kilogram body weight. Analysis of HPLC aliquots treated with and without glucuronidase treatments quantified naringenin and its glucuronides after 2 h at 0.012–0.37% and 5.0–57%, respectively, of the molar naringin dose. No naringin nor naringin glucuronides were found.

In additional investigations, low concentrations (<4 μM) of naringenin glucuronides, but neither naringin nor naringenin, were found in plasma samples from other subjects given grapefruit juice. The metabolism of naringin to naringenin occurred during 24 h of incubation (37°) in three of five feces samples tested. The data suggest that cleavage of the sugar moiety, presumably by intestinal bacteria, is the first step of naringin metabolism. Naringenin formation is thought to be the crucial step in determining the bioavailability of naringenin, which undergoes rapid glucuronidation. The pronounced interindividual variability of naringin kinetics provides a possible explanation for some of the apparently contradictory results of drug interaction studies with grapefruit and naringin (Fuhr and Kummert, 1995).

Taking the oral immunosuppressant drug cyclosporin with grapefruit juice is said to increase the levels of the drug in blood (Yee, 1995). The researchers believe the flavonoid compounds in grapefruit juice inhibit the cytochrome P450 oxidase system in the intestines, which degrades cyclosporin. Similar inhibitors used are expensive drugs such as diltiazem or ketoconazole, so grapefruit juice provides a least expensive regimen. This treatment cuts in half the almost $7000 yearly cost of cyclosporin for organ transplant patients who take the drug (Anon., 1995). Naringin had a marked protective effect against carbontetrachloride induced hepatotoxicity in rats. This was manifested by low serum values of γ-glutamyl transferase (GGT), alanine aminotransferase (ALT), aspartate aminotransferase (AST) and bilirubin. Futhermore, the histopathological changes induced by CCl_4, which showed a ballooning of the hepatocytes and accumulation of lipids in the cytoplasm, were markedly ameliorated by treatment with naringin (Badria et al., 1994). Also in isolated rat hepatocytes, the protein phosphatase inhibitor okadaic acid exerts a strong inhibitory effect on autophagy, which can be partially overcome by certain protein kinase inhibitors, like the isoflavone genistein. From 55 tests, naringin and several other flavonone and flavone glycosides (prunin, neoeriocitrin, neohesperidin, apiin, rhoifolin, kaempferol-3-rutinoside; naringenin, eriodictyol, hesperitin, apigenin) offered virtually complete protection against the autophagy-inhibitory effect of okadaic acid.

Some dentifrice compositions have been identified with rutin or its glycosides and essential oils for treatment of periodontal disease. They suppress active oxygen formation in the body upon infection with causative bacteria and prevent degradation of collagen and worsening of inflammation. Rutin (50 μM) and limonene (1 μM) showed 60–80% inhibition of active oxygen formation by PMA-stimulated polymorphonuclear leukocyte of rats vs. 38.3% by single treatment with rutin (50 μM) (Okada, 1995). Citrus natural products may be useful tools to study intracellular protein phosphorylation and may be suggested for potential therapeutic value as protectants against pathological hyperphosphorylation, environmental toxins, or side effects of chemotherapeutic drugs (Gordon et al., 1995).

Oral treatment in rats of naringin (400 mg/kg), 60 minutes before absolute ethanol was the most effective antiulcer treatment. Subcutaneous administration of indomethacin (10 mg/kg) to the animals treated with naringin partially inhibited gastric protection, but prostaglandin E2 determination did not show any increase in prostanoid levels. The contents of gastric mucus and total proteins were not significantly modified. These results show that naringin has a cytoprotective effect against ethanol injury in the rat, but this property appears to be mediated by nonprostaglandin-dependent mechanisms (Martin et al., 1994).

Flavonoids and limonoids of therapeutic interest may be recovered from buntan (*Citrus grandis* Osbeck) peel-boiled water. Firstly, the flavonoids are crystallized by cooling to 2° and separated, then purified by chromatography on HP-20, XAD-2 and silica gel C-300 with gradient elution using methanol, methylene chloride and ethyl acetate. Naringin was obtained in 98% purity in one fraction. Limonoids in another fraction included deoxylimonin, ichangin, deacetylnomilin, limonin, nomilin and obacunone (Cai et al., 1995).

The research of Hasegawa et al. (1988) at the USDA in Pasadena, California, and other Japanese colleagues predict that biotechnology will solve the bitterness problem posed by anticancer limonoid compounds in citrus juices. They plan to create citrus plants that will accumulate the levels of nonbitter, health-promoting limonoids by inserting an alternative biosynthetic pathway in the fruit tissues. At least two specific genes have been identified for genetic manipulation (*Food Chemical News,* 1992), and a review by The Kuchinotsu Branch, Fruit Tree Res. Station in Nagasaki, Japan, is available in an abridged English version. The 52 reference review is on physiology and genetics of bitterness of citrus fruits, e.g., freezing-induced bitterness and on (an improved) analytical method for (identifying) flavonone glycosides, especially flavonone neohesperiodosides (Matsumoto, 1995).

According to trade sources, bitter foods are a key ingredient for health and overall well-being (Anon., 1988). This activation of basic metabolic responses has prompted NatureWorks, Inc. of Agoura Hills, California, to develop Swedish Bitters Herbal Supplement Capsules. This "precisely formulated herbal supplement provides the benefits of bitter foods without any bitter taste in an easy-to-use form." It has no sugar, artificial color, nor preservatives. Thus, perhaps it makes healthy sense to taste-mask the bitter perception.

Humulone, one of the bitters in hops, was isolated from the female flowers of *Humulus lupulus*. This substance has inhibitory activity against 12-*O*-tetradecanoylphorol-13-acetate (TPA)-induced inflammation. At 1 mg/mouse, humulone inhibited markedly the tumor-promoting effect of TPA (1 μg/mouse) on skin tumor formation following initiation with 7,12-dimethylbenz(a)anthracene (50 μg/mouse). Furthermore, humulone inhibited arachidonic acid-induced inflammatory ear edema in mice (Yasukawa et al., 1995).

Hop bitter acids, humulones and lupulones, were shown to have potent DPPH radical scavenging activity (RSA) and lipid peroxidation inhibitor activity (LIA). Yet, 5-acetyl lupulones and 4-methyl lupulones had more potent LIA than native lupulones but no RSA. This result indicates that the β,β''-triketone moiety of the lupulones has LIA (Tagashira et al., 1995).

Phenolic substances are found to be responsible for some bitterness in foods and beverages. Their sources, chemistry, pharmacological effects and nutritional applications have been reviewed (Shahidi and Naczk, 1995). The bitter substances in a wide variety of food plants and wine have also been reviewed (Herrmann, 1992). Catechins and other polyphenolics possess both astringent and bitter tastes. Epigallocatechin gallate has a threshold of 0.011% (110 ppm) (Oguni and Kozuka, 1995). These said health-promoting substances have been implicated for the lower rates of cardiac disease mortality among people drinking wine regularly in certain European populations and referred to as the "French paradox." The polyphenolics in red wines have been shown to prevent copper-catalyzed oxidation of human low-density lipoprotein (LDL), which transports plasma cholesterol. This antioxidant activity was related to the major phenolic compounds such as gallic acid, catechin, myricetin, quercetin, caffeic acid, epicatechin, rutin, cyanidin and malvidin but not to resveratrol analyzed by HPLC and GC-MS. Red wines inhibited oxidation by 46–100%, whereas white varieties achieved only 3–6% inhibition. The active ingredients are catechin and quercetin and work synergistically with the other flavonoid and polyphenolics found in wines. Wine consumption in moderation (1–2 glasses per day) is purported to be a preventive atherosclerosis treatment (Frankel et al., 1995).

There are some species of alleged medicinal plants in Thailand and Tanzania, it is said, whose shoots apparently sick adult, wild chimpanzees have been seen eating. Subsequently, a female adult's health returned to normal. It was assumed that the bitter principles from *Vernonia extense* D.C. and *Vernonia amygdalina* possess physiological activity. The latter species is a shrub widely distributed over savanna woodland in tropical Africa. The shrubs have known antitumor and antifeedant properties as well as being a tonic food in west Africa. Sterol glucosides such as vernoniosides were isolated and found to be of a bitter taste threshold compared to quinine sulfate (Ponglux et al., 1992; Ohigashi et al., 1991).

A review discusses the structures and antiparasitic activity of the sesquiterpene lactones and steroid glycosides, site-specific localization and seasonal change in their amounts and significance of ingestion (Ohigashi, 1995). Ten of the sesquiterpene lactones found in *Vernonia amygdalina* were tested for their antitumoral and antibacterial activities in vitro against leukemia cells (P-338 and L-1210) and against gram-positive bacteria (*Bacillus subtilis, Micrococcus lulea*). Several of the compounds exhibited the desired activity (Jisaka et al., 1993). Antiparasitic activity tests of the sesquiterpene lactone constituents (vernodalin, vernolide, hydroxyvernolide) and new stigmastane-type steroid glucosides (bitter-tasting vernonioside A1-A4 nonbitter-tasting vernonioside B1-B3) supported the hypothesis that Mahale chimpanzees control parasitic-related diseases by ingesting the pith of the plant. The bitter constituents may play an important role as signals to the ingester, guiding their choice of the appropriate plant, plant part, and possibly also as signals that help to control the amount of intake (Koshimizu et al., 1994).

Bitter substances such as sesquiterpenes, diterpenes, pregnane esters, lanosterols, triterpenes and iridoids of plant origin can be used in cosmetics with medicinal properties for antibacterial skin preparations (Nemet and Then, 1994). Bioflavonoids (naringin, hesperidin and neohesperidin) appear to improve mechanical and cosmetic properties of keratin fibers for hair, eyelashes and nails (Dubief et al., 1995).

BITTER AS DETERRENT

Entomologists are fairly certain that bitter substances in nature act as feeding deterrents. Some compounds that are bitter-tasting to humans, both alkaloidal (quinine, quinidine, atropine and caffeine) and nonalkaloidal (denatonium benzoate, sucrose octaacetate and naringin) deterred feeding and oviposition by *Heliothis virescens* in laboratory and field cage experiments. Preliminary electrophysiological studies of gustatory sensilla on the ovipositor of the species provided evidence of three neurons, one of which is responsive to sucrose. Response of this neuron may be inhibited by quinine and denatonium benzoate (Ramaswamy et al., 1992).

Denatonium benzoate, the most bitter substance known to man, is currently recognized as a means to prevent ingestion of ethyl alcohol intended for industrial use. The addition of 30 ppm denatonium benzoate (DB) to engine coolant (ethylene glycol) and windshield wash (methanol) is also being considered as a deterrent for human ingestion of the substances. At this level, a human taste panel assessed the products as intolerable. Despite international attempts to improve the safety of these products through better labeling and packaging, accidental and intentional ingestions continue as a source of worldwide poisoning and DB is a viable taste deterrent approach to safety (Jackson and Payne, 1995).

The editor does not recommend ingestion of potently bitter denatonium benzoate for any benefit. The bitter aversity was measured in Wales, where school children were administered Bitrex up to 10 ppm in orange juice. This substance does have potential in denaturing poisons. Bitrex, MacFarlan Smith Ltd. (Scotland; Robeco Chem. Inc., NYC, distributor) brand of denatonium benzoate may prevent the >750,000 nondrug poisoning cases in children under age six per year. Ethyl alcohol is denatured at one ounce per 1600 gallons of alcohol. Carol Berning of Procter and Gamble demonstrated among 28 children (18–23) and 80 children (2–4) that Bitrex in dilute dishwasher detergent is so aversive as to "significantly reduce the likelihood of an ingestion involving multiple swallows." A more bitter substance has been identified. Atomergic Chemetals Corp. in Plainview, Illinois, has found Vilest™ (as in "too vile") to be five times the bitter taste of DB or 1 part per 100 million. The substance is claimed to be a denatonium saccharide. Current applications have included warding off foraging animals, rat, dog and wolf (Blum and Hollander, 1988).

Many plant defensive chemicals are bitter to humans. Because of this taste characteristic, and because bitter compounds are often toxic, such substances and the plants that contain them are regarded as generally unpalatable to wildlife. To test the hypothesis that herbivores are indifferent to bitter tastants, the responsiveness of guinea pigs (*Cavia porcellus*) to denatonium benzoate, denatonium saccharide, limonene, L-phenylalanine, naringin, quebracho, quinine and sucrose octaacetate was investigated. Only quinine and sucrose octaacetate, slightly but significantly, reduced feeding. The findings are inconsistent with the notion that herbivores generally avoid what humans describe as bitter taste (Nolte et al., 1994).

BITTER PRINCIPLES AND SOURCES

BEER

Beer bitters have been allegedly ascribed to hop bitter acids. Data are presented from various breweries in Germany for the bitterness of light-colored German beers. Bitterness does not always correlate with the content of bitter acids, and furthermore, bitterness intensity increases during the summer until the spring of the following year or later (Weyh, 1990). The Hans Pfuelf Institute of Hop Research in Wolznach, Fed. Rep. Germany, has reviewed the development of new hop varieties and their bitter acid and flavor compound composition (Maier and Narziss, 1991). The factors affecting beer bitter taste and texture are reviewed (Narziss, 1995).

First wort hopping beers in two breweries scored higher in tastings and exhibited higher bitter yields, even though they were milder in taste. Thus, it is important to monitor bitter substance quality, rather than aim at reducing the amounts of α-acids present. Wort and beer aroma spectra showed differences between conventional hopping and first wort hopping, with the latter exhibiting lower content of hop aroma compounds linalool, terpineol, geraniol and humulene epoxide (Preis and Mitter, 1993).

Hop bitter acids (*cis*- and *trans*-isomers of isocohumulone and isohumulone) were separated by preparative liquid chromatography and their organoleptic impact was determined. The relationship between perceived bitterness and hop bitter acid concentration can be described by an exponential function, the value of the exponent varying from taster to taster. Some hop bitter acids differ significantly in their bitterness intensity not explained on the basis of differences in temporal behavior. Cyclodextrin complexation of hop-derived compounds offers a convenient way to introduce hop character into beer with reduced sensory impact (Hughes and Simpson, 1994).

Centrifugal partition chromatography was used for the preparative separation of bitter acid from a crude supercritical carbon dioxide extraction of hop cones. The main α-acids, humulone, cohumulone and adhumulone were obtained pure in one chromatographic run with the system toluene-0.1 M triethanolamine-HCl pH 8.4 in water. The two-phase system was optimized for pH and the effect of ethylene glycol on the separation was investigated (Hermans-Lokkerbol and Verpoorte, 1994).

Since nonisohumulone bitter compounds (NIBC) are largely not extracted with supercritical carbon dioxide during the manufacture of hop aromas, and may thus represent a loss to the development of a full bitter flavor when such extracts are employed in beer brewing, their effect on beer physiochemistry and organoleptic properties was studied in various experimental beers. As expected, the deployment of aroma or aged hops resulted in an increase in beer NIBC levels compared to CO_2 extracts. The bitter properties of beers prepared with CO_2 extraction residues were only marginally different from those prepared with CO_2 extracts, however, indicating the NIBC was of little significance for product bitterness. This was further confirmed in experiments in which NIBC and isohumulones (IH) extracted from beers prepared with aroma hops were independently added to beers prepared without hop aromas: NIBC exhibited only a very weak bitter potential. Contrary to experiments with IH, NIBC exhibited no positive influence on beer head quality (Wackerbauer and Balzer, 1992). The production of hop-derived bittering agents, the relative advantages of the agents, their bitterness, their analysis in beer and wort and their effects on foam stability and light stability of beer are reviewed (Goldstein and Ting, 1994).

Alcoholic and low-alcohol beverages develop bitter tastes by oxidation during malt roasting. In high-alcohol beverages, the bitter taste is largely masked by the alcohol content, but in low-alcohol brews this benefit is unmasked. Antioxidants such as ascorbic acid (Vitamin C) at 20 ppm may be added to the boiling wort, cooled and yeast added to develop a beer flavor without the production of alcohol. The yeast is then removed from the wort (Schur and Sauer, 1990). A review describes alcoholic beverages and their bitter substances (Moriwaki, 1993).

The established technique of time-resolved fluorescent spectroscopy has been applied to the rapid measurement of beer bitterness. This novel application utilizes the unique long-lived fluorescence properties of the lanthinide, europium. Europium ions have the ability to selectively chelate β-tricarbonyl structures such as the isohumulones. The resultant complex is measured by irradiating the sample at a specific wavelength and measuring the intensity of the long-lived emission. The bitterness method measures the total contribution of iso-alpha acids and related congeners. A linear relationship between standard iso-alpha acid solutions and emission intensity was found. However, in beer the effects of color and other interfering compounds were significant. Decolorization of the beer did not improve the correlation with bitterness units (Tomlinson et al., 1995).

No change in conductometric values for bitterness during seven days of storage of ethanol-pure resin hops extracts at 40° was observed. The content of α-bitter acids is described, where that of iso-α-bitter acid content increased. Total α-bitter acid content analyzed by HPLC was stable during one to four days and decreased by 0.4% after seven days (Biendl, 1995).

Four proposed precursors of hop bitter acids (alkylacyl, aryltriols) were synthesized and used in development of TLC, HPLC and GC chromatograms to prove their presence in cone extracts of several hop cultivars (Fung et al., 1994). The organoleptic characteristics of beer are mainly determined from the bitter-tasting iso-α-acids, which in the brewing process are formed from the α-acids occurring in hops. Yet, co- and isohumulones exhibit higher solubilities in worts and beer compared to other α-acids and iso-α-acid homologs. Thus, *n*-, iso-, and adhumulone levels in sediment and yeast increased during the brewing process, leading to an enrichment of soluble isocohumulones and iso-α-acids in the worts and beer. Experiments with pure bitter acid preparations confirmed that cohumulones give a better bitter yield than *n*- and adhumulones (Wackerbauer and Balzer, 1992).

The history and analytical chemistry of beer bitter acids is reviewed (De Keukeleire et al., 1992). Quantification of the individual iso-α-acids is not straightforward, but recent results obtained by liquid chromatography and micellar electrokinetic chromatography look promising. Supercritical fluid extraction (SFE) is evaluated and optimized for the enrichment and fractionation of the essential oil and the bitter principles of hops (*Humulus lupulus*), both of which contribute to the flavor of beer. The bitter principles, the humulones and lupulones, are analyzed by miniaturized liquid chromatography and micellar electrokinetic chromatography (Verschuere et al., 1992).

Czech research has investigated how the sensorial properties of beers were influenced by the mode of hopping, particularly with regard to aromatics and bitter substances in the hops. In addition to bitterness, effects on fullness and sweetness of the finished beers are discussed (Kozak et al., 1994). Oligopeptides of bitter taste (Trp-Phe, Trp-Pro and Leu-Pro-Trp) were identified as isolated from the hydrolyzate of beer yeast residue (Matsushita and Ozaki, 1993).

Sanyo (1985) has a patent masking the bitterness of monoglycerides with enzymic “essence” from glutathione rich enzymes.

WINE

The “French paradox” is founded from an in-depth study of wine components. Polyphenols are important to fruit products in many ways. The mouth-feel of fruit beverages is due largely to polyphenols, particularly to procyanidins at around their threshold levels. Bitterness and astringency are also associated with procyanidins, the balance between these sensations being a function of molecular weight. Color in apple juices and white wines is due to phenolic oxidation; again, the procyanidins are primarily involved, although phenolic acids are important as coupled oxidants since polyphenoloxidase will not act on procyanidins directly (Lea, 1992).

Trained judges rated intensities of astringency and bitterness of four phenolic substances in white wine by scalar and time-intensity methods. By scalar rating, both tastes increased as a linear function of concentration of catechin, gallic acid, grape seed tannin and tannic acid. Thus, monomeric tannins, gallic acid and catechin, are bitter, whereas the polymeric tannic acid and grapeseed tannin were judged to be more astringent than bitter. Similar results were obtained for maximum intensity ratings by time-intensity methods. A slightly longer time to maximum perception was observed for bitterness than for astringency (Robichaud and Noble, 1990).

Benzoic acid derivatives with varying positions of hydroxyl substitution have remarkably differing taste profiles and attributes. When overall maximum intensities were measured for principal taste components of bitterness or sourness, and then prickling or not, stimulus summaries were pictorialized. Bitterness places itself between astringency and sourness, with sweetness being very polarized away from the other taste modalities (Peleg and Noble, 1995) (Figure 2).

An extensive review of the sensory research has been done to characterize the flavan-3-ols role in eliciting bitterness and astringency in wine and the chromatographic techniques used to separate and isolate these compounds (Thorngate, 1993, and Chapter 7 in this book). A lengthy review discusses the phenolic reactions responsible for must and wine discoloration and the effects of phenols on wine flavor (Chevnier, 1990).

The bitter tastes of isoamyl alcohol and propyl alcohol were studied by sensory tests using model white wines. A strong bitter taste was evident in wine with a low concentration of isoamyl alcohol (200 mg/L). In addition, wine containing > 11 mg/L of propyl alcohol showed a significant increase in body, but also had an unpleasant taste because of excess propyl alcohol (Iino and Watanabe, 1994).

The contribution of ethanol to bitterness and of flavonoid phenols to astringency and bitterness in wines has been thoroughly reviewed and Chapter 7 on the topic appears in this book (Noble, 1994). During the storage and maturation of rice wine, there were increases of sweet amino acids

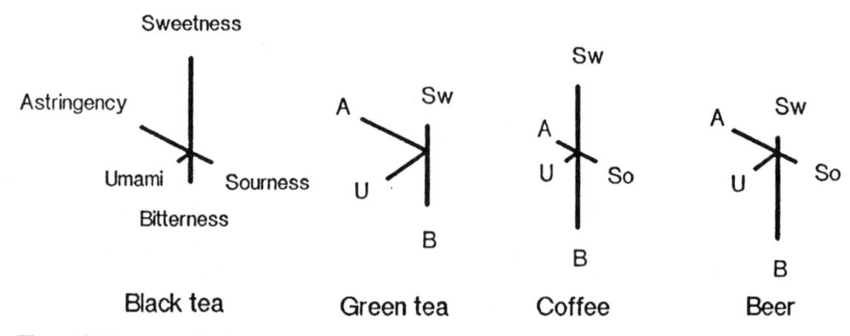

Figure 2 Taste magnitudes as lengths of line in image patterns of black and green tea, coffee, and beer.

and decreases of bitter amino acids. Hence, the pleasant taste of seasoned rice wine may be correlated with its free amino acid concentrations (Wu, 1993).

MISCELLANEOUS BITTER PRINCIPLES AND SOURCES

Bitterness among triterpenes seems almost the normal characteristic and sweetness the exception (Belitz and Wieser, 1985). Triterpenes found in plants are, on occasion, the basis for antifeedant activity to pests. For example, the common cutworm larvae is averted by the presence of amygdalin, phlorizin and sinigrin (Arai et al., 1994).

Structure-taste activity relationships of some triterpenes, the cucurbitane glycosides, show that functional group modifications of the glycosides can change the taste from sweet to bitter or tasteless (Kasai et al., 1987). The large number of tasteless compounds, especially carnosifloside, could have bitterness inhibition or reduction potential (in the same realm as rhoifolin is to naringin), but this test remains to be done. Numerous sweet substances of this family are known. Datiscacin, a cucurbitacin 20-acetate, and other 2-*O*-β-D-glucopyranosyl cucurbitacins have been isolated from the fruits of *Luffa echinata* (gourd of Curcurbitaceae) for the first time (Ahmad et al., 1994). Isolation of the bitter substances 1-tricontanol and celsianol (stigmasta-delta5,9(11)-dien-3-β-ol) from the chloroform extract of the fruits of the bitter variety of *Luffa cylindrica* is reported (Sutradhar et al., 1994). Three new bitter saponins, named mabiosides, from the bark of *Colubrina elliptica* (Rhamnaceae), have been characterized by proton and carbon-13 assignments by 2D NMR spectroscopy (Tinto et al., 1993).

Five bitter-tasting compounds were isolated from cultured oyster, *Crassostrea gigas*, in Korea. NMR and MS spectra suggested cyclic hexa- to heptapeptides including possibly a sulfur containing amino acid. Some nonessential amino acids were thought to be present. Val, Leu and Ile were the most common amino acids present. They are nontoxic to mice at 100 μg/20 mice i.p. and nonantibacterial to *Asp. niger* and *B. subtilis* (10 μg/disk). Other biological studies are in progress (Lee, 1995).

Bitterness of naturally occurring substances seems to be due to a balance between the "bitter unit" and the hydrophobic portions of the molecule. A study based on *Rabdosia diterpenes* indicates that the bitter unit consists of a proton donor DH group and a proton acceptor A group, and perhaps an A-A unit. The bitter substances may approach the bitter receptor site with the bitter unit end oriented into the aqueous phase by hydrogen bonding and the hydrophobic portion aligned into the lipid phase by dispersion forces (Kubo, 1994).

Samples of red-colored lettuce and chicory were assessed for bitterness by sensory analysis and compared with conventional green-colored varieties. The amounts of the sesquiterpene lactones lactucin, 8-deoxylactucin and lactucopicrin and their glycosides were determined using enzyme hydrolysis and high-performance liquid chromatography (HPLC). The level of each of these compounds was compared with the bitterness score found for each sample. The resulting correlation data suggested that the increasing level of one of these compounds, lactucin glycoside, was closely related to the increase in bitterness (Price et al., 1990). Keep in mind, the food pyramid suggests several servings per day of vegetables. How many have you had today?

The bitterness of canned pepper increased significantly when pepper samples were acidified to pH 3.5 and sterilized consecutively. When pepper was processed and sterilized at 100° for 1 h, the bitterness increase represented about 50% in the green mature fruits. The additive heat-treatment at 100° and 2 h caused a slight decrease of bitterness. Storage of sliced pepper fruits at 20° for 24 h did not cause any bitterness increase. The kinetics of bitterness compound production in pepper depending on the medium pH value after heat treatment at 100° for 1 h rose linearly up to a 69% value compared to the control sample as a consequence of the pH value lowering from 6.2 to 2.5 ($R = 0.986$) (Pribela et al., 1994). Additional bitter compounds are created during heat treatment of green sweet peppers and dependent upon the variety and maturity. Immature fruit was exceptionally bitter and compared to quinine hydrochloride as a standard. Treatment of green sweet peppers at 85–100°C for 1 h gave 20% increase of bitterness with subsequent heating to 2 h reducing the bitterness when compared to quinine standards (Pribela et al., 1995).

The main bitter principle of palmyrah (*Borassus flabellifer* L.) fruit pulp has been tentatively identified as a steroidal saponin, which is a tetraglycoside (flabelliferin II) containing two glucose and two rhamnose residues. Bitterness can be removed by the action of naringinase both on the crude bitter principle extracts (containing flabelliferins I and II) and natural fruit pulp. Nariginase released glucose and rhamnose to produce other flabelliferins, two of which occur naturally in palmyrah tuber (Jansz et al., 1994). A bitter principle, tamarindienal, isolated from the tamarind fruit pulp of *Tamarindus indica*, was assigned the structure of 5-hydroxy-2-oxo-hexa-3,5-dienal (Imbabi et al., 1992).

The levels of nutrients, saponins, and bitter substances were compared in wild yam, African yam, cultivated short yam and cultivated long yam. With regard to bitter substances, the levels of crude saponins in cultivated yams were almost similar, but wild yam was more abundant in saponins than cultivated yams. African yam contains high levels of crude saponins; however, it did not contain "β" saponin, and "e"-saponin was higher compared to that in cultivated yams. The bitter taste of African yam was associated more with an ethyl acetate extracted fraction than with a *n*-butanol extract (Im et al., 1995).

Four bitter compounds named lophirosides were isolated from the bark of *Lophira alata* and characterized by chemical and spectral methods. They are rare types of cyanoglucosides containing a cyanomethylene group in the benzoylated and/or cinnamoylated aglycons. The threshold amount for bitterness of each compound was established as 10 mg by a filter-paper tasting method (Murakami et al., 1993).

Cassava parenchyma extraction and chromatography has aided in the identification of a new compound, isopropyl-β-D-apiofuranosyl-(1-6)-β-D-glucopyranoside (IAG). Cassava is a cyanogenic dicotyledon angiosperm of Euphorbiaceae. Linamarin, lotaustralin, citrate, malate and various sugars were also identified. The threshold levels of bitterness of aqueous solutions of linamarin, lotaustralin and IAG were determined, and IAG found to be the major contributor to bitterness. Cassava tubers also possess a sour taste attributed to linamarin with citrate and malate (King and Bradbury, 1995).

A new acetylated ent-atisene glycoside, stevisalioside, has been isolated as a bitter-tasting principle from *Stevia salicifolia* roots. This is the first report of the occurrence of an atisane-type diterpene from the genus *Stevia* (Mata et al., 1992).

The relationship between the content of 6-methoxymellein (6-MM) and the taste of winter carrots harvested in the middle of February was investigated. Sensory evaluation indicated the thresholds of 6-MM for off-flavor and bitter taste were 60 and 100 mg/mL, respectively. An HPLC method was developed to clearly separate 6-MM from other carrot extract components. Although 6-MM was detected in almost all of the carrots tested, the content was lower than the threshold for bitterness. The results indicated that 6-MM has little effect on the taste of winter carrots freshly harvested (Yoshino et al., 1993).

Partially processed carrots of 1/4 lengthwise cut sections were sealed in low-density polyethylene or polypropylene bags into which 100 ppm ethylene was injected and stored at 20°C. Ethylene did not induce bitterness (by isocoumarin) in carrots in either bag during 10 days storage. It is possible that the decrease of oxygen concentration and the increase of carbon dioxide concentration in the bags were responsible for inhibiting the induction of the bitter compound formation by ethylene (Abe and Yoshimura, 1993).

A review of other vegetable and fruit products covers their bitter principles, flavor development, sweetness, acidity, color development, fruit structure and texture, juice and paste production in canned tomatoes and tomato powder, cucumber (*Cucumis melo*) and bur gherkin (*C. anguria*) (Goodenough, 1990).

Sorghum has been found to be a suitable alternative to barley in beer production, but one of the major problems in its use is that of microbial contamination. Bitter leaf (*Vernonia amygdalina)* methanol extract imparts bitterness, exhibits antimicrobial activity and is edible, and it is considered as a possible alternative to hops. Separate steeping of sorghum grains in water, 500 ppm formaldehyde, and 30 mg/L bitter leaf extract for 18 h reduced the bacterial load about 60-, 193- and 849-fold, respectively. Fungal population was virtually unaffected by the bitter leaf extract. Major specific and unique properties of the sorghum malt, wort and beer were not significantly affected by the bitter leaf extract (Okoh et al., 1995).

Stigmasaffron spice is the dry stigmas of *Crocus sativus*. HPLC is the most suitable method for the determination of the bitter principle, picrococin (Sujata et al., 1992). The bitter principle of the toadstool *Tricholoma lascivum* (Agaricales) has been identified as lascivol by chemical and spectroscopic studies (Eizenhoefer et al., 1990).

Certain vegetable products may in fact have demonstrated current toxicity. Consult a reputable edible plants encyclopedia before ingestion of wild products.

Prunasin has been isolated from the stone-fruits of *Prunus jamasakura* and (-)-epicatechin, mandelic acid and new phenylpropanoid sucrose esters with several acetyl groups from those of *P. maximowiczii*. The esters and prunasin were responsible for the very bitter taste of the fruits (Shimazaki et al., 1991).

One triterpenoid and two xanthones were isolated from *Swertia punicea* (Gentianaceae roots). The structures of the triterpenoid proved spectroscopically and chemically to be oleanolic acid and those of the xanthones were decussatin and gentianacaulein. Swertiamarin, sweroside and gentiopicroside, bitter components of *Swertia punicea*, were found in the flowers of the plant. However, amaroswerin and amarogentin, the most bitter components of *Swertia punicea*, were not found by HPLC (Kanamori and Sakamoto, 1993). Unambiguous assignments of the HPLC as well as NMR signals (^{1}H, ^{13}C by 1D and 2D methods) characterized sweroside, amaroswerin, oleanolic acid and β-sitosterol-3β-D-glucoside. Sweroside is now assigned chemically and spectrally (Chaudhuri and Daniewski, 1995).

The bitter taste of amarogentin was evaluated in relation to its use in place of quinine for soft drinks. Based on a recipe for a light bitter lemon beverage, a comparison of amarogentin and quinine bitterness revealed amarogentin to be far more sensitive to changes in sweeteners (Na-saccharin/fructose concentration), acidity (citric acid concentration), lemon flavor and its own concentration. Identical bitter notes for each substance were unattainable. However, amarogentin was evaluated as having a well-rounded and pure bitter taste for the formulation (Busch-Stockfisch and Domke, 1993).

The seeds of olive *Olea europaea* contain the known bitter glucosides salidroside, nuezhenide and a nuezhenide oleoside, together with two new secoiridoid glucosides of uncertain structure containing tyrosol, elenolic acid and glucose moieties (Maestro-Duran et al., 1994).

The relationships of chromatograms of olive oil and the total phenol content, the oil oxidative stability and some of the perceived properties of the virgin olive oil flavor (bitter and astringent) were statistically examined. Solid Phase Extraction (SPE) with adsorbent materials provided a rapid recovery of the phenolic fractions for virgin olive oils originating from Greece, Italy and Spain (Favati et al., 1995).

An assay for the evaluation of the bitter taste in virgin olive oil was developed based on extraction with methanol-water (1:1) of the bitter constituents, including the iridoid, oleuropein. Measurement of the absorbance at 225 nm gave a significant correlation with intensity of bitterness determined by a taste panel on the oil (Rosales et al., 1992).

An automated version for bitterness determination in virgin olive oil is based on flow-injection analysis (FIA) principles and is implemented by coupling a FIA manifold to a solid-liquid retention unit in which the sample matrix is retained on a C18-bonded silica-packed minicolumn, the analytes being monitored by UV spectrophotometry and the retained matrix being eluted in the opposite direction to detection, so it never reaches the detector. The automatic method clearly surpasses its manual counterparts in terms of solvent savings, analytical time (5 vs. 50 min) and labor requirements (Garcia-Mesa et al., 1992). A robotic station has also been proposed, employing monitoring of the absorbance at 225 nm for the separation of the analytes by sorbent extraction (Garcia-Mesa et al., 1993).

REFERENCES

Abe, K. and Yoshimura, K. 1993. Studies on physiological and chemical changes of partially processed carrot. II. Effect of ethylene on the quality and occurrence of bitterness in partially processed carrot. *Nippon Shokuhin Kogyo Gakkaishi, 40*(7), 506–12, in Japanese; *Chem. Abstr.*, 119: 158747j.

Ahmad, M. U., Huq, M. E. and Sutradhar, R. K. 1994. Bitter principles of *Luffa echinata. Phytochemistry, 36*(2), 421–23.

Anon. 1988. Product Alert, Marketing Intelligence Service Ltd., November 14.

Anon. 1995. *Chemical & Engineering News,* May 1, 1995, p. 32.

Arai, N., Hirao, T. and Takano, K. 1994. Plant preferences in the common cutworm larvae, *Spodoptera litura fabricius. Nogyo Seibutsu Shigen Kenkyusho Kenkyu Hokoku, 9,* 221–50.

Badria, F. A., El-Gayar, A. M., El-Kashef, H. A. and El.-Baz, M. A. 1994. A potent hepatoprotective agent from grapefruit. *Alexandria J. Pharm. Sci.*, *8*(3), 165–9.

Belitz, H.-D. and Weiser, H. 1985. Bitter compounds: occurrence and structure-activity relationships. *Food Rev. Intl.*, *1*(2), 271–354.

Biendl, M. 1995. Stability studies on ethanol-pure resin extracts at higher temperatures. Part. 1. *Brauwelt, 135*(18), 872–5.

Blum, M. and Hollander, G. T. 1988. Denatonium saccharide compositions as animal repellent. In US Patent No. 4,661,504 to Atomergic Chemetals Corp.

Busch-Stockfisch, M. and Domke, A. 1993. Sensory evaluation of the bitter taste of amarogentin and its possible exchange for quinine in soft drinks. Part 2. Influence of saccharin-fructose mixture and citric acid. *Z. Lebensm.-Unters. Forsch., 196*(3), 255–8, in German; *Chem. Abstr.*, 119: 179598b.

Cai, H., Hashinaga, F., Ueno, H. and Watanabe, Y. 1995. Recovery of bitter compounds from buntan peel-boiled water. *Nippon Shikuhin Kagaku Kogaku Kaishi, 42*(10), 802–7.

Chaudhuri, P. K. and Daniewski, W. M. 1995. The major principle of *Swertia chirata* . . . and characterization of other constituents. *Pol. J. Chem., 69*(11), 1514–19, in English.

Cheynier, V. 1990. Effect of phenolic compounds on the organoleptic properties of wines. *Bull. Liaison—Groupe Polyphenols, 15,* 275–84, in French; *Chem. Abstr.*, 115: 134022k.

De Keukeleire, D., Vindevogel, J., Szucs, R. and Sandra, P. 1992. The history and analytical chemistry of beer bitter acids. *Trends Anal. Chem., 11*(8), 275–80.

Dubief, C., Braida-Valerio, D. and Cauwet, D. 1995. Use of bioflavonoids as keratinous fiber protecting agents. In Eur. Pat. Appl. EP 0681827 to Oreal S.A.

Eizenhoefer, T., Fugmann, B., Sheldrick, W. S., Steffan, B. and Steglich, W. 1990. Lascivol, the bitter component of *Tricholoma lascivum* (Agaricales). *Liebigs Ann. Chem.*, (11), 1115–18.

Favati, F., Caporale, G., Monteleone, E. and Bertuccioli, M. 1995. Rapid extraction and determination of phenols in extra virgin olive oil. *Dev. Food Sci., 37A*, 429–52.

Food Chemical News. September 14, 1992, p. 23.

Frankel, E. N., Waterhouse, A. L. and Teissedre, P. L. 1995. Principal phenolic phytochemicals in selected California wines and their antioxidant activity in inhibiting oxidation of human low-density lipoproteins. *J. Agric. Food Chem., 43*, 890–4.

Fuhr, U., Klittich, K. and Staib, A. H. 1993. Inhibitory effect of grapefruit juice and its bitter principle, naringenin, on CYP1A2 dependent metabolism of caffeine in man. *Br. J. Clin. Pharmacol., 35*(4), 431–6.

Fuhr, U. and Kummert, A. L. 1995. The fate of naringin in humans: a key to grapefruit juice-drug interactions? *Clin. Pharmacol. Ther. (St. Louis), 58*(4), 365–73.

Fung, S. Y., Brussee, J. van der Hoeven, R. A. M., Wilfried, W. M. A., Scheffer, J. J. C. and Verpoorte, R. 1994. Analysis of proposed aromatic precursors of hop bitter acids. *J. Nat. Prod., 57*(4), 452–9.

Garcia-Mesa, J. A., Luque de Castro, M. D. and Valcarcel, M. 1992. Direct automatic determination of bitterness in virgin olive oil by use of a flow-injection-sorbent extraction system. *Anal. Chim. Acta, 261*(1–2), 367–74.

Garcia-Mesa, J. A., Luque de Castro, M. D. and Valcarcel, M. 1993. Determination of bitterness in virgin olive oil by using a robotic station. *Lab. Rob. Autom., 5*(1), 29–32.

Goldstein, H. and Ting, P. 1994. Post kettle bittering compounds: analysis, taste, foam, and light stability. *Monogr.-Eur. Brew. Conv.*, 22, 141–64.

Goodenough, P. W. 1990. Tomato, cucumber, and gherkin. *Dev. Food Sci.*, *3C*(Food Flavors, Pt. C), 327–50.

Gordon, P. B., Holen, I. and Seglen, P. O. 1995. Protection by naringin and some other flavonoids of hepatocytic autophagy and endocytosis against inhibition by okadaic acid. *J. Biol. Chem.*, *270*(11), 5830–8.

Hasegawa, S., Herman, Z., Bennett, R. D., Fong, C. H. and Ou, P. 1988. Limonoid glucosides in citrus and their possible role to human nutrition. Abstr. from *196th Natl. ACS Meet.*, Los Angeles.

Hermans-Lokkerbol, A. C. J. and Verpoorte, R. 1994. Preparative separation and isolation of three bitter acids from hop, *Humulus lupulus* L., by centrifugal partition chromatography. *J. Chromatogr.*, *A*, *664*(1), 45–53.

Herrmann, K. 1992. Bitter compounds in fruits and vegetables. *Ernaehr.-Umsch.*, *39*(1), 14–19, in German; *Chem. Abstr.*, 116: 213047w.

Hughes, P. S. and Simpson, W. J. 1994. Sensory impact of hop-derived compounds. *Monogr.-Eur. Brew. Conv.*, *22*, 128–40.

Iino, S. and Watanabe, M. 1994. Bitter taste of higher alcohols in wine. *Nippon Jozo Kyokaishi*, *89*(12), 996–8, in Japanese.

Im, S.-A., Kim, Y.-H., Oh, S.-H., Ha, T.-I. and Lee, M.-J. 1995. The study on the comparisons of ingredients in yam and bitter taste materials of African yam. *Han'guk Yongyang Siklyong Hakhoechi*, *24*(1), 74–81, in Korean.

Imbabi, E. S., Ibrahim, K. E., Ahmed, B. M., Abulfutuh, I. M. and Hulbert, P. 1992. Chemical characterization of tamarind bitter principle, tamarindienal. *Fitoterapia*, *63*(6), 537–8.

Jackson, M. H. and Payne, H. A. S. 1995. Bittering agents: their potential application in reducing ingestions of engine coolants and windshield wash. *Vet. Hum. Toxicol.*, *37*(4), 323–6.

Jansz, E. R., Nikawela, J. K., Gooneratne, J. and Theivendirarajah, K. 1994. Studies on the bitter principle and debittering of palmyrah fruit pulp. *J. Sci. Food Agric.*, *65*(2), 185–9.

Jisaka, M., Ohigashi, H., Takegawa, K., Huffman, M. A. and Koshimizu, K. 1993. Antitumoral and antimicrobial activities of bitter sesquiterpene lactones of *Vernonia amygdalina*, a possible medicinal plant used by wild chimpanzees. *Biosci., Biotechnol., Biochem.*, *57*(5), 833–4.

Kanamori, H. and Sakamoto, I. 1993. The components of *Swertia punicea*. *Hiroshima-ken Hoken Kankyo Senta Kenkyu Hokoku*, *1*, 13–16, in Japanese; *Chem. Abstr.*, 121: 200871p.

Kasai, R., Matsumoto, K., Nie, R.-L., Morita, T., Awazu, A., Zhou, J. and Tanaka, O. 1987. Sweet and bitter cucurbitane glycosides from *Hemsleya carnosiflora*. *Phytochemistry*, *26*(5), 1371.

Kasai, R., Matsumoto, K., Nie, R.-L., Zhou, J. and Tanaka, O. 1988. Glycosides from Chinese medicinal plant, *Hemsleya panacis-scandens*, and structure-taste relationships of cucurbitane glycosides. *Chem. Pharm. Bull.*, *36*(1), 234.

King, N. L. and Bradbury, J. H. 1995. Bitterness of cassava: identification of a new apiosyl glucoside and other compounds that affect its bitter taste. *J. Sci. Food Agric.*, *68*(2), 223–30.

Koshimizu, K., Ohigashi, H. and Huffman, M. A. 1994. Use of *Vernonia amygdalina* by wild chimpanzee: possible roles of its bitter and related constituents. *Physiol. Behav.*, *56*(6), 1209–16.

Kozak, J., Koran, M., Totis, T. and Haximo, R. 1994. Effect of hop bitter principles on sensorial properties of beers. *Kvasny Prum.*, *40*(10), 300–5, in Czech.

Kubo, I. 1994. Structural basis for bitterness based on Rabdosia diterpenes. *Physiol. Behav.*, *56*(6), 1203–7.

Lam, L. K. T., Li, Y. and Hasegawa, S. 1989. Effects of citrus limonoids on glutathione S-transferase activity in mice. *J. Agric. Food Chem.*, *37*, 878.

Lea, A. G. H. 1992. Flavor, color, and stability in fruit products: the effect of polyphenols. *Basic Life Sci.*, *59*(Plant Polyphenols), 827–47.

Lee, J-S. 1995. Isolation and some properties of bitter taste compounds from cultured oyster, *Crassostrea gigas. Han'guk Susan Hakhoechi*, *28*(1), 98–104, in Korean.

Maestro-Duran, R., Leon-Cabello, R., Ruiz-Gutierrez, V., Fiestas, P. and Vazquez-Roncero, A. 1994. Bitter phenolic glucosides from seeds of olive (*Olea europaea*). *Grasas Aceites (Seville)*, *45*(5), 332–5, in Spanish.

Maier, J. and Narziss, L. 1991. New hop varieties. Part I. New breeds of aroma varieties from the Hans Pfuelf Institute of Hop Research in Huell. *Brauwelt*, *131*(13/14), 497–502, 515, in German; *Chem. Abstr.*, 114: 244232r.

Martin, M. J., Marhuenda, E., Perez-Guerrero, C. and Franco, J. M. 1994. Antiulcer effect of naringin on gastric lesions induced by ethanol in rats. *Pharmacology*, *49*(3), 144–50.

Mata, R., Rodriguez, V., Pereda-Miranda, R., Kaneda, N. and Kinghorn, A. D. 1992. Chemical studies on Mexican plants used in traditional medicine, XXV. Stevisalioside A, a novel bitter-tasting ent-atisene glycoside from the roots of *Stevia salicifolia. J. Nat. Prod.*, 55(5), 660–6.

Matsumoto, R. 1995. Studies on genetics of bitterness of Citrus fruit and application to breeding of bitterless Citrus cultivars with special reference to the bitterness caused by flavonone glycosides. *Kaju Shikenjo Hokoku, Extra 6*, 1–74, in Japanese/English; *Chem. Abstr.*, 1996: 647890.

Matsushita, I. and Ozaki, S. 1993. Bitter peptides from beer yeast extracts. *Pept. Chem.*, *31st*, 165–8.

Moriwaki, M. 1993. Alcoholic beverages and bitter substances. (Saneigen FFI K.K.), *Foods Food Ingredient J.*, *155*, 64–75, in Japanese; *Chem. Abstr.*, 121: 155776y.

Murakami, A., Ohigashi, H., Tanaka, S., Hirota, M., Irie, R., Takeda, N., Tatematsu, A. and Koshimizu, K. 1993. Bitter cyanoglucosides from *Lophira alata. Phytochemistry*, *32*(6), 1461–6.

Narziss, L. 1995. Beer flavor and the influence of raw materials and technological factors on it. *Brauwelt*, *135*(45), 2286–2301.

Nemet, A. and Then, M. 1994. Bitter substances as new plants agents in medicinal cosmetics. *Olaj, Szappan, Kozmet.* (Spec. Issue), 49–51, in Hungarian.

Noble, A. C. 1994. Bitterness in wines. *Physiol. Behav.*, *56*(6), 1251–5.

Nolte, D. L., Mason, J. R. and Lewis, S. L. 1994. Tolerance of bitter compounds by an herbivore, *Cavia porcellus. J. Chem. Ecol.*, *20*(2), 303–8.

Oguni, I. and Kozuka, H. 1995. Utilization of tea catechins as functional food material. Taste threshold of (-)-epigallocatechin gallate. *Tokubetsu K.H.S-kD.T.D.* 1993 (Publ. 1995), 57–62, in Japanese; *Chem. Abstr.*, 124: 54073y.

Ohigashi, H., Jisaka, M., Takagaki, T., Nozaki, H., Tada, T., Huffman, M. A., Nishida, T., Kaji, M. and Koshimizu, K. 1991. Bitter principle and a related steroid glucoside from *Vernonia amygdalina*, a possible medicinal plant for wild chimpanzees. *Agric. Biol. Chem.*, *55*(4), 1201–3.

Ohigashi, H. 1995. Medicinal use of plant diets by chimpanzees in the wild. *Nippon Shokuhin Kagaku Kogaku Kaishi, 42*(10), 859–68, in Japanese.

Okada, T. 1995. Dentifrice compositions containing rutin or its glycosides and essential oils for treatment of periodontal disease. In JP 07,165,547 to Lion Corp.

Okoh, I. A., Babalola, G. O. and Ilori, M. O. 1995. Effect of methanol extract of *Vernonia amygdalina* on malting and brewing properties of sorghum. *Tech. Q.-Master Brew. Assoc. Am., 32*(1), 11–14.

Peleg, H. and Noble, A. C. 1995. Perceptual properties of benzoic acid derivatives. *Chem. Senses, 20,* 393–400.

Ponglux, D., Wongseripipatana, S., Aimi, N., Oya, N., Hosokawa, H., Haginiwa, J. and Sakai, S. 1992. Structure of two new bitter principles isolated from a Thai medicinal plant, *Vernonia extense* D. C. *Chem. Pharm. Bull., 40*(2), 553–5.

Preis, F. and Mitter, W. 1993. Rediscovery of first wort hopping. *Brauwelt, 133*(42), 2137–8, 2140–4.

Pribela, A., Karovicova, J., Kovacova, M. and Michnya, F. 1994. Influence of pH on bitter compounds production in canned pepper (*Capsicum annum* L.). *Potravin. Vedy, 12*(6), 437–44, in English.

Pribela, A., Piry, J., Karovicova, J., Kovacova, M. and Michnya, F. 1995. Bitterness of sweet pepper (*Capsicum annum* L.). Part 1: bitter compounds production during heat-treatment of sweet pepper. *Nahrung, 39*(1), 83–9.

Price, K. R., DuPont, M. S., Shepherd, R., Chan, H. W. S. and Fenwick, G. R. 1990. Relationship between the chemical and sensory properties of exotic salad crops—colored lettuce (*Lactuca sativa*) and (*Cichorium intybus*). *J. Sci. Food Agric., 53*(2), 185–92.

Ramaswamy, R. S., Cohen, N. E. and Hanson, F. E. 1992. Deterrence of feeding and oviposition responses of adult *Heliothis virescens* by some compounds bitter-tasting to humans. *Entomol. Exp. Appl., 65*(1), 81–93.

Robichaud, J. L. and Noble, A. C. 1990. Astringency and bitterness of selected phenolics in wine. *J. Sci. Food Agric., 53*(3), 343–353.

Rosales, F. G., Perdiguero, S., Gutierrez, R. and Olias, J. M. 1992. Evaluation of the bitter taste in virgin olive oil. *J. Amer. Oil Chem. Soc., 69*(4), 394–5.

Sanyo. 1985. Masking bitterness, irritating taste, and odor of mono-glycerides by adding enzymic essence obtained from glutathione rich enzymes to monoglycerides emulsifying agent, Sanyo Kokusaku Pulp, Japanese Patent JP 198648.

Schur, F. and Sauer, P. 1990. US Patent 4,971,807.

Shahidi, F. and Naczk, M. 1995. *Food Phenolics: Sources, Chemistry, Effects Applications,* Technomic Publishing Co., Inc., Lancaster, PA.

Shimazaki, N., Mimaki, Y. and Sashida, Y. 1991. Prunasin and acetylated phenylpropanoic acid sucrose esters, bitter principles from the fruits of *Prunus jamasakura* and *P. maximowiczii. Phytochemistry, 30*(5), 1475–80.

Sujata, V., Ravishankar, G. A., Venkataraman, L. V. 1992. Methods for the analysis of the saffron metabolites crocin, crocetins, picrocrocin and safranal for the determination of the quality of the spice using thin-layer chromatography, high-performance liquid chromatography and gas chromatography. *J. Chromatogr., 624*(1–2), 497–502.

Sutradhar, R. K., Huq, M. E. and Ahmad, M. U. 1994. Chemical constituents of the fruits of *Luffa cylindrica* (bitter variety). *Bangladesh Chem. Soc., 7*(1), 87–91.

Tagashira, M., Watanabe, M. and Uemitsu, N. 1995. Antioxidative activity of hop bitter acids and their analogs. *Biosci., Biotech., Biochem.*, 59(4), 740–2.

Thorngate, J. H., III. 1993. Flavan-3-ols and their polymers. Analytical techniques and sensory considerations. *ACS Symp. Ser. 536* (Beer and Wine Production), 51–63.

Tinto, W. F., Reynolds, W. F., Seaforth, C. E., Mohammed, S. and Maxwell, A. 1993. New bitter saponins from the bark of *Colubrina elliptica:* proton and carbon-13 assignments by 2D NMR spectroscopy. *Magn. Reson. Chem., 31*(9), 859–64.

Tomlinson, J. B., Ormrod, I. H. L. and Sharp, F. R. 1995. A novel method for bitterness determination in beer using a delayed fluorescence technique. *J. Inst. Brew., 101*(2), 113–18.

Verschuere, M., Sandra, P. and David, F. 1992. Fractionation by SFE and microcolumn analysis of the essential oil and the bitter principles of hops. *J. Chromatogr. Sci., 30*(10), 388–91.

Wackerbauer, K. and Balzer, U. 1992. Hop bitter principles. Part I. Changes in their composition during brewing process. *Brauwelt, 132*(5), 152–5, in German; *Chem. Abstr.*, 117: 68658k.

Weyh, H. 1990. Consistency of beer bitters. *Brauwelt, 130*(46), 2179–82, 2184–8, in German; *Chem. Abstr.*, 114: 60389r.

Wu, Z. 1993. Investigation of free amino acids in rice wine affecting on its taste style. *Shipin Kexue (Beijing), 165*, 20–1, in Chinese; *Chem. Abstr.*, 120: 76014z.

Yasukawa, K., Takeuchi, M. and Takido, M. 1995. Humulone, a bitter in the hop, inhibits tumor promotion by 12-*O*-tetradecanoylphorbol-13-acetate in two-stage carcinogenesis in mouse skin. *Oncology, 52*(2), 156–8.

Yee, G. C. 1995. *Lancet, 345*, 955.

Yoshino, N., Kawaguchi, T., Tokuoka, K., Ishitani, T. and Hirata, T. 1993. 6-Methoxymellein levels in fresh carrots in relation to the sensory quality. *Nippon Shokuhin Kogyo Gakkaishi, 40*(1), 17–21, in Japanese; *Chem. Abstr.*, 118: 253712g.

CHAPTER 5

General Ingredient or Process Approaches to Bitterness Inhibition and Reduction in Foods and Beverages

GLENN ROY[1]

RECENT growth in the market for low-sugar and low-fat foods sometimes revealed bitter attributes previously masked by the sugar and fat content of the products. Although consumers were once comfortable with the use of sugar and fat as a means of masking bitterness, the market for high-sugar/high-fat foods is more and more limited. Fortunately, most of the revealed bitterness with the sugar and fat substitutes is masked by their mouth-feel characteristics. Increasing the viscosity of a product can decrease the rate of diffusion of bitter substances from the food to the taste buds. A final product can be formulated with natural ingredients that confer increased viscosity or encapsulation opportunities, e.g., gelatin, starch, gelatinized starch, lecithins, chitosan, cyclodextrins, liposomes, surfactants and other polymeric materials. Basic taste responses may also be confused with the acidic amino acids, flavors and salts to improve taste.

As was evident in previous discussion on the similarity of sweetness to bitterness, artificial sweeteners are very popular in formulations. Evaluation of the effects of product formulations on bitterness perception will become an increasingly important aspect of the evaluation of new formulations, as certain consumers are averted from certain products with a bitter taste. We have seen an evolution in processing cheese or protein hydrolysate products with purified enzymes, exported enzymes by certain bacteria, ultrafiltration, or adsorbents. This is a commercially proven and effective industry. In the following discussion then, bitterness is reduced or inhibited by the use of ingredients and/or process including natural and artificial sweeteners, flavors, amino acids, certain bacteria, enzymes,

[1]Pepsi-Cola Co., 100 E. Stevens Ave., Valhalla, NY 10595, U.S.A.

cyclodextrin, starch, gelatin, gelatinized starch, PVP, fat, ultrafiltration, adsorbents and inorganic salts.

INGREDIENT APPROACHES

ADDITION OR MODIFICATION OF SWEETENERS AND INHIBITORS

Bitterness inhibition in taste perception research has become a serious consideration worldwide. The tools of research include molecular modeling, sensory analysis, and synthetic chemistry to recognize molecular attributes of some, but certainly not all, bitter compounds within sweetness receptor models. Natural sweeteners such as sugars have long been used for taste masking. To prepare a low-calorie sugar-based sweetener, the sugar is normally diluted with some substance known as a bulking agent. Such substances include cellobiulose, which possesses a bitter taste. However, cellobiulose prepared by the isomerization of cellobiose in water at pH 11 and 80–90°C for 15 minutes as sugar (four parts) and cellobiulose (one part) results in a nonbitter formulation useful for coffee or tea (Ozawa and Hino, 1993). Other artificial sweeteners have a bitter taste (e.g., glycyrrhizin and stevioside, saccharin and acesulfame-K). Some discussion will address that problem.

Pleasant-tasting flavor enhancers based on monoammonium glycyrrhizinate products used at levels of 20–1000 ppm are already available in the form of the "MAGNASWEET" series (MacAndrews & Forbes Co.). Cyclodextrin glucosyltransferase may also be used to improve the sweetening properties of stevioside (Tanaka et al., 1991).

Debittering extracts of *Hydrangea macrophylla* is achieved with γ-cyclodextrin to form glycoside inclusion compounds or by treating the extract with α-glucosyltransferase in the presence of starch. The resultant product is useful for diabetes treatment (Anon., 1995). The presence of rebaudioside A, which is 1.5 times sweeter than stevioside, can mask the bitter taste of stevioside. Generally, the extraction of *Stevia rebaudiana* leaves will give a mixture of diterpene glycosides; these act synergistically to improve the overall sweetening properties of the extract (Crammer et al., 1990). The bitter substances in steviosides may be removed by steam distillation for 30–60 min without degradation of the sweetness of the steviosides (Zhou and Wen, 1994).

The taste-enhancing properties and ability to mask bitterness and saltiness of neohesperidin dihydrochalcone (NHDHC), an artificial sweetener with a long-lasting sweet taste, have been reviewed (Baer et al., 1990). Neohesperidin dihydrochalcone and hesperidin dihydrochalcone 4′-β-D-glucoside have the ability to reduce the perception of bitterness in citrus beverages as well as pharmaceuticals by virtue of the NHDHC lingering (long ET), cooling aftertaste and sweetness (Horowitz and Gentili, 1969, 1977, 1986; Horowitz, 1978). The characteristic lingering cooling aftertaste has a stability optimum from pH 2–6, but increasing temperature promotes hydrolysis. The established use level without health risk is up to 5 mg/kg body wt (Borrego and Montijano, 1995).

The most comprehensive structure-taste activity relationships are found in reviews (Belitz et al., 1979; Belitz and Weiser, 1985), but more trends cannot be easily rationalized. The continuum of review attests to the tremendous variety of bitter compound classes. The degree of bitterness and human selectivity to its perception also varies, and may be compared generally to quinine, with a threshold of 0.0010%, whereas casein, soybean, zein and gliadin proteins have a threshold of 0.0005 to 0.0008%; naringin 0.002 to 0.005%; neohesperidin 0.05%; and limonin 0.0001%. Humans have an evident genetic predisposition to certain bitter substance perception as was discussed. For the present though, consider sweetness and bitterness to be recognized on similar regions of our taste buds.

Competition for the "active" site(s) of a taste bud (more specifically a stimulus receptor) has been fully recognized as the source of bitter taste inhibition specificity. Sweetness perception, in general, also works in this manner. From 1983 to the present, several patents claim sweetness inhibition of all sweeteners. Most notably, Tate and Lyle's Lactisole ((-)-2-(4-methoxyphenoxy)propanoic acid)) is available with Domino sugar for commercial baking applications under the brand name CYPHA. The significantly potent inhibition capability permits use of sugar for the colligative and bulking properties but without sweetness. Lactisole is touted as "nature-identical" since having been found naturally in roasted Colombia Arabica coffee beans (Rathbone et al., 1989a, 1989b). Structurally similar analogs occur in many other foodstuffs and could contribute to their palatability. For example, 2-phenylpropanoic acid (wine, beer); 3-phenylpropanoic acid (wine, grape strawberry and soya bean); 4-methoxyphenylacetic acid (cocoa); 4-methoxybenzoic acid (currants, cocoa and aniseed); 3-(2-methoxyphenyl)propanoic acid (cinnamon). General Foods Corporation has also patented several similar chemical classes of compounds as sweetness inhibitors (Barnett and Yarger, 1985a, 1985b, 1986, 1987; Vellucci et al., 1987) (Figure 1).

The efficacy of any natural sweetness or bitterness inhibitor remains to be substantiated in a food or beverage product of choice. Recent studies suggest the commercial potential for Lactisole as a bitterness inhibitor is limited by the higher levels required for its effectiveness as they exceed the levels permitted by FEMA (Flavour and Extract Manufacturer's Association) for commercial products. Additionally, no bitter-masking potential was observed in trained sensory panels at the lower, permitted levels of 150

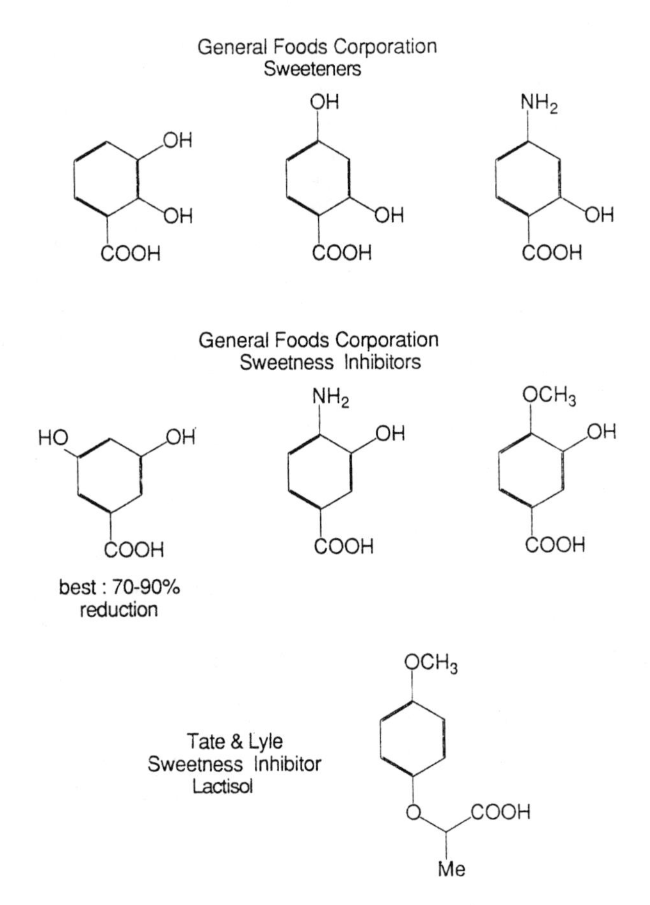

Figure 1 Aryl hydroxy acids as inhibitors of sweet taste.

ppm, with some results suggesting increased bitterness perception! The increase in perceived bitterness intensity and persistence is in agreement with the expected release from mixture suppression. Hence, at levels permitted for food use, a bitter inhibitor application is unlikely. Lactisole is, however, an effective sweetness inhibitor at low levels for both carbohydrate and artificial sweeteners in baking applications. These and other results support the existence of two distinct receptor sites/loci in sweet and bitter chemoreception (Johnson et al., 1994).

The Japanese claim the use of Lactisole and analogs as a method of increasing the bitterness, astringency, sourness and pungency in foods. Evidently, it is possible to reduce the level of use of the bitter, sour or pungent ingredient. Thaumatin and certain lactisole analogs are shown to improve the taste of bitter foods containing salts or fermented foods (Oonishi et al., 1995a, 1995b).

The supply of natural thaumatin is limited as isolated from the fruit of the tropical plant *Thaumatoccus danielli.* The desirable lingering sweetness masks bitterness. Thaumatin can be expressed via transgenic organisms albeit in low yields at present. Ideally, thaumatin could be engineered directly into selected fruit and vegetable crops to improve their flavor and sweetness (Zemanek and Wasserman, 1995). The Japanese have patented somatin, as a thaumatin sweetener, for masking the bitter aftertaste in dipeptide sweeteners (Anon., 1984). The solubility, stability, taste modification and masking, enhancement and synergism, regulatory status and "naturalness" of thaumatin have been reviewed (Anon. 1996).

ADDITION OF FLAVORS

The recent surge in sports drink consumption spawned many new products. Flavors have a major role in sports drink palatability. The flavors can provide a masking effect, which can contribute to increased overall consumption of the beverage. Mineral salts are probably the most difficult to mask due to their astringency, bitterness and saltiness, and with any vitamins also contributing significantly to bitterness. In general, citrus flavors tend to be acceptable with the mineral salts if acidity is adjusted, which tends also to offset the salty character (Peterson, 1992).

The bitter taste of Vitamin B1 derivatives at 0.001–0.05% by wt in beverages is controlled by adding 0.001–10.0% by wt extracts of medicinal plants such as *Epimedium grandiflorum,* Astragali, *Lycium rhombifolium* and ginseng. It is said the bitterness of Vitamin B1 is replaced by the bitterness of the medicinal extract and palatability is still achieved (Yano et al., 1995).

The bitterness of potassium chloride-based salt substitutes can be substantially masked by the addition of 0.1–0.8 mol% sclareolide, a natural diterpenoid flavoring available from International Flavors and Fragrances Inc., Dayton, New Jersey [Figure 2(a)]. Sclareolide is shown to reduce the bitterness of black coffee; 0.01% sclareolide in an aqueous water/ethanol solution was sprayed onto roasted coffee beans to give a final concentration of 0.001 ppm in the dried coffee grounds. The bitterness of products containing citrus fruits, caused by the presence of naringin, can also be reduced by 100 ppm sclareolide (Buckholz and Farbood, 1991). International Flavors and Fragrances also reports the addition of aconitic, gluconic and/or succinic acid, optionally with sclareolide, to improve the or-

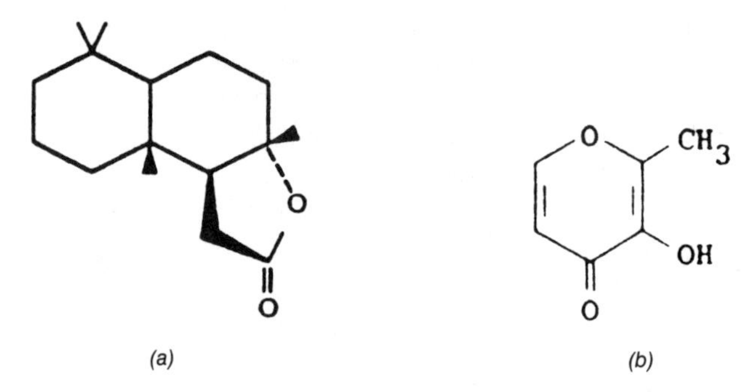

Figure 2 Bitterness modifiers: (a) sclareolide and (b) maltol.

ganoleptic properties of foods, especially by imparting a umami effect and full mouthfeel in the absence of amino acids, sulfur containing amino acids and their salts, pyrrolidonecarboxylic acids and their salts and nucleotides (Buckholz, 1994).

Methyl and ethyl maltol, mouthfeel-enhancing ingredients with a "fresh-baked" odor or creamy, cotton candy flavor (3-hydroxy-2-alkyl-4-pyrone), mask the bitter taste of metal salts such as potassium chloride in seasoned foods [Figure 2(b)]. Cod roes (1 kg) were seasoned with an aqueous solution of sodium chloride (40 g), potassium chloride (40 g), methyl maltol (0.4 g) and water (40 g) and the taste of the KCl was masked (Kato and Yamabe, 1991). Commercial food-grade maltol is available as Veltol™ from Pfizer Food Science. This product would be useful in sports drinks containing vitamins and minerals. Currently, lemon-lime, orange and fruit punch flavors are marketed for improved palatability. Bitterness modifiers as liquid flavors are available from Bush Boake Allen, Inc. They are recommended for use at 0.6–0.8% in the finished product (natural flavor #15555) and 0.1% in finished product (artificial flavor #36734). Both compositions are Trade Secrets and the formula withheld as per 29 CFR 1910.1200 [Bush Boake Allen, Inc., 2711 W. Irving Park Rd., Chicago, IL 60618-9931; Phone (312) 463-7600].

ADDITION OF OTHER TASTANT COMPOUNDS

Trisodium citrate, which alone has a sour taste, is added to a seasoning salt composed of sodium chloride, potassium chloride, calcium chloride, magnesium chloride and ferrous sulfate to mask the bitterness of the potassium chloride (Akai, 1991). KCl bitterness in a 0.5% solution was reduced 78% when mixed with 0.05% calcium lactate as judged by panelists (Nakagawa, 1992a). A beverage containing KCl and NaCl is treated with a mushroom extract that masks the flavor of KCl (Ishida et al., 1993). A salt substitute containing autolysed yeast and ammonium chloride is not very bitter and has an intense salty flavor (Murray and Shackelford, 1991). It is possible that debittering peptides are present in the exogenous enzymic treatment to make the autolysed yeast that reduces the perceived bitterness of the ammonium chloride.

Amino acids and organic acids are also said to reduce the bitterness of KCl when precipitated from aqueous solution (Omari, 1992). A low-salt product available as Cardia™ salt from Applied Microbiology, Tarrytown, New York, was acquired in 1995 from Oriola Oy of the Orion Group, the largest pharmaceutical company in Finland. It was originally marketed as Pansalt in Finland and Japan. A recent controlled clinical trial demonstrated reduced blood pressure in hypertensive patients by ingesting potassium and magnesium salts. The traditional bitter, metallic taste of potassium chloride and magnesium salts in the product is improved by addition of lysine. Sodium chloride is present in low amounts. The addition of 0.025% lambda-carrageenan to an aqueous 0.5% potassium chloride solution reduced the bitterness by 46%. Organic calcium and/or magnesium acid salts are also said to reduce bitterness (Nakagawa, 1992a, 1992b). Nutritional supplements may be improved by converting the bitter-tasting amino acids of the formulation to a palatable acidic salt. Arginine, valine and other polyamines are converted to salts of food-grade acids such as citric or phosphoric acids. Taste tests showed the palatability was greatly improved by acidic pH (Greenberg et al., 1993). Astringent peptide additives such as γ-Glu-Gly, γ-Glu-Asp, γ-Glu-Ala and γ-Glu-Glu were said to mask bitter taste (Kato et alo., 1988).

A delicious tetrapeptide was isolated from beer yeast seasoning and identified as Asp-Asp-Asp-Asp by Edman degradation. Larger quantities made by α-chymotrysin induced oligomerization synthesis (pH 7.5–8.5 at $-25°C$) were further tested to show significant ability to mask bitter taste (Matsushita and Ozaki, 1995). The bitter taste of L-valine, L-tryptophan, L-trileucine and a tryptic casein hydrolysate was diminished to 17–35% of its original strength by the addition of 0.6–14 M excess of glutamic acid. The bitter taste of naringin and limonin (citrus bitters) was not affected by the glutamic acid additions. Quinine and caffeine were reduced 71 and 77% of their original bitter taste, respectively (Warmke and Belitz, 1993). Taurine can reduce bitterness by 50% when added at a concentration of 300 mM to a series of bitter peptides. The use of taurine did not give the sour taste obtained when other acidic amino acids were used (Tamura et al., 1990b). The addition of 1% taurine to soy sauce improved the flavor while also taking advantage of taurine's ability to suppress increases of blood pressure and serum cholesterol levels. Highly salted soy sauces tend to put consumers at risk of hypertension (Moriguchi, 1990). Caffeine at

130 ppm imparts bitter taste with higher temperature, higher pH and polyphenols increasing the bitter perception. Addition of amino acids reduces the bitter taste (Chen et al., 1992). For example, the results of both iodometric and HPLC determination of caffeine in tea infusions were compared. The bitterness of tea infusions was tested by sensory analysis and no significant correlation was found between the content of caffeine and the degree of bitterness, probably owing to the interfering effect of other substances (Lai et al., 1987). The bitterness of kojic acid (an antioxidant) is masked with the addition of amino acids (Wakayama, 1992).

Sodium hexametaphosphate (Sporix™ brand imported from Korean food processors) was reported to debitter tomato juice with added KCl. Sporix also provides buffering and chelating functions and masks other bitter flavor of low sodium products (*Food Eng.*, 1984a). A 50/50 solution of the purine-5′-ribonucleotides, inosine monophosphate (IMP) and guanine monophosphate (GMP) reduces the perception of bitter and sour tastes (Oberdieck, 1987). However, IMP has thermal instability. The difference threshold of umami taste of 0.005% IMP solution in the presence of a 0.05% MSG solution was 0.002%. The taste threshold in sensory evaluation of a 0.005% IMP solution decreased about one-half when heated at 95° for 15 h. Inosine, one of the main products of thermal degradation of IMP, had a bitter taste and the detection varied widely among panelists. Therefore, umami taste used to reduce bitter taste may be prone to develop bitter taste upon heat treatment in the formulation (Kuchiba-Manabe et al., 1991).

Confectionery and nonfrozen foods are bitter with greater than 5 w/w% of added glycerine. Removing bitterness from those food items is accomplished by addition of 0.3 to 1 w/w% sodium acetate (Anon., 1988). The taste of glycerin is improved by the addition of calcium carbonate. Bean jam (500 g) was mixed with glycerin (50 g) and calcium carbonate (5 mg) to provide a mild taste even after freezing and thawing (Narubayashi, 1991).

The addition of persimmon tannin has been reported to improve the normally bitter taste of coffee and to suppress the odors of various food components even though, paradoxically, persimmon itself has a bitter taste (Izumitani and Kazuhiko, 1991). Interestingly, the bitter taste of persimmon can be eliminated by wrapping the fruit in a plastic bag and increasing the temperature to accelerate respiration, while maintaining a high carbon dioxide concentration and a low oxygen concentration in the bag (Kitagawa, 1991). Persimmon placed with ethanol in a plastic container with moisture permeability ranging from 35–500 g/m^2 per 24 h and higher ethanol permeability removes the bitter taste (Koyakumaru and Ono, 1995). An apparatus is described to treat persimmons in gaseous ethanol to eliminate bitter taste (Kiguchi, 1995).

PROCESS APPROACHES

ENZYMIC TREATMENTS TO CONTROL BITTERNESS IN PEPTIDE, PROTEIN AND CARBOHYDRATE OF VEGETABLE AND ANIMAL ORIGIN

Considerable applied research on bitterness reduction and inhibition has taken place in the area of food proteins. New applications of proteins in foods and beverages have presented problems relating to bitterness. Protein hydrolysates find use as a means of nutritionally fortifying cereal foods and as low-cost flavors. Proteins as a source of flavor have also been reviewed (Weir, 1992). Hydrolyzed proteins find many uses in the food industry for dehydrated soups, sauce flavoring, bouillon, processed foodstuffs and animal feed. Seasoning protein powders with sweeteners are commercialized in Japan (Fujii et al., 1995). The utilization of Flavourzyme (an *Aspergillus niger* extract) in the manufacture of flavors from plant and animal proteins is reviewed. The effects of pH, time, temperature, combination with Alcalase (Novo Nordisk A/S), substrate pretreatments and hydrolysate posttreatments are discussed (Nielsen and Jimenez, 1995). Protein sources include casein, whey, vegetable, gluten, zein, grains, fish, meat and tannery waste. The protein hydrolysis process may include acids and/or enzymes. The harsh acid process, while reagent economical and used since the early 1950s, may result in the loss, partial destruction, or racemization of essential amino acids, undesirable side reactions with nonprotein components and formation of toxic by-products such as chloropropanols.

Some of the off-flavors and organochlorine compounds from soybean meal hydrolysates are said to be diminished by adjusting the hydrolysate to pH 7–9 and heating from 70–100°, then acidified to the prior pH. Storage of treated samples versus untreated samples had a pronounced principal-component sensory difference in favor of alkaline treatment (Velisek and Dolezal, 1994). Recent progress in enzymatic processing offers a valuable tool for avoiding the unhealthy disadvantages of acid hydrolysis. The mild conditions and specificity of enzyme processing provide retained nutritional quality, desirable functional properties, easy digestibility, high water solubility and resistance to heat denaturation. Bitterness buildup may be controlled.

Odor development is another problem. Protein hydrolytic products from gelatin, albumin, collagen, keratin, elastin, fibroin, wheat peptide, soybean peptide and casein generally have undesirable odors. The unfavorable odor may be removed by supercritical CO_2/ethanol extraction at above room temperatures without causing heat denaturation (Imamura and Takeuchi, 1994).

The control of bitterness development arises from an understanding of amino acid compositions or Fischer ratio. For example, enzymic hydrolysis may improve the Fischer ratio (the molar ratio of the branched-chain amino acids—leucine, isoleucine and valine—to the aromatic amino acids—phenylalanine and tyrosine) of a peptide fraction. The bitterness intensity of protein hydrolysates has been quantified against principles of hydrophobic parameters (Mogenson and Adler-Nissen, 1988). Bitter-tasting peptides exhibit a high content of hydrophobic amino acids. In 1971 Ney established a Q-rule to predict whether a hydrolysate peptide is bitter or not with its Q value, an average of the hydrophobicity of the side-chain amino acids involved. This rule is said to apply to peptides up to 6000 Da because larger peptides are generally not bitter. Though the Q-rule is reputed to be an accurate predictor of bitterness, the control of bitterness is agreed to be dependent entirely on the proteolytic process choice. Some post-processing may also alleviate bitter taste in the hydrolysates. The bitter flavor that is sometimes generated during hydrolysis may be avoided by using suitable proteases (Kilara, 1985). N-terminal hydrophobic amino acids of peptides are generally the bitter causatives in food and beverage products. Traditional methods of bitterness control include masking, removal of bitter peptides, or prevention by limiting the degree of protein hydrolysis. Masking methods include the addition of polyphosphate, gelatin, glycine or glutamic acid. Enzymatic treatments can provide some control of bitterness by selectively clipping away at peptide chains, and numerous reviews have discussed the enzymatic processes that are in use (*Food Eng.*, 1989).

Sunflower protein, partially hydrolyzed first with kerase (CEPA S.A., Madrid), membrane fractionated and then with certain fractions treated with actinase and further chromatography, gave a marked reduction in the aromatic amino acid content (Bautista et al., 1996). Thus, ultrafiltration may provide useful nutrient materials for patients with liver diseases. Zein was hydrolyzed, first with an alkaline proteinase and then with actinase to liberate the aromatic amino acids, which were then removed by gel permeation chromatography. The hydrolysate was bitter, but further process modification with transglutaminase decreased the bitterness. The process resulted in a product with a Fischer ratio of 20.0 and a yield of 56%, suggesting that it may be economically feasible to use the method on an industrial scale (Tanimoto et al., 1991). Zeins are first hydrolyzed with an alkaline protease to expose the terminal aromatic amino acids, this is followed by hydrolysis with Actinase E to free those terminal aromatic amino acids yielding peptides that were further treated with bacterial transglutaminase to remove the bitterness (Watanabe et al., 1992).

The proteolytic enzyme named molsin has been claimed to yield soybean hydrolysates of low bitterness (Roozen, 1989). Partial hydrolysates of both milk and soy proteins without a bitter taste have been prepared using partially inactivated aminopeptidases (from *Aspergillus sojae*) that had been heated to 50–80°C at a pH of 5–9 in the presence of a polyol such as glycerol or sorbitol (Sproessler and Plainer, 1990). Hydrolysis of milk protein with the untreated enzyme resulted in the preparation becoming bitter at only 4.2% hydrolysis; addition of the partially inactivated aminopeptidase delayed the onset of bitterness to 11.4% hydrolysis. A freeze-dried mixture of proteinase and peptidase isolated from the culture medium of *Rhizopus pseudochinesis* IMA6042 has also been used to produce nonbitter peptides in the enzymatic degradation of acidic casein and soy proteins (Sproessler and Plainer, 1990). Hydrolysis of casein by proteases from *Penicillium citrinum, Aspergillus* spp. and *Rhizopus* spp. results in mildy bitter but acceptable dipeptides (Nakamura et al., 1991). Bitterness scores for a tryptic hydrolysate of casein decreased during incubation with immobilized cells of *Erwinia ananas,* an ice-nucleating bacterium. The aminopeptidase activity of *E. ananas,* combined with its effectiveness at low temperatures, may offer possibilities for use in processes that require low temperatures (Watanabe et al., 1990).

Deltown Specialities (Greenwich, Connecticut, U.S.A.) manufactures protein hydrolysates for use as an energy source and to provide body and desirable mouth-feel characteristics to low-calorie drinks. The bitterness commonly associated with protein hydrolysates is significantly reduced by the chocolate, coffee, vanilla and tomato flavors added to the beverages (Deltown, 1991). Plant protein hydrolysates comprised of peptides of molecular weight 200–4000 that contain aromatic amino acids of 3.6 mol% are prepared and used to manufacture beverages for athletes. The protein hydrolysates that do not have bitter taste may be prepared in various forms such as tablet, granule, powder and foaming agent. Corn protein suspension obtained from a wet milling operation was digested stepwise with a starch degradation enzyme, an alkaline protease of *Bacillus* and a neutral protease, followed by chromatography and freeze drying. The molecular ratio of aromatic and branched amino acids was 1:38.3. A composition for a beverage with good flavor and further fortified with vitamins, minerals, sugars and flavoring agents is disclosed (Kori et al., 1991).

A bitter-like flavor may be noticed when tasting starch. Starch is an all-encompassing term for amylose, amylopectin, starch hydrolysates (maltodextrins, gelatinized starch), cross-linked starches (esters, ethers). Sometimes, carbohydrate residue is sourced from gluten (protein) remaining within a carbohydrate matrix. For those gluten-free and bitter-tasting starches with other oligopeptides (typically <0.5 wt%), debittering is accomplished by incubating the starch with an enzymatically active, protease-containing cell preparation for 6–12 h at 37–40°C. This process provides substantial enzymatic peptidolysis of residual oligopeptides and reduced bitterness. The protease-containing cell preparation is a crude cell lysate extract used without purification from *Lactococcus lactis* (a form of Debitrase™ from Imperial Biotechnology, UK) or *Aspergillus* sp. (Corolase 7093™ from Roehm, Germany) to provide primarily exopeptidase activity. Incubation is stopped before any residual amylase activity can substantially degrade the starch (Haring et a., 1993).

Out of thirteen commercial protease evaluated, a debittering enzyme was recently found to also provide hypoallergenic, palatable protein. Alcalase 2.4L (Novo Nordisk A/S), Papain W-40 and Proleather were effective in decreasing the antigenicity of whey protein concentrates. The hydrolysate obtained with Papain W-40 was not bitter and, therefore, holds promise for preparing hydrolysates for hypoallergenic infant formula (Nakamura et al., 1992, 1993).

Hydrolysis of β-lactoglobulin with porcine trypsin (a neutral protease) for 90 min, ultra-filtered and freeze-dried, provides allergy-free hydrolysates useful for foods and nutrient supplements. The low antigenicity was determined by rat-passive cutaneous anaphylaxis reaction (Kaneko et al., 1993a, 1993b).

Avonmore Foods plc provides a whey protein hydrolysate with a very low bitterness profile and suitable in hypoallergenic infant formula and clinical nutrition products [brand name, Nutravon 181; Avonmore Ingredients, Inc., 307 11th St., Monroe, WI 53566-0453; Phone (608) 328-5700].

Fish and Meat Protein and Hemoglobin

Alcalase or Actinase (alkaline proteases) treatments folowed by adsorptive purification were effective in producing fish protein hydrolysates that are soluble and highly nutritious. Alkaline proteases were more effective than neutral or acidic proteases in the treatment of defatted *Sardinops malanostictus* (Sugiyama et al., 1991).

It was reported that a simplified synthetic extract of scallop muscle satisfactorily reproduced the taste of natural extract. Also, neither the natural nor synthetic extract was bitter, despite their having the arginine concentration approximately six times as high as threshold concentration (50 mg/100 mL). Those results suggest that one or more of the extractive components mask the bitterness of 0.3% arginine. Sensory tests were conducted and determined arginine did not impart significant bitterness to the extract. Yet the Kyoritsu Women's University in Tokyo, Japan, identified common table salt as an effective substance for masking bitterness of 0.03% arginine in a designed, synthetic extract of scallop (Michikawa and Konosu, 1995). Bitter taste of protein hydrolysates is controlled by treating hydrolysates with acidic carboxypeptidase derived from squid liver or this enzyme in combination with exopeptidase derived from animal, plant and/or microorganisms (Gocho et al., 1995).

Fish protein hydrolysates were prepared using minced fillets with Alcalase and papain treatments, from raw herring *(Clupea harengus)* and herring defatted by ethanol extraction, cooking and pressing. Physiochemical, sensory, storage properties and molecular weight estimations were evaluated in spray-dried hydrolysates. Fat extraction before hydrolysis reduced the degree of hydrolysis. Alcalase hydrolyzed samples to a higher degree than papain, and the hydrolysates with papain were more bitter than those made with Alcalase. Ethanol extraction reduced fishy odor to barely detectable levels (Hoyle and Merritt, 1994).

Liquid condiments are manufactured by first incubation of the hot water extract of the residues of dried bonito (kezuribushi) with an endo-protease for 2 h at 50° in neutral conditions for solubilization of proteins, and then with an immobilized exo-protease at pH 3 for decomposition of the bitter ingredients and increase of the umami ingredients. The glutamic acid content was 193.5 mg/mL vs. 7.2 mg/mL for a control condiment without exoprotease treatment (Hiraoka, 1993). The preparation of angiotensin-converting enzyme-inhibiting proteins without bitterness is achieved by extracting dried bonito in hot water to remove soluble proteins, and the insoluble proteins then digested with thermolysin. The hydrolysate is then passed through a styrene divinylbenzene copolymer column to remove 90% of bitter taste (Yasumoto, 1994). The structural requirements for sweetness or decreasing the bitterness of angiotensin-converting enzyme-inhibitory peptides were described from structure-taste relationship with Leu-Lys-Tyr analogs (Kawakami et al., 1995b). Waste meat from lobster (*Panulirus* spp.) was optimally hydrolyzed by a fungal protease and progress measured by liberation of tyrosine. The products hold potential as flavorants (Vieira et al., 1995). A process has been reported recently in which the fish mince is digested by bacterial enzymes without the addition of water, thereby reducing the cost of drying after hydrolysis (Hirano et al., 1991).

Many flavor peptides in raw pork meat and dry-cured ham are generated as a consequence of muscle protein degradation. Low-molecular mass peptides (Mr <3000 Da) of the water-soluble extract from both pork meat and dry-cured ham were fractionated by gel filtration chromatography. The fractions were found to have a wide range of flavors as revealed by sensory testing. Bitterness was detected in the earlier-running fractions (below 1800 Da) and a slightly acid taste in the latter-running fractions (below 1000 Da). A savory ham flavor when assaying dry-cured ham extracts and a brothy/umami flavor in the case of meat were detected at the molecular mass range 1500–1700 Da. Peptide profiling of these desirable fractions by reversed-phase HPLC and free solution capillary electrophoresis showed that they were composed largely of compounds thought to be hydrophilic peptides (Aristoy and Toldra, 1995).

A little-studied source of protein is bovine slaughterhouse blood, which is naturally high in nutritional value, cost competitive, and available in large quantities. INSA of Toullouse, France, has found hemoglobin protein is three times higher in histidine than casein, an essential amino acid for infant formulas. Alcalase (Novo) or Pepsin (Merck) enzyme proteolysis of hemoglobin provides a decolorized permeate. Pepsin-hydrolyzed hemoglobin exhibits a distinct bitter taste and unpleasant odor. The bitter taste is not due to heme or its derivatives, as a decolored hemoglobin hydrolysate has the same flavor. Different separation processes were developed to isolate the bitter fractions using ultrafiltration, reversed-phase chromatography, and organic solvent extraction. Bitter peptides were concentrated with an ultrafiltration YM5 membrane (cutoff 5000, Amicon Grace) to afford permeate. Bovine hemoglobin hydrolysate was eluted with 30–40% acetonitrile from a C18 bonded silica column (Pep RPC, Pharmacia LKB) and Lichroprep (Merck). Bovine hemoglobin hydrolysate was selectively extracted by 2-butanol or adsorbed by a Superose 12 gel filtration column (Pharmacia LKB) to represent a chromatographic peak appearing after total column volume.

Chromatography on Superose 12 is said to constitute an interesting analytical method for detection of bitterness in hydrolysates. The bitter peptide that corresponded to VV-hemorphin 7, i.e., the fragment 32–40 of the β chain of bovine Hb, is first generated during proteolysis, then hydrolyzed by pepsin. The isolated substance exhibited a strong bitterness at 0.25 mM equivalent to 0.073 mM quinine sulfate or 21 mM caffeine (Aubes-Dufau et al., 1995a, 1995b). Porcine blood plasma treated with Alcalase at 73°C for 30 min, then treated with activated carbon, filtered and the filtrate passed through HP 20 resin prepares odorless, bitterness-free, salt-free peptides. The peptides are useful as antiallergenic agents for foods and cosmetics (Kawakami, 1993).

Studies on the relationship between peptide concentrations in beef soup stock and the taste revealed that the amounts of glutamic acid are responsible for good taste, alanine and glycine for sweetness and histidine in peptides of mol wt 10,000–100,000 for bitterness. The components increased as the cooking time from 30 to 120 min at 92° was prolonged (Mega, 1994).

Low-lipid animal proteins may be hydrolyzed and said to provide low dichloropropanol and monochloropropanol content by a combination of hydrolysis with HCl and proteinase. For example, a concentrated pig bone extract containing less than 0.5% lipids was treated with Alcalase in water at pH 8.0 and 55°C for 6 h and then treated with HCl in water at 100°C for 15 h to give a hydrolysate containing less than 50 ppb of dichloro- and 80 ppb monochloropropanol content (Fujii and Kai, 1994).

In practice then, value-added products are derived from low-value substrates normally regarded as waste products. Thus, proteolytic enzymes offer a valuable tool for modifying proteins to improve their functionality, palatability and nutritional properties. The reader is encouraged to explore enzyme ingredients discussed in Chapter 11 by Imperial Biotech. Future innovations, both by the enzymologist and the food technologist, will continue to produce novel uses for enzymes in reducing the bitter taste during food processing.

Legume and Grain

Bitterness- and pungency-free protein hydrolysates from soybean 7S globulin are prepared by hydrolysis of the proteins with trypsin at a 6.5–7.5 pH and not at the optimal pH 8.0. The products are useful as foaming agents and emulsifiers (Uesugi et al., 1994). Wheat gluten was hydrolyzed by six proteases, pepsin, rapidase, actinase, bromelain, bioprase and papain. Peptide fractions possessing molecular weight between 500 and 10,000 were prepared from the gluten hydrolysates with ultrafiltration. Except for Actinase, hydrolysis by the other five proteases elicited strong bitterness. The Actinase hydrolysate showed slight bitterness and umami. Deamidation of the latter peptide fraction in 0.5 N HCl at 120°C for 15 minutes further decreased its bitterness and increased umami. This deamidated fraction was separated into two further fractions with molecular weight 500–1000 and molecular weight 1000–10,000, with the former adding umami to a niboshi soup stock (Ishii et al., 1994a). The addition of the deamidated peptide fraction with an umami flavor to sour and bitter solutions was found to suppress the sour and bitter taste perceptions (Ishii et al., 1994b).

Tartarian buckwheat *(Fagopyrum tataricum)* flour is commonly eaten in East Asian countries. The flour contains high amounts of rutin and is antihypertensive, but its characteristic bitter taste prevents its acceptance in Japan. Bitter components (mainly located to the outside of the grain) could be extracted from the flour by washing with 75 v/v% aqueous ethanol solution. Quercetin and two unknown compounds were found in the bitter ethanolic fraction. These components could also be prevented by heating the flour at >70°. Because these two treatments were determined not to decrease the rutin content, debittered Tartarian buckwheat flour may be useful as a functional food for improving hypertension (Kawakami et al., 1995b). The low digestibility and poor taste of buckwheat protein may be improved by using a protein extraction procedure consisting mainly of temperature-controlled globulin isolation by isolectric precipitation. The resultant protein from tartary buckwheat had no bitter taste (Kawakami et al., 1994). Buckwheat protein is said to be the best known source of high biological value proteins in the plant kingdom and to have amino acid compositions nutritionally superior to most cereals.

Pretreatment by spraying adzuki bean with aqueous neutral polyphenol oxidase or protease at 50°C for 2 h and subsequent pressure cooking removed bitterness (Okazawa et al., 1993).

Cheese, Milk

A review outlines the role of proteolytic enzymes in cheese ripening and flavor development (Bockelmann, 1992). Some emphasis is placed on the proteolytic enzymes from starter bacteria because these, in particular, have been the subjects of recent and current research (Visser, 1993). A hypothesis to account for bitter development in cheddar cheese flavor was advanced many years ago (Lowrie and Lawrence, 1972). Cultures of starter *Streptococci* produce low-molecular-weight bitter peptides by the rennet-induced degradation of casein. The choice of proteinase governs the activity and provides a pool of peptides, some bitter and some not, that over time cascade to a pool of more nonbitter peptides and amino acids. The compositions are extremely complex and structural identifications rare. For example, BPII (cyclo(-Trp-Leu-Trp-Leu-)) is a bitter peptide from protein hydrolysate later found to be cyclo(-Trp-Leu-) (Minamiura et al., 1972; Shiba and Nunami, 1974). Other examples of some of the bitter linear peptides identified in protein hydrolysates and cheese products are represented.

Soy globulin hydrolyzed with pepsin:	Arg-Leu; Gly-Leu; Leu-Lys; Phe-Leu; Gln-Tyr-Phe-Leu; Ser-Lys-Gly-Leu
Zein hydrolyzed with pepsin:	Ala-Ile-Ala; Gly-Ala-Leu; Leu-Val-Leu; Leu-Pro-Phe-Ser-Gln-Leu
Casein hydrolyzed with trypsin:	Phe-Ala-Leu-Pro-Gln-Tyr-Leu-Lys; Gly-Pro-Phe-Pro-Ile-Ile-Val
Cheddar cheese:	Pro-Phe-Pro-Gly-Ile-Pro; Pro-Phe-Pro-Gly-Pro-Ile-Pro-Asn-Ser; Leu-Val-Tyr-Pro-Phe-Pro-Gly-Pro-Ile-Pro

A rapid procedure for preparative isolation of the carboxyl-terminal fragment 193–209 of β-casein, the bitter C-peptide, is described. The C-peptide was preferentially cleaved from β-casein by chymosin, and subsequently purified by ultrafiltration after acid precipitation of residual large β-casein fragments. The C-peptide was strongly hydrophobic but did not interact with other hydrophobic parts of β-casein. Ultrafiltration and dialysis experiments showed that the peptide was probably oligomeric in aqueous solutions with oligomers in equilibrium with lower-molecular-weight species (Vreeman et al., 1994).

A preparative-scale procedure to isolate bitter peptides in cheddar cheese includes water extraction, membrane ultrafiltration, and reversed-phase chromatography. Most of the isolated bitter peptides contained a high level of the hydrophilic amino acids, glutamic acid, glutamine and serine. Hydrophobic amino acids such as phenylalanine and tyrosine were said to be absent in all isolated bitter peptides. The preparative fractions were reincorporated into a full fat and reduced-fat cheese matrix for sensory evaluation. Bitterness perception was more pronounced in the reduced-fat cheeses (Lee and Warthesen, 1996).

Carbon dioxide extraction of an aqueous extraction of Comte cheese removed numerous bitter-tasting N-acylamino acids, diketopiperazines, and nonpeptide compounds such as xanthines (Roudot-Algaron et al., 1993).

The activity of the purified aminopeptidase from *Pseudomonas fluorescens* ATCC 948 on synthetic bitter peptides, bitter hydrolysate of UHT milk proteins, and on the ripening of Italian Caciotta-type cheese indicated nearly complete hydrolysis of the bitter pentapeptide H-Leu-Trp-Met-Arg-Phe-OH and liberated valine from the bitter tetrapeptide H-Val-Pro-Leu-Leu-OH. No cleavage of the proline bond was observed. In UHT milk, in which bitter protein hydrolysate was caused by externally added proteinase, the aminopeptidase produced a concentration of free amino acids about eight times higher than those determined on UHT unhydrolyzed nonbitter milk (211.3 vs. 25.3 μg/mL). Glutamic acid, leucine, methionine, tryptophan, and valine were the most abundant amino acids. The enzyme showed 44% of its maximum activity at the storage temperature of UHT milk (20°). The aminopeptidase was stable during the ripening of Caciotta-type cheese: no activity was lost during two months. After 45 days of ripening, the cheese containing aminopeptidase had a higher amino acid level (990.2 μg/mL) than the untreated control (591.1 μg/mL). The amino acid profile of the treated cheese also reflected the aminopeptidase activity. The specific hydrolysis of peptidic bonds involving the amino acids (leucine, tryptophan and valine), usually identified as major components of bitter peptides, probably indicates a debittering activity of this aminopeptidase (Gobbetti et al., 1995).

Debittering mechanisms have been proposed, whereby wheat carboxypeptidase may be useful for debittering protein hydrolysates in food applications. Imperial Biotechnology Inc., a fermentation company based at London's Imperial College, markets a product called Debitrase™ as a mixture of exopeptidase enzymes that can be classified as typical α-aminoacylpeptide hydrolases (Pawlett and Fullbrook, 1988). The enzyme system has been cleared for use in food products by the British Ministry of Agriculture, Fisheries and Food (MAFF). The system is capable of debittering protein isolates of soy, casein, gluten and whey. Integrated Ingredients (Alameda, California), in partnership with Imperial Biotech Ltd., has introduced the product Accelase DB for direct addition to cheese milk while working synergistically with starter-culture enzymes. A low-fat cheddar cheese product without the enzymes synergy was rated moderate to high in bitterness. Bitterness was rated low with the enzyme synergy. Lower perceived bitterness permits stronger consumer perception of natural cheese flavor.

Sixteen experimental Camembert cheeses were prepared with four strains of *Penicillium camemberti* used alone or in mixed culture with three strains of *Geotrichum candidum.* They were tasted by a panel of 18 selected and trained judges. Nitrogen fractions of these cheeses were analyzed (total N, soluble N, at pH 4.6, soluble N in phosphotungstic acid, ammonia). Proteolytic and aminopeptidase activities were also measured. Bitterness of cheeses is correlated with the concentration of soluble nitrogen at pH 4.6 or concentration of peptides. The ammonia taste is also correlated with the free ammonia concentration. Cheeses inoculated with *Penicillium camemberti* were judged to be more bitter. They also had the highest concentrations of soluble nitrogen and peptides at pH 4.6. *Geotrichum candidum* has a higher aminopeptidase activity than *Penicillium camemberti;* it may be the pathway by which *Geotrichum candidum* would decrease bitterness. Total N concentration and soluble N in phosphotungstic acid were not significantly different among the 16 cheeses (Molimard et al., 1994). The mixed-culture strain products had enhanced dairy product flavors and decreased rancid, moldy, cardboard and plastic flavor notes (Molimard et al., 1995).

A proteinase line of products effective at pH 3–11, Corolase™ from Rohm Tech Inc. (Malden, Massachusetts), was introduced high in endopeptidase and exopeptidase activity for debittering (*Food Eng.*, 1989). The molecular distribution patterns of a whey protein hydrolysate as obtained by gel permeation chromatography show differences between purely endoproteolytic (Corolase N = bacterial proteinase) and endo/exoproteolytic (Corolase PP = pancreatic enzyme) breakdown. The degree of hydrolysis has been adjusted to 15%. Corolase N attacks the β-lactoglobulin fraction less strongly than Corolase PP. The combined endo/exo activity additionally causes a shift of the fractional distribution pattern to the low molecular weight range. The aromatic amino acids detected at 280 nm, which account for part of the hydrophobic amino acids, have increased, whereas the number of peptide bonds, measured at 215 nm, has decreased. This means that potential bitter peptides were hydrolyzed and individual amino acids formed. The hydrolysate obtained with Corolase N tasted bitter, whereas Corolase PP had markedly reduced the bitterness. The exo-activity for aminopeptidase is high specificity for hydrophobic amino acids (Leu, Phe, Tyr), which is conducive to the controlled elimination of bitterness. This hydrolytic specificity is also found with the pure exopeptidase Corolase 7093, which as a heat-stable enzyme can be obtained from *Aspergillus oryzae* by thermal inactivation of the endopeptidase—a patented process—and thus is available in technical quantities. Combinations of exoproteinase hydrolysis with endoproteinase must be controlled in stoichiometry such that the proteinase activity does not gallop ahead of the aminopeptidase activity. The commercial range of enzymes may be applied to proteins of vegetable and animal origin: soy, gluten, potato, yeast, meat, fish and milk such as casein and whey (Moll, 1990). They are distinguished by their high peptidase activity for the hydrophobic amino acids Phe, Leu, Tyr, Pro. The bitter peptides from milk casein are cleaved, and bitterness can be decreased (Umetsu et al., 1983).

Papain and neutral *Bacillus subtilis* proteinase treatments of α-casein make a bitter protein digest whose individual hydrophobic bitter fractions may be tasted as separated by HPLC. When the digest is conducted with food-grade fungal peptidase *(Aspergillus oryzae)*, a change in the HPLC profile was observed, indicating reduced formation of the hydrophobic bitter peptide fractions and more formation of hydrophilic ones. The hydrophilic peptide fractions are said to provide possible novel savory flavors (Gallagher et al., 1994a). Pineapple stem bromelain and a *Bacillus* protease act on casein to produce hydrolysates of very different peptide composition. The *Bacillus* protease action produces a greater variety of hydrophobic hydrolysate containing a greater number of low molecular weight peptides ($<$ 10 kDa), while bromelain action gives hydrolysate with higher molecular weight peptides ($>$ 10 kDa), normally associated with flavor fractions (Gallagher et al., 1994b).

The changes of osmolality and other properties of tryptic and peptic whey protein hydrolysates with various enzyme/substrate ratios were studied. The degree of hydrolysis for the peptic hydrolysate was lower than for the tryptic ones. At pH 4.6, nitrogen solubility index values were much lower and more differentiated than at pH 7.0. The osmolality in all analyzed hydrolysates was maintained below physiological blood plasma osmolality. The hydrophobicity of protein hydrolysates calculated on the basis of amino acid analysis was 4839 J/mol. The bitterness in all hydrolysates was not strong, probably because of the masking effect of salts formed during the hydrolysis. The results indicated a high potency of the hydrolysates for the production of specific dietetic formulas or for protein fortification of acidic beverages (Ziajka et al., 1994).

Two fungal lipases (Palatase 200 from *Mucor miehei* and Palatase 750L from *Aspergillus niger*) alone and combined with a fungal protease from

Aspergillus oryzae (MKC fungal protease) and a bacterial neutral protease from *Bacillus subtilis* (neutrase) to determine the practical application of these enzymes in the maturation of Manchego type cheese. Combinations of proteases and lipases increased proteolysis as measured by indexes of soluble nitrogen and HPLC of free amino acids. There was selectivity for the intense breakdown of β-caseins but not of α-caseins. The exclusive addition of lipases decreased proteolysis and increased soapiness due to excessive release of long-chain fatty acids. Only the cheeses containing *B. subtilis* protease developed bitterness and sticky and crumbly texture because of the intense breakdown of β-casein (Fernandez-Garcia et al., 1994).

The bitterness of casein proteins is allegedly removed by complete hydrolysis of the protein with an immobilized composite exopeptidase of carboxypeptidase and/or aminopeptidase from bovine or porcine pancreas (Ge and Zhang, 1993).

The origin of bitterness in β-casein has been systematically investigated by CD spectra and sensory evaluation of synthetic peptides H-Arg-Gly-Pro-X-Pro-Ile-Ile-Val-OH, where X = L-Phe, D-Phe, LLys, Gly, Glu, L-pyrenylalanine. The results indicate that the location of a hydrophobic amino acid with the L-configuration between two proline residues should be important for this series of peptides producing a bitter taste (Nakatani et al., 1994). Similar findings were reported in German research. Trypsin-digested β-casein A2 from a homozygous cow was separated into 18 peptide fractions representing 97% of the protein source. Only three peptides had a bitter taste and were labeled I49-N68 (recognition threshold 1 mg/mL, 0.45 mmol/L), I49-N97 (recognition threshold 1.5 mg/mL, 0.28 mmol/L), G203-V209 (recognition threshold 0.175 mg/mL, 0.23 mmol/L). Their contribution to the overall bitterness of the casein hydrolysate (2.67 mg/mL) was about 11, 21 and 60%, respectively. Based on structural evaluations of other peptide fragments, it was concluded that in the case of larger peptides neither hydrophobicity nor size is responsible alone for bitter potency, but conformational parameters must be of great importance. Only a part of the structure is actually in contact with the receptor (Bumberger and Belitz, 1993).

Ras cheese samples (Cephalotery) from a local Egyptian market were analyzed for bitter peptides, amino acids, biogenic amines, glycerides and fatty acids in order to evaluate the relation of these components to the detected bitter taste. All cheese samples, either bitter or nonbitter, contained four bitter peptides. Quite bitter and extremely bitter cheeses contained high amounts of the amino acids proline and valine. The biogenic amine tyramine was found in all cheese samples, while some bitter cheeses contained histamine. Extremely bitter cheese contained low amounts of triglycerides and high amounts of free fatty acids, mono- and diglycerides and short-chain fatty acids (Darwish et al., 1994a). Ras cheese samples from the local Egyptian market were analyzed for compositional and microbiological qualities in order to evaluate the relationship between such properties and the detected bitter taste (Darwish et al., 1994b).

NONENZYMIC TREATMENTS TO CONTROL BITTERNESS IN PEPTIDE- AND PROTEIN-CONTAINING FOODS

Extraction

It is suggested that phenolic acids are responsible for the sour, bitter and astringent flavors found in many vegetable proteins. Ethanol washing, heat treatment and acid washing have been used to remove threshold taste levels of phenolic acids, such as ortho-coumaric (20 ppm), para-coumaric (48 ppm) and ferulic acids (90 ppm), thereby reducing the perceived bitterness of corn germ protein flour (Huang and Zayas, 1991). Soy sauce lees (sediments) develop a bitter flavor, when in the brewing of soy sauce glycosides are enzymatically converted to their aglycones. Ethanol washing removes the bitter components, mainly aglycon saponins, allowing the use of the lees as flour substitutes for use in bread formulations (Okubo and Otomo, 1991).

The bitterness of waste hop foliage can be completely removed by solvent extraction to provide a material 23% in proteins. Ten different organic solvents were utilized and the efficiency found to be a function of solid/liquid phase ratio and the initial humidity of the meal. The extract concentrate contains hop resins, waxes, essential oils and dyes, which upon further processing may be of value as well (Grilc and Dabrowski, 1994).

General Foods eliminates the harsh bitter flavor associated with bran cereal by contacting the bran cereal with dried citrus peel (Triani and Meczkowski, 1987). A traditional Egyptian practice for debittering lupin seed flour involves a saline steeping to remove alkaloids. The resulting flour did not show trypsin and α-amylase inhibitor activity or hemaglutinin activity. Lupin seed flour did not show caseinolytic activity after debittering. The in vitro digestibility of debittered lupin seed proteins with pepsin/pancreatin was slightly higher (Rahma and Rao, 1984). Lupin *(L. mutabilis)* debittering may be accomplished by 60–87% isopropanol extraction (Guerrero et al., 1991). In one process, lupin protein blends had protein efficiency ratios equal to those of casein (Univ. of Giessen, 1983). The contribution of saponins to the bitter of dried pea *(Psium sativum)* flour is discussed (Price et al., 1985). Extrusion is a method for debittering partially hydrolyzed pea-milling products (Maack et al., 1984).

Seven sensory characteristics (bitter, green, beany, metallic, astringent, floury and woody) were identified for a slurry of eight samples of milled lupin seed in water. This quantitative assay was applied to the seed of five species of lupin, which included three sweet and five bitter varieties. The levels of both total and individual alkaloids and of total tannin and catechin were determined by the chemical analysis of the seed and related to sensory characteristics. A positive correlation was observed both between the bitter, metallic and woody tastes and between these tastes and total alkaloid content. Individual alkaloids, catechin and total tannin levels were not related to any of the sensory characters. Floury taste was inversely correlated to bitter, metallic and woody tastes, while astringent, green and beany tastes appeared to be independent of the other sensory characteristics (DuPont et al., 1994).

Heat treatment of soybean meal is intended to debitter the beans and eliminate inhibitory substances (Kurz, 1986). A process for producing defatted and debittered soybean meal is described (Steinkraus, 1984). A refined debittered full-fat soy flour (Nurupan) has been applied in various types of chocolate, sugar and baked confection (Anon., 1986). Debittered soybean meal and other components are suitable as substitute for 60% of the cocoa constituents of a chocolate crumb product (Turos, 1982; Miller, 1981). CocoNo™, a caffeine-free chocolate and cocoa pound-for-pound substitute, made from carob and debittered brewer's yeast *(Saccharomyces uvarum)* was introduced by Coors Food Products Company (*Baker's Digest,* 1984; *Food Eng.,* 1984a). Test marketing in Boise, Wichita and Fargo was unsuccessful. However, in large school districts "the kids never knew the difference." Cocoa bitterness is due to an interaction of theobromine and cyclic peptides (Pickenhagen et al., 1975; Ney, 1986). Dried, debittered brewer's yeast from the Pyramid Beer Co. has been used to enrich bread (Alian et al., 1983). The utilization of spent grains from brewery by-products is illustrated as bakery additives and fodder after debittering (O'Rourke, 1980; Kann et al., 1982). The bitter taste of beer yeast is removed by treatment with aqueous alkaline pH 9 to 11 solution (Anon., 1982). White bean flour (25 to 40% protein) is deflavored, debittered and intended for use in bakery products, snacks, snack replacers, and pasta foods (Andres, 1981). The bitterness of quinoa, a hardy and nutritious South American food crop, can be reduced by nonenzymatic means. Quinoa seeds have a higher nutritive value than most cereal grains, and the protein has a high biological value (Meyer et al., 1990). However, before consumption, the bitter and toxic saponins must be removed from the seeds. A "foam height analytical method" is used to determine the saponin content, and therefore the efficacy of the nonenzymatic debittering process involving abrasive dehulling (polishing) followed by aqueous extraction of the seeds (Koziol, 1991). Green banana treated with ethanol, placed in a contained bag and sealed, eliminates the bitter taste (Kitagawa and Urase, 1995).

The bitter principles of fish protein hydrolysate have been removed by extraction (Chakrabati, 1983). Blanched and water-washed (pH 5) sugar beet pulp (fiber) is useful as a noncaloric, noncoloring, bland ingredient in foodstuffs. The process also includes alcoholic extraction to remove bitter-tasting flavor and residual color. The resulting tissue-milled fiber was much improved in a comparison with other nontreated flours, oat flour and α-cellulose used in a 5–6.7% fiber-containing muffin mix, 10–20% fiber-containing cereal, and 1.5% fiber-containing diet drink. Bitterness and texture appear to be the largest components impacting the overall preference (McGillivray et al., 1993).

The main chemical, physical, microbial and sensorial characteristics of a powdered dietary fiber obtained from white seedless Marsh grapefruit were reviewed. The product had a high level of total dietary fiber (65.6%) which is a satisfactory result considering that the sample showed an equilibrated composition of soluble and insoluble fractions. The water-holding capacity was high (9.8 g/g) and the caloric value was low (0.78 cal/g). The microbiological quality of the product was satisfactory and with regard to the sensory quality, the bitter taste was moderate and the color a light brown (Fernandez et al., 1993).

INCLUSION COMPLEXES, CHELATING AGENTS, POLYSACCHARIDES AND ION-EXCHANGE MEMBRANES

Potentially useful spheres of spray-dried starch granules with small amounts of proteins or water-soluble polysaccharides have been reviewed for controlled release of flavor (Zhao and Whistler, 1994).

Frozen or salted herring eggs soaked overnight in an aqueous solution of sodium chloride containing branched cyclodextrin were not bitter (Hosokawa et al., 1993). Soaking herring eggs in aqueous solutions of chelating agents such as EDTA, HEDTA or DTPA for more than 5 h removed the bitterness (Nagabori et al., 1993). Sake is subjected to electrodialysis with cation- and anion-exchange membranes to manufacture bitterness- and astringency-reduced sake. The contents of the sake (ethanol, isomaltose, glucose, galactose, glycerin, propyl alcohol, isobutyl alcohol, isoamyl alcohol, isoamyl acetate and ethyl caproate) are not significantly changed while the contents of succinic acid, lactic acid, malic acid, sodium ion, potassium ion, tryptophan, arginine, lysine, putrescine and cadaverine are reduced (Iwase et al., 1992).

Treating protein hydrolysates with a cross-linked styrene and divinyl benzene adsorptive resin removes color components, odor components, bitter components and certain amino acids (Garbutt, 1993). A strongly basic ion-exchange resin was used to reduce the bitter taste of the fat and sugar substitute polydextrose (Bunick and Luo, 1991). The Pfizer Food Science Group (Groton, Connecticut), now Cultor Food Science, manufactured a nominally palatable polydextrose several years ago, but the sensory properties of the newly marketed version of polydextrose (Litesse™) are well accepted. Low color, low bitterness, high viscosity and low hygroscopicity of polydextrose are achieved by treatment with glucose oxidase, ion-exchange resins, and optionally followed by catalytic hydrogenation (Duflot, 1995).

Carob has been used as a solid ligneous adsorbent of plant origin to debitter juices (Farr and Magnolato, 1981) and whey protein (Farr and Magnolato, 1983; Magnolato, 1983). The bitter taste of certain pharmaceuticals or foods used to control diabetes can be masked by the addition of chitosan, which is derived from the polysaccharide chitin found in the skeletons of crustaceans (Ikezuki, 1990). A therapeutic composition of polyphenol containing tea extracts is useful for renal failure treatment. The bitter taste of the tea extract is masked when the extracts are chitin-impregnated (Yoshimura et al., 1993). Gelatinized starch has been recommended for the debittering of peptides; gentle heating of a starch solution is believed to encapsulate the bitter substances in the gel matrix (Tamura et al., 1990b). Similarly, the bitter taste of aqueous 10% casein hydrolysates can be removed by spray drying with cyclodextrin (Saito and Misawa, 1990). The bitterness and "antisweet" character of gymnemic acid have been successfully eliminated by treating a mixture of starch and gymnemic acid with cyclomaltodextrin glucosyltransferase. The bitterness disappeared, and "antisweet" activity was reduced 15-fold. The addition of γ-cyclodextrin to gymnemic acid was also effective in removing the bitterness and "antisweet" character (Nagaoka et al., 1990). Cyclodextrins have yet to be approved for use as food additives in the USA; however, pharmaceutical applications for the hydroxypropyl cyclodextrins have been approved. A taste-modifying preparation from *Curculigo latifolia,* a Malaysian herb containing curculin, salts, carbohydrates, organic acids and proteins, has been used to mask and potentiate flavors (Kurihara et al., 1990).

CHEMICAL MODIFICATIONS OF BITTER SUBSTRATES

CARBOHYDRATES, PROTEIN, PEPTIDES

Chemical modification of simple sugars and glycosides results in products that are bitter, sweet and occasionally tasteless. Birch found that bitterness in all these molecules appears to be the result of polar as well as nonpolar (lipophilic) molecular features (Birch and Lee, 1976; Birch et al., 1972). Unlike sweetness, which seems primarily associated with the third and fourth hydroxyls of glucopyranosyl structures, bitterness is chiefly associated with hydroxyl groups on carbon atoms 1, 2 and 6, and with the ring oxygen atom. Goodwin of the USDA found tasteless glucosides with extended alkylene side chains (Goodwin and Hodge, 1981). The bitterness of glycyrrhizin-containing solutions (potently sweet licorice extracts) may be reduced by the covalent addition of sugar moieties to glycyrrhizin, using the α-transglucosidase activity of cyclodextrin glucosyltransferase; the glucose residues are donated by starch. The reaction products have only 20% of the bitterness of the initial extract (Shidehara et al., 1990). Various sugar alcohol additions are patented to reduce bitter taste (Anon., 1986).

A number of papers have examined practical treatment strategies of various peptides with a bitterness equivalent to that of 10 mM caffeine (Noguchi et al., 1975; Tamura et al., 1990a, 1990b; Ishibashi et al., 1987, 1988a, 1988b; Shinoda et al., 1987). For example, acetylation of simple bitter hydrophobic amino acids, or increasing their hydrophilicity by the addition of or derivatization with acidic amino acids can be effective (Shinoda et al., 1987). Some mechanisms for reducing bitter taste perception in peptides have been proposed (Ishibashi et al., 1988a, 1988b; Umetsu et al., 1983; Arai et al., 1970) and were previously discussed. Studies of bitter reduction resulted in the assignment of bitterness-producing units composed of stimulating and binding units. The attachment of Gly-Gly residues to the C- or N-terminal end of some bitter peptides (e.g., Arg-Pro-Phe-Phe, 25 times as bitter as caffeine) reduced bitterness in the C-terminal case by one-third to one-half.

This capping technique has been recommended for use in debittering foods, particularly in manufacturing processes, where hydrolysis of proteins releases hydrophobic bitter substances. Unfortunately, BPIa (bovine peptide Ia) from casein hydrolysate (Arg-Gly-Pro-Phe-Ile-Val, 20 times caffeine) was unaffected in taste by Gly-Gly residues placed at the N- or C-terminal or both. BPIc (Val-Tyr-Pro-Phe-Pro-Pro-Gly-Ile-Asn-His, 20 times caffeine) was used to build molecular modeling bitter determinants but not "capped" and tasted. Another bitter bovine β-casein C-terminal octapeptide (Arg-Gly-Pro-Phe-Pro-Ile-Ile-Val) provided a good model to study the altered bitterness/sweetness taste profile in a series of peptides. Synthesis of the octapeptide with modification of the sequence provided a 1000-fold sucrose sweetness with bitterness when the amino acid Glu was substituted for Gly, while Glu substitution for Phe gave only slight sweetness with bitterness. Furthermore, an analog with both Arg and Phe substituted by Glu residues produced sweetness that was 667-fold stronger than sucrose and had no bitterness. These results have provided a model for sweet and bitter taste recognition by a single taste receptor (Takahashi et al., 1995).

The "capping" could prove useful to a quality-minded producer of cheese, for example. The introduction of hydrophilic peptides into the C-terminus of bitter peptides (e.g., endopeptidase-catalyzed reverse reaction, the plastein reaction) was effective for decreasing bitterness (Arai et al., 1970). Yet, the caffeine-equivalent bitterness of Phe-Phe was not reduced in Phe-Phe-Glu-Glu (Ishibashi et al., 1988a).

Chemical acetylation of soy protein hydrolysates on lysines is effective in the reduction of hydrolysate bitterness. Isolated soy protein (ISP) was treated with *N*-acetylimidazole for acetylation of lysine and tyrosine. *O*-acetyl tyrosine was deacetylated by subsequent treatment at pH 11. The lysine-acetylated soy protein and control protein were hydrolyzed by bromelain to the same degree of 10% hydrolysis, and then the bitterness was evaluated. Sensory analysis indicated that bitterness of the lysine-acetylated hydrolysates decreased in comparison with hydrolysates of the control ISP. Surface hydrophobicity of hydrolysates of the lysine-acetylated ISP slightly increased, and they had fewer lysine residues at the C-terminal region than hydrolysates of the control ISP (Yeom et al., 1994).

Simple alteration of chemical structure in saccharin removes the bitter aftertaste associated with the sweetener. By incorporating nitrogen into the aryl ring of the sweetener saccharin, bitterness is reduced. E. I. du Pont reported research on pyridine analogues of saccharin said to be sweet to the taste, not to possess the bitter aftertaste, and to have an approximate lethal dose in rats of 11 g/kg (Chiang, 1991).

SPECIFIC APPLICATIONS OF BITTERNESS INHIBITION AND REDUCTION

ENZYMIC TREATMENT OF CITRUS BEVERAGES

The most abundant examinations of debittering are found in the citrus industry. The bitterness of citrus juices was found to be attributed mostly to naringin, limonin, and nomilin. Robert M. Horowitz passed away at age 71 with a lifelong interest in natural product chemistry, specifically in the chemistry of citrus flavonoids. He was the Director of the U.S. Department of Agriculture's Fruit and Vegetable Laboratory in Pasadena, California. With Bruno Gentili, he isolated more than 20 new flavonoids and phenols from lemons. To determine the structure of these complex molecules, Horowitz devised new ultraviolet and nuclear magnetic resonance spectroscopic techniques. His pioneering work on the relationship between the structure and taste of flavonoids established the importance of the disaccharide in the bitterness of flavonones. The flavonoid neodiosmin was seen to be a potentially useful suppressor of bitterness.

As early as 1969, Robert Horowitz studied dihydrochalcones (DHCs) as debittering agents and recogr'zed 2-0-α-L-rhamnosyl β-D-glucosides as the cause of bitter in DHCs, while β-rutinose(6-0-α-L-rhamnosyl β-D-glucose) appended to the same DHC provided a tasteless compound (Horowitz and Gentili, 1986). The rhamnosyltransferase catalyzing the production of the bitter flavanone-glucosides, naringin and neohesperidin, is purified to homogeneity and commonly called naringinase. The enzyme catalyzes the transfer of rhamnose from UDP-rhamnose to the C-2 hydroxyl group of glucose attached via C-7-*O*- of naringenin or hesperitin. This first complete purification of rhamnosyltransferase was accomplished from young pummelo leaves to a >2700-fold purity and specific activity of >600 pmol/min/mg of protein by sequential column chromatography on Sephacryl S-200, reactive green 19-agarose, and Mono-Q. The enzyme was selectively eluted from the green dye column with only three other proteins by a pulse of the substrate hesperitin-7-*O*-glucoside followed by UDP. The rhamnosyltransferase is monomeric (~52 kDa) by gel filtration and electrophoresis. The enzyme rhamnosylates only with UDP-rhamnose. Flavonoid-7-*O*-glucosides are usable acceptors, but 5-*O*-glucosides or aglycones are not. The rhamnosyltransferase is inhibited by 10 μm UDP, its end product, but not by naringinin or neohesperidin. Several flavonoid-aglycons at 100 μm inhibited the rhamnosyltransferase; UDP-sugars did not. The K_m for UDP-rhamnose was similar with prunin (1.3 μM) and hesperitin-7-*O*-glucoside (1.1 μM) as substrate. The affinity for the natural acceptor prunin (K_m = 2.4 μm) was much higher than for hesperitin-7-*O*-glucoside (K_m = 41.5 μM). The isolation of the gene may enable its use in genetic engineering directed at modifying grapefruit bitterness while preserving the consumption of healthy flavonoids (Bar-Peled et al., 1991).

Debittering of juices may simply be done by splitting the glycosides in a fermentation process (Burkhardt, 1971). The bitter taste of citrus peel, orange rind, and immature orange extract is reduced by treating these with naringinase enzyme, and then with oligosaccharides, carboxylic acids and gelation agents to produce materials useful in making jellies or beverages (Kumai, 1995). Bitterness perception was studied with marmalade prepared from a 1:1 mixture of two kinds of citrus peel. The amount of flavonoids was decreased markedly by heating, indicating they were not responsible for the bitterness. While limonin was also decreased by boiling, the amount of nomilin increased, suggesting that it was responsible for the bitterness. Yet, fruit containing naringin tasted good, but fresh fruit containing a large amount of limonin was not favorable. Mixture suppression of bitterness may account for the improvement of the flavor of the marmalade (Morishita, 1994).

Receptor recognition of sweet and bitter occurs in many classes of structurally similar triterpene substances. Sweetness in a saponin bis-glycoside, osladin, is evident if one of the sugar moieties is a 2-0-α-L-rhamnopyranosyl β-D-glucopyranosyl residue. The linkage between C-26 and the rhamnose unit in osladin is probably the same sugar residue that is responsible for sweet taste in neohesperidin dihydrochalcone (Horowitz and Gentili, 1986). Japanese researchers also understood the positional effects of sugars (Esaki et al., 1983; Konishi et al., 1983). Kimball reviewed the research and quality control aspects of citrus processing, including a discussion of debittering technology (Kimball, 1991).

Multifaceted approaches have been employed to understand the bitter and sweet perceptions of flavone and flavanones. Bitter flavonoids such as naringin and neohesperidin become intensely sweet when reduced to their dihydrochalcone derivatives. Yet, phyllodulcin is as sweet as the dihydrochalcone although its structure more closely resembles that of bitter flavanones or flavones. X-ray structure analysis, energy calculation, and structure comparison attempt to clarify the structure-taste correlations in these classes of compounds. In the crystal, naringin dihydrochalcone assumes a J-shaped conformation with a fully extended dihydrochalcone moiety, while neohesperidin dihydrochalcone assumes the same overall conformation but with a partially extended moiety. A 2D conformational energy map of dihydrochalcone obtained using molecular mechanics revealed nine local minimum. The pseudoequatorial and pseudoaxial forms of phyllodulcin have the same AM1 energies with a low energy barrier between them. The partially extended form of dihydrochalcone and the pseudoequatorial form of phyllodulcin, which are the maximally superimposable conformations, are proposed to be the active conformers. The major differences between the structures of flavone (bitter) and phyllodulcin (sweet) is not in the overall planarity but in the relative orientation of the pyrone and the phenyl ring systems (Shin et al., 1995).

The screening of microorganisms (bacteria, yeasts, molds) from rotten fruit and soil yielded eight molds that degraded >80% limonin, whereas 13 molds and eight yeasts were capable of degrading 60–85% limonin (Hashinaga et al., 1984). Cells capable of metabolizing limonoids have been prepared. The cells were immobilized on acrylamide gel and used to successfully debitter juices. No deacidification effect was found. The microorganism methodology was particularly effective toward nomilin. Nomilin acetyl-lyase, further purified from cell-free extracts of *Corynebacterium fascians,* was responsible for the conversion of bitter nomilin to nonbitter obacunone (Herman et al., 1985; Hasegawa et al., 1985). Obacunone is also found in phellodendron bark (the oxidized limonoid found in species of *Rutaceae, Simaroubaceae, Meliaceae*). The conversion of bitter citrus chemicals into obacunone is intriguing. Just 0.1 to 5 mg/kg per day of the limonoid, obacunone, as a botanical treatment has recently been found to be a central-nervous activator, thereby enhancing our mental faculty to ameliorate drowsiness or malaise. Obacunone is said to impart a refreshingness, vitality and cleanness to the inherent taste of foods (Masanobu and Keiji, 1994). In these times of bringing more functionality to beverages, perhaps orange juice could be made a more stimulating drink than our morning coffee.

Tsen utilized enzymic technology by immobilizing naringinase on chitin and treated grapefruit juice. The technique failed to debitter the juice and it was determined that naringinase was inhibited by some noncompetitive inhibitors (glucose and fructose) in the juice (Tsen, 1984). The most effective method to remove the major bitter component of grapefruit juice, naringin, is to hydrolyze with naringinase. The major debittering activity of naringinase lies in its α-rhamnosidase activity. The specific activities of α-rhamnosidase from *Pencillium decumbens* and *Aspergillus niger* were increased 2.16- and 5.12-fold, respectively, after repurification through DEAE-cellulose and Sephadex G-200 column chromatography. The optimum pH was determined to be 3.7 and 4.5, respectively, with the former having better heat stability. Neither naringinase is a metal-requiring enzyme. Except for the liberated rhamnose, the glucose and fructose were found to have little effect on either naringinase's activities (Hsieh and Tsen, 1991).

Screening of matrices for immobilization of naringinase demonstrated that 2% sodium alginate was an optimal matrix. Upon immobilization, 30 U of naringinase gave 82% naringin hydrolysis in 3 h. Broadening the pH optima has attributed desirable flexibility for debittering kinnow juice of varying pHs. Temperature profiles indicated an improved thermostability, which could be handy during the reduction in cost of debittering. Alginate permitted attainment of equilibrium readily with no hindrance in the inflow of naringin and outflow of naringenin/prunin. Adequate mechanical stability is indicating the feasibility of the process for commercial exploitation. The application of kinetic parameters optimized with pure naringin to kinnow juice resulted in 60% debittering. Ultrafiltration to minimize product inhibition and maximize naringin hydrolysis is also reported with fortification of the permeate at the end of debittering to formulate sweetened kinnow juice with nourishing and natural characteristics (Puri et al., 1996).

The incubation of 5 v/v% *Rhodococcus fascians* with kinnow mandarin juice for 48 h at 25°C and pH 4 afforded optimal 60% limonin degradation. Additional alkaline treatment increased the degradation to 78% (Marwaha et al., 1994). The removal of limonin by free and immobilized cells of *Rhodococcus fascians* was studied in a model buffer system and in the real system. The microorganism was immobilized by adsorption on chitin. The effect of pH on the chemical conversion of limonin to its acid forms was studied together with its subsequent biodegradation. In the real system of orange juice at pH 3.5 employing 1.5–4 g of wet biomass, no significant degradation was observed. In the buffer systems with 24.0 ppm of limonin and different concentrations of dry cells, at 2 mg/mL for the free cells and at concentrations 10 times lower for the immobilized ones, the amount of limonin removed at pH 7 was 13.1 and 12.7 ppm, respectively. It may consequently be assumed that the biodegradation of limonin by *R. fascians* depends largely on limonin conversion to its acid forms and, therefore, on the pH value (Bianchi et al., 1995).

A packed bed reactor of naringinase (from *Penicillium* sp.) immobilized on glycophase-coated porous glass was utilized for continuous operation in debittering fruit juice. In 1987, this industrially optimized process with an active form of naringinase was developed by the Spanish (Jimeno et al., 1987; Manjon et al., 1985). Naringinase (from *Penicillium* spp.) immobilized on cellulose triacetate in a bioreactor was effective (Tsen, 1990; Hoo, 1991). An alternative approach is the adsorption of the bitter substances directly onto cellulose triacetate powder, where 70% of limonin and 8% of naringin were removed from orange juice using 2% cellulose triacetate at 15°C for 60 minutes (Johnson and Chandler, 1985a, 1985b, 1986).

Recent research demonstrates that cellulose monoacetate gel beads are as effective as the cellulose triacetate fibers. Sugar components, total organic acids and turbidity were not affected and the enzyme column could be regenerated with warm water without operational instability. The method could be considered for industrial use (Tsen and Yu, 1991). A hexameric unit cell-crystallized immobilization matrix of cell surface layer (*Thermoanaerobacter thermohydrosulfuricus* L111-69) with a monolayer at the outermost surface of nariginase (EC 3.2.1.40) shows 60–80% of original activity of immobilized enzyme correlating with the extent of interactions with the matrix. Some smaller enzymes can penetrate the central funnel shape and lose activity, other larger enzymes may covalently bind on the surface (Kuepcue et al., 1995).

Chinese citrus research has studied five methods of extracting naringin from grapefruit peel. Treatments included lime solution, water, pectinase and water, pectinase and aluminium chloride and water and pectinase and *Saccharomyces cerevisiae* and water. Lime treatment resulted in the highest recovery rate (49%), but the purity of the product was the lowest (72%) and its residues could not be reused. Treatment with yeast gave the product of highest purity, resulted in more rapid precipitation and the highest recovery rate, and peel residues could be used further for pectin extraction (Wu et al., 1991).

NONENZYMIC TREATMENT OF CITRUS BEVERAGES

The USDA in 1985 reported the use of β-cyclodextrin-bound polymer in debittering juices (Swientek, 1988; Shaw and Wilson, 1983, 1985). The process was clever, but claimed to be costly for a 50% bitter reduction. Sixteen bed volumes of juice were bitter-reduced with the polymer that could be regenerated 19 times. The Chinese also report the reduction of bitterness with β-cyclodextrin, which awaits food use approval in the US (Chiu et al., 1988).

It has been reported that the bitterness due to limonoids of citrus products decreased by addition of β-cyclodextrin (β-CD). However, in the case of canned Iyo orange segments in syrup, the addition of β-CD caused heavy white precipitation in the syrup. The cause of the precipitation was concluded to be the formation of inclusion compounds of β-CD with ingredients of juice sac membrane of Iyo orange and the poor solubility of those inclusion compounds in the syrup. More recently, the addition of maltosyl-β-CD formulations, which have 150 times the solubility of β-CD formulations alone, showed no cloud or precipitate with 0.5% maltosyl-β-CD added. The sensory evaluation for limonin bitterness of the Iyo orange samples was reduced remarkably (Kodama, 1992).

Treatment of citrus fruit juice with immobilized naringinase (purified from *Aspergillus niger*) at 40–45° for 120 min removed ~50% of naringin in the juice. Soaking of the fruit in 2000 ppm ether for 60 min, then storing the fruit for 3–5 days before extraction of fruit juice, reduced naringin content by 49% and limonin content by 45%. Addition of 0.3–0.5% β-cyclodextrin reduced free naringin by 24–28% and free limonin by 49%. The combined use of β-cyclodextrin and ether treatment for reduction of the bitter compounds is recommended (Xu and Liu, 1992).

Membrane extraction is popular for selective removal of bittering agents from citrus juices. One type of apparatus describes a membrane extraction of citrus juice with a hydrophobic bittering agent as extraction fluid. In the other type, a juice permeate is directed to a membrane containing hydrophobic bittering agents within the pores of the microporous polymeric membrane to facilitate transport of the bittering agents across the membrane (van Eikeren and Brose, 1993).

Perhaps the cheapest citrus-debittering method was reported by the Australians from a program begun in 1974. Johnson and Chandler (1986) describe the utility of resins in both debittering and acid-reducing citrus juices for a cost estimated to be about 1 cent per liter. Novel applications of cellulosic ion-exchange resin filtration media are reported (Hughes, 1985). Grapefruit juice and grapefruit pulp wash were ultrafiltered and debittered with a pilot system containing polysulfone ultrafiltration cartridges and a nonionic polystyrene-divinylbenzene resin debittering column. Limonin was removed from both juice sources independent of temperatures from 13–48°, but the resins were exhausted more easily for naringin, narirutin, hesperitin and neohesperitin particularly at 13°. Taste panels suggested the flavor was appreciably increased (Hernandez et al., 1992). The specific surface area of the resin appeared to be the major factor in the debittering process. Regeneration with 95% ethanol was more efficient than warm water (Manlan et al., 1990). Columns of acrylic resin and styrene resins in series are effective for adsorbing naringin and hesperidin. Debittering efficiency was confirmed by organoleptic assessment (Tateo and Caimi, 1993). Limonin and naringin are economically removed from kinnow mandarin juice with Amberlite XAD-16 resin. The treatment did not affect the quality of the juice except for a small loss of titratable acidity. Based on recurring expenses, resin treatment showed the least cost of production among various treatments (Premi et al., 1995). Styrene-divinylbenzene and acrylic resins were found to be limited in adsorption by sucrose content. In model systems of naringin, sucrose and citric acid, the amount of adsorbed naringin decreased as the content of sucrose increased. Adsorption equilibrium was reached within 90 min in the solution containing 12% sucrose at pH 3.2 (Inoue et al., 1989). Removal of the bitter substances with Diion HP20 was demonstrated combined with sodium hydroxide regeneration (Terayama, 1991). Another review discusses similar treatment schemes (We et al., 1992).

The bitterness of hassaku juice due to limonin was decreased 90% by adsorption on styrene-vinylbenzene copolymer beads. Adsorption equilibrium was established within 60 minutes with 0.5% dose of the copolymer. Adsorption was 1.2 times more efficient at 20° than 5° and the limonin could be removed with 96% efficiency from the resin with acetone-hot water (Inoue and Sakamoto, 1993; Li, 1992). Supercritical CO_2 was effective in reducing limonin an overall 25% and reviewed (Kimball, 1987; Dekker, 1988).

Aerobic processing at 30°C of grapefruit juice increases the bitter taste. Different processing methods can reduce delayed bitterness during fruit expression. Grapefruit (*Citrus paradici* MACF) was cut and the juice expressed and filtered in an experimental design of aerobic and anaerobic conditions at 30°C and 5°C. Limonin and naringin contents were measured analytically by HPLC and bitter taste by ranking taste evaluation. The limonin content was highest in juice aerobically treated at 30°C. Naringin contents were unchanged across all conditions. Nomilin and obacunone contents were also estimated; they remained unchanged. It is suggested that the preservation conditions of the expressed juice affected the biosynthesis of limonin and thereby increased the bitter taste of the juice (Morishita, 1993).

Optimum conditions of fruit maturity and processing for improved quality of Thai tangerine juice were evaluated. Limonin and naringin components causing bitterness, acidity, total soluble solids and Vitamin C were quantified in specified fruit settings and processing conditions. Higher limonin contents were observed in tangerines harvested early in the season of 1989, whereas naringin contents gradually decreased with maturity. The optimum harvesting time for Thai tangerines, which meets worldwide quality indicators of extracted juice, was nine months after fruit set. Low-temperature storage of tangerine juice was only effective in delaying limonin formation if not pasteurized, which resulted in higher limonin concentrations at the start of the storage period. Naringin concentrations of tangerine juice were not affected by storage conditions and the pasteurization process. Lower juice extraction pressure resulted in low limonin and naringin concentrations (Noomhorm and Kasemsuksakul, 1992).

Hesperidin and naringin were extracted from orange peel and grapefruit peel, respectively, by treating the peel with $Ca(OH)_2$. The highest extraction yields and purity were obtained by recycling the extracting liquor from orange peel extracted in the early season. The highest yield of hesperidin was 15.5 g/2 kg peel, and the highest naringin was 12 g/2 kg peel. A hot-water leach was used to recover naringin and pectin successively from grapefruit peel by using hydrochloric acid and alcohol precipitation for the pectin. The hot-water leach resulted in greater recoveries of naringin than did the control, a caustic leach similar to commercial practice. The highest yields of naringin were obtained using a 1:3 ratio of peel to leach water at 88–90°C for five minutes. A threefold concentration of leach water increased the yield of isolated naringin by more than 25% (El-Nawawi, 1995).

A solution of naringin containing a large amount of rhoifolin is less bitter than naringin alone (Kuang-chih and Hua-zhong, 1987). This suggests that rhoifolin is a poor competitor with naringin for sites on the taste receptors, although rhoifolin does not produce a taste response of its own. USDA research in 1976 utilized neodiosmin to suppress the bitter taste of limonin and naringin in citrus juices (Guadagni et al., 1976). At 40 ppm in water solution, neodiosmin is essentially tasteless, but as little as 10 ppm in water increased the threshold for naringin and limonin 3.5- and 4.0-fold, respectively. In other terms, the addition of 60–100 ppm of neodiosmin to orange juice containing 10 ppm of added limonin reduced the limonin bitterness to the equivalent of 4–5 ppm of limonin in orange juice. This ineffective use of the debittering agent reflects the non-competitive nature, in receptor terms, of neodiosmin.

Different concentrations of ethephon (ethylene) affected the levels of the sequiterpene, nootkatone, and the flavanones, naringin and narirutin, in grapefruit. Nootkatone synthesis and/or accumulation was stimulated by all the concentrations of ethephon assayed, while the levels of naringin and narirutin in the rind diminished. These results open up new perspectives concerning the possible regulation of the secondary metabolism of plants (Garcia Puig et al., 1995).

The use of plant growth regulators such as gibberellic acid has merit for reducing the seasonal content of bitter agents in fruits. Limonin and naringin content decreased with maturity in some Marsh grapefruit albedo-treated hybrids over four seasons (Shaw et al., 1991). Chinese cabbage, radish and spinach showed enhanced growth as a result of treatment with naringin, naringenin and naringin glucosides (Komai, 1994).

A mixture of L-threonine, alanine, glycine, amino acids, nucleic acid-oligosaccharide complexes, vitamins, auxins and cytokinin applied to roots and leaves of tea and leafy vegetables improved the yield as well as taste of the plants (Handa, 1994).

OTHER BEVERAGE JUICES

More than 650 metric tons of bitter apricot seeds are produced in Turkey per year as a by-product of the fruit canning industry. Chang was capable of debittering Japanese apricot juice *(Prunus mume)* with an aqueous extract of its own seed kernel (Chang and Wu, 1985). The seed kernels have been previously evaluated as unconventional protein sources (Gabrial et al., 1981). The agent held responsible for the bitter taste was identified as amygdalin (later attempted as a cure for cancer under the name of laetrile). Amygdalin is a cyanogenic glycoside occurring, among others, in almonds and bitter apricot seeds with interesting levels of dietary protein. Acids or enzymes can decompose amygdalin to prussic acid and benzaldehyde, but prussic acid is toxic. An acidic (pH 5–6) soaking solution at room temperature and exchanged twice a day (3 × volume of raw material) is said to reduce bitterness and toxicity (Zhu et al., 1993). Utilization of seeds for human or animal nutrition requires adequate detoxification.

Of microbial degradation studies with selected filamentous fungi *(Mucor circinelloides, Penecillium nalgiovenese)* and yeasts *(Hanseniaspora valbyensis, Endomyces fibuliger)*, only *Endomyces fibuliger* showed in situ ability to decompose amygdalin. This study showed autoclaved bitter apricot seeds *(Prunus armeniaca)* were detoxified from 30 μmol/g dry matter to less than 1 μmol/g dry matter after 48 h of incubation at 27°C (Nout et al., 1995). However, Chinese researchers report that treatment of bitter apricot seeds *(Prunus armeniaca)* with boiling water, steam, storage or baking did not significantly decrease a 5% amygdalin content

(Gao and Jin, 1992). Grinding, cooking and soaking must be combined with microbiological methods (Tuncel et al., 1995).

The chemical composition of bitter and sweet varieties of apricot *(Prunus armeniaca)* kernel were found to contain more oil in the sweet variety (53 g/100 g) and less soluble sugars (7 g/100 g) than bitter kernels (43 and 14 g/100 g, respectively). No significant differences in the protein content were found in either variety. Oleic acid and linoleic acid are approximately 92 g/100 g of total fatty acids. Pectic polysaccharides, cellulose, and hemicelluloses (in decreasing amounts) were inferred to be the main-component polysaccharides. Essential amino acids constitute 32–34 g/100 g of the total amino acids determined. Amygdalin content was very high (5.5 g/100 g) in bitter cultivars and was not detected in the sweet variety (Femenia et al., 1995). Gelatin and sodium or potassium citrate additives effectively remove the bitterness and astringency of tannins found in cranberry juice (Shimazu et al., 1992).

Five commercial pectinases were used to investigate the efficacy for improvement of juice yield and quality from Stanley plums. Pectinases, to a varying degree, improved the yield, color (assayed as release of anthocyanins) and clarity of the juice. A significant increase in the effectiveness of pectinases was observed as the concentration was increased from 0.01–0.60% v/w. However, at concentrations >0.20% they tended to impart a bitter flavor in the juice, Clarex L at 0.20% is recommended (Chang et al., 1995).

For the practical use of danewort *(Sambucus ebulus)* containing anthocyanin colorants at 3.5–5.2 g/kg, it is necessary to remove an unpleasant bitter taste, a mousy smell and to stabilize anthocyanins in a suitable form. Treatment of danewort juice with cellulase and adsorption on corn starch was an effective way to improve the quality of its anthocyanins. The resulting colorants could be used as a substitute for red colorants in clear lemonades (Kovacova et al., 1993).

Vegetables are not eaten as frequently as nutritionists recommend. New debittering techniques applied to vegetables could provide more palatable food products for both children and adults. Undesirable odor or bitter taste of fruit or vegetable juice is controlled by adding >0.01 w% glycerol diester of C6–C22 fatty acid or diglyceride or sucrose fatty acid ester (So et al., 1995). Protease treatment with Pancidase NP-2 has been used to eliminate the bitterness from sweet pepper juice (Hoshikawa and Kikuchi, 1991). A palatable juice without unpleasant bitter aftertaste can also be prepared from green vegetables such as spinach, celery, lettuce, parsley and cabbage by extracting the juice with supercritical carbon dioxide (Osaka, 1990). Bitterness of vegetable or fruit juices is removed by treatment with silica gel. For example, the bitterness of spinach juice was removed by treatment with 0.5 weight percent silica gel (Hoshikawa and

Kikuchi, 1993). Stable and bitterness-free beverages are prepared by using a base of soyasaponin (0.1 g) dissolved in 2 mL of 70% ethanol and mixed with 100 mL of water and γ-cyclodextrin (0.07 g). The solution is heated with 3% citric acid at 80°C for 30 minutes without precipitation (Murakami and Minami, 1992).

Protamines and/or their salts may be added to beverages such as coffee, tea, grapefruit juice, and wine to remove bitter taste. Protamines are a group of simple proteins that yield basic amino acids upon hydrolysis and occur combined with nucleic acids in many species of fish. Tyrosine and tryptophan are always absent in the hydrolyzates. The protamines are useful for suppressing astringent or bitter-tasting foods and beverages such as those containing 25 wt% saponins with only 0.03 wt% protamines added (Sugakawa and Masuda, 1993, 1994).

BITTERNESS REDUCTION AND INHIBITION IN CARBOHYDRATE PRODUCTS

Aloe vera is traditionally known as a wound healer, for burns specifically. Recent clinical data have suggested soothing effects to mucous membranes can result from drinking aloe beverages. However, expressed aloe juices contain anthraquinone substances, which exhibit a laxative effect and possess irritating cathartic properties. A properly prepared aloe juice should contain no anthraquinone substances and maximized contents of the beneficial active ingredients known as mucilaginous polysaccharides. Generally, the processing destroys mucilaginous polysaccharides, also known as acemannan. Acemannan has a physiological activity as an antiviral agent, demonstrates immuno-modulating activity, reduces opportunistic infections and stimulates healing processes in gastric ulcers.

Acemannan at 500–1500 mg/L in beverages generally possesses a bitter taste even with various flavoring agents. So a need exists for a standardized acemannan beverage that does not have a bitter taste and is anthraquinone-free. A palatable drink has been prepared with alcohol-precipitated mucilaginous polysaccharides. The process involves taking the clean inner gel of aloe vera leaves without the rind. The filtered gel is acidified to pH 3.2 with dilute hydrochloric acid. The acidified gel is then extracted for 4–5 h with four volumes of 95% ethanol at room temperature. This extraction provides a precipitate that is collected by centrifugation. The solid material is now free of most organic acids, oligosaccharides, monosaccharides, anthraquinones and inorganic salts. Bitter taste is essentially removed. This solid aloe vera product is tradenamed Mucipol. Reconstitution of a freeze-dried solid by homogenization in water at 0.1–0.5% provides a palatable drink. Buffer addition enhances beverage stability at pH

4.5. The authors prefer glycine as sweetener and sodium metabisulfite as antioxidant (Moore and McAnalley, 1995).

The bitterness of β-glucooligosaccharides is removed by the reduction of the oligosaccharides. The products are useful as sweeteners and improvers for intestinal *Bifidobacterium* growth. Gentio-oligosaccharide syrup containing gentiobiose (73.7%), gentiotriose (20.2%) and gentiotetraose (4.3%), manufactured from glucose with β-glucosidase, was hydrogenated at 130° over Raney-nickel catalyst for 4 h to provide less bitterness (Okada et al., 1992).

The polycondensation of starch hydrolysates with sugar alcohols, catalyzed by weak inorganic acids, results in high molecular weight bitter polysaccharides. Subsequent enzymatic degradation produces nonbitter polysaccharides. For example, sorbitol, phosphoric acid and anhydrous glucose polymerized at 130°C and heated slowly to 170°C is treated with glucoamylase overnight at 55°C to give a sweet product with no bitter taste (Yoshida et al., 1990).

Liquid-liquid partitioning has also been successfully used to remove bitter compounds from glycolipid extracts. Solvent extraction of the lipids from the flower bud tips of *Lycopersicon pennellii* (tomato) yields a green lipid extract that has a bitter taste and an unpleasant odor. This extract can be separated into colorless, odorless, bitter and nonbitter glycolipid fractions by liquid/liquid partitioning; the method has been successfully scaled up for the production of bitter and nonbitter glycolipids (Nair et al., 1990).

STIMULANT BEVERAGES AND BITTER SWEETENERS

Xanthines in Tea, Coffee, Cocoa and Chocolate

Xanthines are a family of nitrogenous natural substances largely responsible for bitterness in tea, coffee, cocoa and chocolate. Caffeine is one xanthine very familiar to all coffee drinkers. The routine methods of decaffeination are well known and were briefly reviewed (Roy, 1994). It is disputable that supercritical fluid extraction with carbon dioxide is a welcomed improvement to the established aqueous adsorption method because of cost for SCFE. And, both abate health concerns with the methylene chloride extraction technology for caffeine.

The Merck Index provides some interesting approaches to debittering, utilizing a "salting" of the bitter substrate. For example, theophylline, a bitter xanthine component of tea, is tasteless as a dicalcium-disalicylate salt: $Ca(C_7H_7N_4O_2)_2$ and $Ca(C_7H_5O_3)_2$ (Merck, 1983). It appears that formation of dicalcium-disalicylate salts would be effective in debittering xanthines found in cocoa and coffee. Theobromine, a bitter xanthine component of cacao bean, cola nuts and tea, has a slightly saline taste as a double salt. The taste of caffeine as salicylate salts has not been reported but would be expected by precedence to be tasteless. Salicylic acid, however, may carry some toxicity, but the levels of ingestion would have to be examined for any toxic effects.

LinTech and General Foods prepared salts of aryl hydroxy acids (sweetness inhibitors) (Lindley and Rathbone, 1985; Barnett and Yarger, 1985a, 1985b; Vellucci et al., 1987) for nontoxic applications as sweetness inhibitors. Some of them were found to possess bitterness inhibition qualities (Kurtz and Fuller, 1993). Hypothetically, naturally occurring aryl acids could be formulated as calcium salts for bitterness inhibition and calcium supplementation. Kraft Foods (formerly General Foods) has provided recent patent data to illustrate bitterness reduction using sodium salts of hydroxy and alkoxy-substituted cinnamic acids such as caffeic and ferulic acid. Both *cis*- and *trans*-forms are said to be effective but the *trans*-form is the natural derivative. The sodium salt of ferulic acid (450 ppm) was tested and found to significantly reduce the perceived bitterness in solutions of 500 ppm caffeine (65%), 500 ppm saccharin (37%), 1000 ppm acesulfame-K (55%) and 8 ppm quinine sulfate (37%). Sensory panel evaluation of a black cocoa powder aqueous extract with 100–300 ppm sodium ferulate found it to have 23–35% reduced bitter taste. Additionally, the ferulate salt was tested in an unsweetened, brewed, Columbian roasted and ground coffee at three concentration levels: 100, 250 and 500 ppm. The majority of the eight tasters felt there was significant reduction of the coffee bitterness at all three levels. Some tasters described the higher levels as being flat or noncoffee-like because of the almost total elimination of bitterness. Some off-notes were detected at the highest sodium ferulate level. In these cases of 0.01%–0.1% addition, bitter taste reduction was favorably competitive with sweet-taste reduction (Reimer, 1994). The bitter taste and sweetness of acesulfame-K (0.044%) is improved by the addition of 0.4% DL-alanine, 0.4% glycine (sweeteners). Less bitter but possible disagreeable acid taste was obtained with 0.002% glutamate mixed with nucleotides, 0.04% histidine or 0.02% L-Arg-L-Glu. Compositions are presented for acceptable taste products such as mandarin juice, juice sherberts, apple jam, black tea and ice cream (Nakajima, 1984).

Selecto, Inc. has petitioned the FDA to permit the addition of 5 ppm of the reducing agent, sodium thiosulfate, directly into packages of dry coffee and tea as well as to the coffee filters and tea bags. Evidently, coffee or tea brewed with water being treated with sodium thiosulfate eliminates chlorine, hypochlorites and chloramines found in the potable water that often impart bitter flavor (Levy, 1992). As of June 1993, the FDA has asked for additional data in proof of chlorine elimination from potable water.

Typical aqueous extraction of tea leaves removes the bitterness of catechin and caffeine. Additional healthy polyphenols may then be extracted from the waste residue with ethanol and added as medicinal components for various foods (Sakanaka et al., 1992).

From a sensory point of view, chocolate has been considered to be an ideal substance. The added sugar and natural fat conveniently masks and dilutes the inherent bitterness of the xanthines found in cocoa (Table 1). While the fat content of chocolate is close to the hedonic ideal, the sucrose content must also be high enough to satisfy the consumer's palate without bitterness.

Nestec has patented the "exsorption" method (supercritical fluid extraction and activated carbon adsorption of xanthines) for the debittering of cocoa. Thus, the physiological effects of xanthine stimulants can be eliminated in the treated products. The theobromine content of green, unroasted cocoa nibs can be decreased by 71–89% without loss of fatty material. This technology should permit the formulation of chocolate products with less sugar and, therefore, fewer calories (Margolis et al., 1989). The bitterness of cocoa is due to an interaction of theobromine and cyclic peptides (Pickenhagen et al., 1975; Ney, 1986). The bitter cyclic peptides (diketopiperazines) are representative of those found in cocoa:

Cyclo(-Pro-Leu-), cyclo(-Val-Phe-), cyclo(-Pro-Phe-),
cyclo(-Pro-Gly-), cyclo(-Ala-Val-), cyclo(-Ala-Gly-),
cyclo(-Ala-Phe-), cyclo(-Phe-Gly-), cyclo(-Pro-Asn-),
cyclo(-Asn-Phe-)

The aroma of cocoa is also derived from protein and peptides. Cocoa-specific aroma precursors were obtained by degradation of the cocoa globulin with the aspartic endoprotease from cocoa seeds followed by treatment with carboxypeptidase A from porcine pancreas. Proteolysis products derived from the cocoa globulin by successive digestion with pepsin and carboxypeptidase A also revealed a typical, but less pronounced cocoa aroma upon roasting in the presence of sugars. However, no cocoa-specific aroma precursors were generated by degradation of

TABLE 1. Comparison of the Concentrations of Bitter Methyl Xanthines in Various Products.

Product	Theobromine (%, w/w)	Caffeine (%, w/w)
Cocoq	1.9	0.2
Plain chocolate	0.45	0.07
Milk chocolate	0.16	0.025
Coffee	ND	1–2

ND: not detected.
Taken from Schuman et al., 1987, *J. Nerv. Ment. Dis.*, *175*, 491–495.

cocoa globulin with chymotrypsin and carboxypeptidase A. Therefore, the specific mixture of oligopeptides and hydrophobic-free amino acids required for the formation of the typical cocoa aroma components is not only determined by the structure of the protein substrate but also dependent on the specificity of the endoprotease (Voigt et al., 1994).

Fermentation trials of Amelonado, BAL209, Nanay33 and UIT1 cocoa genotypes, grown in Malaysia, have revealed consistent differences in cocoa flavor intensity, bitterness and astringency relating both to the conditions of post-harvest processing and the polyphenol composition of the fermented and dried beans. Combined liquid chromatography and mass spectrometric analyses showed the procyanidins to be the principal class of compounds responsible for the typically astringent taste of cocoa beans and liquors (Clapperton et al., 1992).

Shelled and degermed cacao beans were soaked in solutions of crude laccase of *Coriolus versicolor* IFO 30388 (a polyphenol oxidase) at 45° for 3 h, dried, roasted and ground. The process improves the flavor and taste of cacao nib and its chocolate products by removing bitterness and other unpleasant tastes (Takemori et al., 1992). Obvious market need questions arise as to whether a bitter inhibition ingredient or process is really a necessity for cocoa or coffee. Kava™, the debittered coffee product, was never successful (. . .CocoNo™).

REFERENCES

Akabas, M. H., Dodd, J. and Al-Awqati, Q. 1988. A bitter substance induces a rise in intracellular calcium in a subpopulation of rat taste cells. *Science, 242,* 1047.

Akai, H. 1991. Japanese Patent JP 03,139,258.

Alain, A., El-Akher, M. A., Abdou, I. and Aid, N. 1983. Enrichment of local bread with dried brewery yeast. 1. Chemical and biological evaluation of dried brewery yeast and flour. *Egypt. J. Food Sci., 11*(1/2), 23.

Andres, C. 1981. High protein flour from white beans. *Food Process, 42*(5), 64.

Anon. 1982. Ebiosu Yakuhin Kogyo KK, Debittering of yeast, Japan Patent No. JP 5,735,631 B2.

Anon. 1984. Dipeptide sweetener improvement with somatin, JP 5,914,764 to San-Ei Chemical Industries, Ltd.; *Chem. Abstr.* 100: 173435g.

Anon. 1986. An interesting raw material. Application of full-fat soy flour. I. *Zucker-Suesswarenwirtschaft, 39*(2), Abstr. 51.

Anon. 1986. Preparation of low calorie polysaccharide with reduced bitterness by reacting lactose with sugar alcohol in presence of (in)organic acid and recovering excess sugar alcohol, Sansho Kogyo KK, JP 134,813, June 12, 1986.

Anon. 1988. Removing bitterness from glycerin-containing foods by addition of sodium acetate, Daiei Shokuhin Kogy JP 096,642.

Anon. 1995. Removal of bitter taste from extracts of *Hydrangea macrophylla,* JP 07,327,602.

Anon. 1996. Thaumatin: the sweetest substance known to man has a wide range of food applications. *Food Technology, 50*(1), 74–5.

Arai, S., Noguchi, M. Kurosawa, S., Kato, H. and Fujimaki, M. 1970. Applying proteolytic enzymes on soybean. VI. Deodorization effect of Aspergillopeptidase A and debittering effect of aspergillus acid carboxypeptidase. *J. Food Sci., 35,* 392.

Aristoy, M.-C. and Toldra, F. 1995. Isolation of flavor peptides from raw pork meat and dry-cured ham. *Dev, Food Sci., 37B,* 1323–44.

Aubes-Dufau, I., Capdevielle, J., Seris, J.-L. and Combes, D. 1995a. Bitter peptide from hemoglobin hydrolyzate: isolation and characterization. *FEBS Lett., 364*(2), 115–19.

Aubes-Dufau, I., Seris, J.-L. and Combes, D. 1995b. Production of peptic hemoglobin hydrolyzates: bitterness demonstration and characterization. *J. Agric. Food Chem., 43*(8), 1982–8.

Baer, A., Borrego, F., Benavente, O., Castillo, J. and Del Rio, J. A. 1990. Neohesperidin dihydro-chalcone: properties and applications. *Lebensm. Wiss. Technol., 23*(5), 371–376.

Baker's Digest. 1984. No caffeine cocoa flavor. Brewer's yeast product earns use in cookies. *58*(4), 300.

Barnett, R. E. and Yarger, R. G. 1985a. European Patent Application EP0132444 (3-hydroxy benzoic acid).

Barnett, R. E. and Yarger, R. G. 1985b. European Patent Application EP0131640 (3-amino benzoic acid).

Barnett, R. E. and Yarger, R. G. 1986. Canadian Patent No. 1208966 (3-hydroxy 4-aminobenzoic acid, 3-hydroxy-4-methoxy benzoic acid and 3,5-dihydroxy benzoic acid).

Barnett, R. E. and Yarger, R. G. 1987. Derivatives useful as sweetness inhibitors, U.S. Patent No. 4,642,240 (3-amino benzenesulfonic acid).

Bar-Peled, M., Lewinsohn, E., Flurh, R. and Gressel, J. 1991. UDP-rhamnose:flavanone-7-*O*-glucoside-2″-*O*-rhamnosyltransferase. Purification and characterization of an enzyme catalyzing the production of bitter compounds in citrus. *J. Biol. Chem., 266*(31), 20953–9.

Bautista, J., Hernandez-Pinzon, I., Alaiz, M., Parrado, J. and Millan, F. 1996. Low molecular weight sunflower protein hydrolysate with low concentration in aromatic amino acids. *J. Agric. Food Chem., 44,* 967–971.

Belitz, H.-D., Chen, W., Jugel, H., Treleano, R., Weiser, H., Gasteiger, J. and Marsili, M. 1979. Sweet and bitter compounds: structure and taste relationship, in *Food Taste Chemistry, Vol. 115,* Boudreau, J. C., Ed., *American Chemical Society Symposium Series, 115,* 93–131.

Belitz, H.-D. and Weiser, H. 1985. Bitter compounds: occurrence and structure-activity relationships. *Food Rev. Intl., 1*(2), 271–354.

Belitz, H.-D., Rohse, H., Stempfl, W., Wieser, H., Gasteiger, J. and Hiller, C. 1988. Schematic sweet and bitter receptors—a computer approach, *Proc. 5th Int. Flavor Conf.* July 1-3, 1987, in Frontiers of Flavor, Charalambous, G., Ed., Elsevier, Amsterdam, 49.

Bianchi, G., Setti, L., Pifferi, P. G. and Spagna, G. 1995. Limonin removal by free and immobilized cells. *Cerevisia, 20*(2), 41–6.

Birch, G. G., Cowell, N. D. and Young, R. H. 1972. *J. Sci. Food Agric., 23,* 1207.

Birch, G. G. and Lee, C. K. 1976. Structure functions and taste in the sugar series: the structural basis of bitterness in sugar analogues. *J. Food Sci., 41*(6), 1403.

"Bitter isn't better . . ." 1994. *Food Business,* June, p. 55.

Bockelmann, W. 1992. The significance of proteolytic enzymes in cheese ripening. *Lebensmittelind. Milchwirtsch., 113*(25), 739–43, in German; *Chem. Abstr.* 118: 21096f.

Bocklet, G. 1980. Nougat and marzipan raw masses. Characterization and technology. *Suesswarentechnik, 24*(10), 18.

Borrego, F. and Montijano, H. 1995. Potential applications of the sweetener neohesperidin dihydrochalcone in pharmaceuticals. *Pharm. Ind., 57*(10), 880–2, in German.

Buckholz, J. L. 1994. Use of aconitic, gluconic and/or succinic acid for improvement of the organoleptic properties of foods, DE 4,339,522 to IFF; *Chem. Abstr.* 121: 178318q.

Buckholz, L., Jr., Farbood, M. I., Kossiakoff, N. and Scharpf, L. G. 1991. U.S. Patent 4,999,207 to IFF.

Bumberger, E. and Belitz, H. D. 1993. Bitter taste of enzymic hydrolyzate of casein. Part 1. Isolation, structural and sensorial analysis of peptides from tryptic hydrolyzates of β-casein. *Z. Lebensm.-Unters. Forsch., 197*(1), 14–19.

Bunick, F. J., Luo, S. J. and Warner-Lambert Co. 1991. European Patent Appl. EP 0,458,748.

Burkhardt, R. 1971. Observations on removal of bitterness from grapefruit and grapefruit juice. *Mitteilungsblatt der Gdch-Fachgruppe Lebensmittelchemie und Gerichtliche Chemie, 25*(12), 354.

Chakrabati, R. 1983. A method of debittering fish protein hydrolysate. *J. Food Sci. Technol, India, 20*(4), 154.

Chang, C. M. and Wu, J. S. 1985. Debittering of Japanese apricot *(Prunus mume)* juice with kernel extract. *J. Chin. Agric. Chem. Soc., 23*(3/4), 282.

Chang, T.-S., Siddiq, M., Sinha, N. K. and Cash, J. N. 1995. Commercial pectinases and the yield and quality of Stanley plum juice. *J. Food Process. Preserv., 19*(2), 89–101.

Chen, Z., Bao, X. and Wang, B. 1992. Taste characterization of caffeine. *Shipin Kexue (Beijing), 145,* 1–2, in Chinese; *Chem. Abstr.* 117: 88928b.

Chiang, G. C. S. 1991. Method of using pyridine analogues of saccharin for sweetening. U.S. Pat. No. 5,043,181 to du Pont de Nemours; *Chem. Abstr.* 116: 40104g.

Chiu, C. P., Lin, T. C., Chang, S. W. and Chuang, W. L. 1988. Studies on functional properties and applications of β-cyclodextrins I. Emulsification, reduction of bitterness, hygroscopicity and foaming. *Chieh Mien K'o Hsueh Chih, 11*(1), 38; *Chem. Abstr.* 109: 34512c.

Clapperton, J., Hammerstone, J. F., Jr., Romanczyk, L. J. Jr., Yow, S., Chan, J., Lim, D. and Lockwood, R. 1992. Polyphenols and cocoa flavor. *Bull. Liaison—Groupe Polyphenols, 16*(Pt. 2), 112–16.

Crammer, B., Ikan, R., Weinstein, V. 1990. Israel Patent IL 81,351.

Darwish, S. M., El-Difrawy, E. A., Mashaly, R. and Aiad, E. 1994a. Compositional and microbiological properties related to bitterness in Ras cheese on the market. *Egypt. J. Dairy Sci., 22*(1), 11–17.

Darwish, S. M., El-Difrawy, E. A., Mashaly, R. and Aiad, E. 1994b. An assay for bitter peptides, amino acids, biogenic amines, glycerides, and fatty acids, in the bitter Ras cheese on local market. *Egypt. J. Dairy Sci., 22*(1), 1–10.

Dekker, R. F. H. 1988. Debittering of citrus fruit juices: specific removal of limonin and other bitter principles. *Aust. J. Biotechnol.*, *2*(1), 65.

Deltown Specialties. 1991. *Food Eng.*, October, p. 37.

Duflot, P. 1995. Process for purifying hypocaloric soluble glucose polymers and the obtained products. Eur. Pat. Appl. EP 0654483 to Roquette Freres.

DuPont, M. S., Muzquiz, M., Estrella, I., Fenwick, G. R. and Price, K. R. 1994. Relationship between the sensory properties of lupine seed with alkaloid and tannin content. *J. Sci. Food Agric.*, *65*(1), 95–100.

El-Nawawi, S. A. 1995. Extraction of citrus glucosides. *Carbohydr. Polym.*, *27*(1), 1–4.

Esaki, S., Tanaka, R. and Kamiya, S. 1983. Structure-taste relationships of flavanone and dihydrochalcone glycosides containing (1 → 2) linked disaccharides. *Agric. Biol. Chem.*, *47*(10), 2319.

Farr, D. R. and Magnolato, D. 1981. Process for debittering a protein hydrolysate and the debittered hydrolysate obtained by this process, U.S. Patent No. 4,293,583.

Farr, D. R. and Magnolato, D. 1983. Method for debittering a protein hydrolysate and the resultant hydrolysate, Nestle SA, Swiss Patent CH 635491 A5.

Femenia, A., Rossello, C., Mulet, A. and Canellas, J. 1995. Chemical composition of bitter and sweet apricot kernels. *J. Agric. Food Chem.*, *43*(2), 356–61.

Fernandez, M., Borroto, B., Larrauri, J. A. and Sevillano, E. 1993. Dietary fiber in grapefruit: a natural product with additives. *Alimentaria (Madrid)*, *247,* 81–3, in Spanish.

Fernandez-Garcia, E., Lopez-Fandino, R., Alonso, L. and Ramos, M. 1994. The use of lipolytic and proteolytic enzymes in the manufacture of Manchego type cheese from ovine and bovine milk. *J. Dairy Sci.*, *77*(8), 2139–49.

Food Eng. 1984a. Healthy cocoa alternative made from yeast and carob. *Food Eng.*, *56*(3), 105.

Food Eng. 1984b. Sodium hexametaphosphate . . . debitters KCl. *Food Eng.*, *56*(3), 96.

Food Eng. 1989. Przybyla, A. E., Ed., Enzymes used for protein hydrolysate . . . debittering. *Food Eng.*, 51–52.

Fujii, K., Masutake, K., Fukumoto, T., Yoshizaki, M. and Nishama, H. 1995. Manufacture of seasoning powder with starch, gums, neohesperitin dihydrochalcone and thaumatins, JP 07,99,916; JP 07,99,918; JP 07,99,919; JP 07,99,920 to Saneigen Efu Efu Ai Kk.

Fujii, M. and Kai, F. 1994. Manufacture of protein hydrolyzate seasonings with low chloro alcohol contents, JP 06,105,665 to Ashai Chemical Ind.; *Chem. Abstr.* 121: 132700e.

Fuller, W. D. and Kurz, R. J. 1991. Ingestibles containing substantially tasteless sweetness inhibitors as bitter taste reducers or substantially tasteless bitter inhibitors as sweet taste reducers, World Patent Cooperation Treaty, WO 91/18523.

Gabrial, G. N., El-Nahry, F. I., Awadalla, M. Z. and Girgis, S. M. 1981. Unconventional protein sources: apricot seed kernels. *Z. Emaehrungswiss.*, *20*(3), 208.

Gallagher, J., Kanekanian, A. D. and Evans, E. P. 1994a. Debittering of α-casein hydrolyzates by a fungal peptidase. *Spec. Publ. Royal Soc. Chem.*, *150* (Biochemistry of Milk Products), 143–51.

Gallagher, J., Kanekanian, A. D. and Evans, E. P. 1994b. Hydrolysis of casein: a comparative study of two proteases and their peptide maps. *Int. J. Food Technol.*, *29*(3), 279–85.

Gamay, A. 1990. World Patent Cooperation Treaty WO 91/17664.

Gao, J. and Jin, C. 1992. Comparison of glucoside content of bitter apricot seeds processed in different ways and stored routinely for one year. *Zhongguo Zhongyao Zazhi, 17*(11), 658–9, in Chinese; *Chem. Abstr.* 118: 56226a.

Garbut, T. 1993. Non-bitter protein hydrolyzates. Grain Processing Corp. U.S. Pat. No. 5,266,685.

Garcia Puig, D., Perez, M. L., Fuster, M. D., Ortuno, A., Sabater, F., Porras, I., Garcia Lidon, A. and Del Rio, J. A. 1995. Effect of ethylene on naringin, narirutin and nootkatone accumulation in grapefruit. *Planta Med., 61*(3), 283–5.

Ge, S. and Zhang, L. 1993. Removal of bitterness of proteins by complete hydrolysis with an immobilized composite exopeptidase, CN 1,078,495; *Chem. Abstr.* 120: 268365m.

Gobbetti, M., Cossignani, L., Simonetti, M. S. and Damiani, P. 1995. Effect of the aminopeptidase from *Pseudomonas fluorescens* ATCC 948 on synthetic bitter peptides, bitter hydrolyzate of UHT milk proteins and on the ripening of Italian Caciotta type cheese. *Lait, 75*(2), 169–79.

Gocho,. S., Kawabata, Y. and Nakanishi, J. 1995. Control of bitter taste of protein hydrolyzates by enzymes, JP 07,115,913 to Hasegawa, T. Co., Ltd.

Goodwin, J. C. and Hodge, J. E. 1981. Sweetness and bitterness of some aliphatic, α,ω-glycol D-glucopyranosides. *J. Agric. Food Chem., 29*(5), 935.

Greenberg, N. A., Kvamme, C. and Schmidl, M. K. 1993. Oral nutritional compositions having improved palatability containing acid salts of amino acids, Eur. Pat. Appl. EP 0567433 to Sandoz Nutrition Ltd.

Grilc, V. and Dabrowski, B. 1994. Utilization of waste hop foliage by solvent extraction. *Bioresour. Technol., 49*(1), 7–12.

Guadagni, D. G., Maier, V. P. and Turnbaugh, J. G. 1976. Effect of Neodiosmin on threshold and bitterness of limonin in water and orange juice. *J. Food Sci., 41*(3), 681.

Guerrero, M., Acuna, P. and Ruales, J. 1991. Bench-scale debittering of chocho *(Lupinus mutabilis)* seeds, isolation of sparteine, and preliminary trials concerning texturization of the protein concentrate. *Politecnica, 16*(2), 49–79, in Spanish; *Chem. Abstr.* 117: 47022h.

Handa, K. 1994. Application of plant hormone mixtures for improving taste of leafy vegetables and tea, JP 06,169,642.

Haring, P. G. M., De Kok, P. M. T., Potman, R. P. and Wesdorp, J. J. 1993. Starches with an improved flavor, U.S. Pat. No. 5,246,718 to Van den Bergh Foods.

Hasegawa, S., Dillberger, A. M. and Choi, G. Y. 1984. Metabolism of limonoids: conversion of nomilin to obacunone in *Corynebacterium fascians. J. Agric. Food Chem., 32*(3), 457.

Hasegawa, S., Vandercook, C. E., Choi, G. Y., Herman, Z. and Ou, P. 1985. Limonoid debittering of citrus juice sera by immobilized cells of *Corynebacterium fascians. J. Food Sci., 50*(2), 330.

Hashinaga. F., Toyofuku, K. and Itoo, S. 1984. Screening of some microorganisms producing limonoid-degrading enzyme used for debittering citrus juices. *Bull. Fac. Agric., 34,* 39.

Hatzold, T., Elmadfa, I., Gross, R., Wink, M., Hartmann, T. and Witte, L. 1983. Quinolizidine alkaloids in seeds of *Lupinus mutabilis. J. Agric. Food Chem., 31*(5), 934.

Herman, Z., Hasegawa, S. and Ou, P. 1985. Nomilin acetyl-lyase, a bacterial enzyme for nomilin debittering of citrus juices. *J. Food Sci., 50*(1), 118.

Hernandez, E., Couture, R., Rouseff, R., Chen, C. S. and Barros, S. 1992. Evaluation of ultrafiltration of adsorption to debitter grapefruit juice and grapefruit pulp wash. *J. Food. Sci.*, *57*(3), 664–6, 670.

Hirano, K., Ito, H., Fujoshi, T., Nakamura, T. and Yadanobe, Y. 1991. Japanese Patent JP 03,123,484.

Hiraoka, Y. 1993. Manufacture of liquid condiments with protease, JP 05,308,922 to Ehime Prefecture; *Chem. Abstr.* 120: 162221w.

Hoo, Y. T. 1991. Japanese Patent JP 03 91,482.

Horowitz, R. M. and Gentili, B. 1969. Taste and structure in phenolic glycosides. *J. Agric. Food Chem.*, *17*(4), 696.

Horowitz, R. M. and Gentili, B. 1977. Flavonoids: uses and byproducts. *Proc. Int. Soc. Citricult.*, 743.

Horowitz, R. M. 1978. Taste and structure relations of flavonoid compounds. *Int. Congr. Food Sci. Technol. Abstr.* 66.

Horowitz, R. M. and Gentili, B. 1986. In *Alternative Sweeteners*, O'Brien Nabors, L. and Gelardi, R. C., Eds., Marcel Dekker, New York, 142.

Hoshikawa, H. and Kikuchi, H. 1991. Japanese Patent JP 03,191,767.

Hoshikawa, H. and Kikuchi, H. 1993. Bitterness removal from juices with silica gel, JP 05,111,371 to Yakult Honsho Kk, 1993; *Chem. Abstr.* 119: 94124u.

Hosokawa, M., Fujiwara, Y. and Toyoda, N. 1993. Improvement of flavor of herring eggs with branched cyclodextrin, JP 05,161,477 to Hasegawa T. Co. Ltd.; *Chem. Abstr.* 119: 138081j.

Hoyle, N. T. and Merritt, J. H. 1994. Quality of fish protein hydrolyzates from herring *(Clupea harengus)*. *J. Food Sci.*, *59*(1), 76–9, 129.

Hsieh, P. C. and Tsen, H. Y. 1991. Purification and characterization of α-rhamnosidase from two fungal naringinase preparations. *Zhongguo Nongye Huaxue Huizhi*, *29*(1), 61–73, in Chinese; *Chem. Abstr.* 115: 181683m.

Huang, C. J. and Zayas, J. F. 1991. Phenolic acid contribution to taste characteristics of corn germ protein flour products. *J. Food Sci.*, *56*(5), 1308–1310, 1315.

Hughes, R. 1985. Resins solve food processing problems. *Food Technol. N. Z.*, *20*(7), 33.

Ikezuki, Y. 1990. Japanese Patent JP 02,291,244.

Imamaura, H. and Takeuchi, N. 1994. Manufacture of odorless protein hydrolytic products as cosmetic materials, JP 06,279,229 to Nippon Oxygen Co., Ltd.

Inoue, A., Sakamoto, K. and Iyama, M. 1989. Adsorption of naringin by styrene-divinylbenzene and acrylic resins. *Hiroshima-kenritsu Shokuhin Kogyo Gijutsu Senta Kenkyu Hokoku*, (19), 15–21, in Japanese; *Chem. Abstr.* 112: 177206d.

Inoue, A. and Sakamoto, K. 1993. Studies on removal of bitter components in the juice of citrus. II. Adsorption of limonin in the juice of Hassaku (*C. hassaku* hort. ex Tanaka) in styrene-vinylbenzene copolymers. *Hiroshima-kenritsu Shokuhin Kogyo Gijutsu Senta Kenkyu Hokoku*, *20*, 33–35, in Japanese; *Chem. Abstr.* 122: 30146g.

Ishibashi, N., Arita, Y., Kanehisa, H., Kouge, K., Okai, H. and Fukui, S. 1987. Bitterness of leucine containing peptides. *Agric. Biol. Chem.*, *51*(8), 2389–2394.

Ishibashi, N., Ono, I, Kato, K, Shigenaga, T., Shinoda, I., Okai, H. and Fukui, S. 1988a. Role of the hydrophobic amino acid residue in the bitterness of peptides. *Agric. Biol. Chem.*, *52*(1), 91.

Ishibashi, N., Kouge, K., Shinoda, I., Kanehisa, H. and Okai, H. 1988b. A mechanism for bitter taste sensibility in peptides. *Agric. Biol. Chem.*, *52*(3), 819.

Ishida, T., Takahura, Y., Kawabe, T. and Morita, H. 1993. Additive for improving taste of potassium salts in beverages, JP 07,115,933 to Takara Shuzo Co.

Ishii, K., Nashimura, T., Ono, T., Hatae, K. and Shimada, A. 1994a. Taste of peptides in wheat gluten hydrolyzates by protease. *Nippon Kasei Gakkaishi, 45*(7), 615–20, in Japanese; *Chem. Abstr.* 121: 203838u.

Ishii, K., Nishimura, T., Ono, T., Hatae, K. and Shimada, A. 1994b. Effect of peptides in wheat gluten hydrolyzates on basic tastes. *Nippon Kasei Gakkaishi, 45*(9), 791–6, 797–801, in Japanese; *Chem. Abstr.* 122: 8399.

Iwase, T., Kayama, K. and Nakahara, H. 1992. Taste-improved sake and its manufacture by electrodialysis, JP 04,141,080 to Toyo Jozo Co., Ltd.; *Chem. Abstr.* 117: 169933b.

Izumitani, M. and Kazuhiko, N. 1991. *New Food Ind.*, *33*(3), 1–6.

Jimeno, A., Manjon, A., Canovas, M. and Iborra, J. L. 1987. Use of naringinase immobilized on glycophase-coated porous glass for fruit juice debittering. *Process Biochem.*, *22*(1), 13.

Johnson, C., Birch, G. G. and MacDougall, D. B. 1994. The effect of the sweetness inhibitor 2-(4-methoxyphenoxy)propanoic acid (sodium salt) (Na-PMP) on the taste of bitter-sweet stimuli. *Chemical Senses, 19*(4), 349–358.

Johnson, R. L. and Chandler, B. V. 1985a. Economic feasibility of adsorptive deacidification and debittering of Australian citrus juices. *CSIRO Food Res. Q.*, *45*(2), 25.

Johnson, R. L. and Chandler, B. V. 1985b. Ion exchange and adsorbent resins for removal of acids and bitter principles from citrus juices. *J. Sci. Food Agric.*, *36*(6), 480.

Johnson, R. L. and Chandler, B. V. 1986. Debittering and de-acidification of fruit juices. *Food Technol. Aust.*, *38*(7), 294.

Kaneko, T., Kojima, T., Kuwata, T. and Yamamoto, Y. 1993a. β-Lactoglobulin hydrolyzates with low antigenicity and no unpleasant taste and their manufacture, JP 05,137,515 to Meiji Milk Prod. Co., Ltd.; *Chem. Abstr.* 119: 115976r.

Kaneko, T., Kojima, T., Kuwata, T. and Yamamoto, Y. 1993b. Protein hydrolyzates free from unpleasant taste and antigenicity and their enzymic manufacture, JP 05,344,847 to Meiji Milk Prod. Co., Ltd.; *Chem. Abstr.* 120: 215760e.

Kann, A. G., Kask, K. A., Annusver, K. Kh. and Rand, T. I. 1982. Possible applications for brewing byproducts. *Tallina Poluetehnilise Instituudi Toimetised, 537,* 21.

Kato, H., Rhue, M. R. and Nishimura, T. 1988. Taste of free amino acids and peptides in foods. *Agric. Food Chem. Sect. of ACS Meeting,* Abstr. #20, Toronto, June.

Kato, T. and Yamabe, K. 1991. Japanese Patent JP 03,206,854.

Kawakami, A., Kayahara, H., Tadasa, K. and Ujihara, A. 1994. Isolation and taste improvement of tartary buckwheat protein by isoelectric precipitation. *Nippon Shokuhin Kogyo Gakkaishi, 41*(7), 481–4, in Japanese; *Chem. Abstr.* 121: 178269z.

Kawakami, A., Kayahara, H. and Tadasa, K. 1995a. Taste evaluation of angiotensin I converting enzyme inhibitors. *Biosci., Biotechnol., Biochem.*, *59*(4), 709–10.

Kawakami, A., Kayahara, H. and Ujihara, A. 1995b. Properties and elimination of bitter components derived from Tartarian buckwheat (*Fagopyrum tartaricum*) flour. *Nippon Shokuhin Kagaku Kogaku Kaishi, 42*(11), 892–8, in Japanese.

Kawakami, T. 1993. Manufacture of odorless peptides with protease, JP 05,244,978 to Matsudaira Tennenbutsu Kenkyus; *Chem. Abstr.* 120: 6933z.

Kiguchi, T. 1995. Removal of bitter taste of persimmon by ethanol, JP 07,107,903.

Kilara, A. 1985. *Process Biochem.*, *20,* 149–158.

Kimball, D. A. 1987. Debittering of citrus juices using supercritical carbon dioxide. *J. Food Sci.*, *52*(2), 481.

Kimball, D. A. 1991. *Citrus Processing, Quality Control and Technology,* Van Nostrand Reinhold.

Kitagawa, H. 1991. Japanese Patent JP 03,119,954.

Kitagawa, H. and Urase, K. 1995. Control of bitterness in banana by alcohol treatment, JP 07 87,885.

Kodama, M. 1992. Bitterness reduction of syruped Iyo orange (Citrus-Iyo Hort ex. Tanaka) segments with addition of branched cyclodextrin (Utilization and processing of middle or late ripening variety citrus fruits). *J. Japanese Soc. Food Science and Technology, 39*(5), 446–450.

Komai, K. 1994. Naringin derivatives as vegetable growth enhancers, JP 06,279,212 to Wakayama Aguri Baio Kenkyu Sen; Agurosu Kk.

Konishi, F., Esaki, S. and Kamiya, S. 1983. Synthesis and taste of some flavanone and dihydrochalcone glycosides in which carbohydrate moieties are located at differing positions of the aglycones. *Agric. Biol. Chem.*, *47*(7), 1419.

Kori, H., Matsubara, M., Fujii, Y. and Ueno, H. 1991. Plant protein hydrolyzates, their preparation, and their use in manufacturing beverages, JP 03,251,161 to Otsuka Pharmaceutical Co., Ltd.; Japan Maize Co., Ltd.; *Chem. Abstr.* 116: 40139x.

Kovacova, M., Pribela, A. and Kozarova, I. 1993. Stability of anthocyanin colorants and elimination of undesirable flavor and odor constituents of danewort. *Potravin. Vedy, 11*(5), 401–8, in Slovakian; *Chem. Abstr.* 120: 105293x.

Koyakumaru, T. and Ono, Y. 1995. Control of bitter taste of persimmon by ethanol, JP 07 99,884 to Rengo Co., Ltd.

Koyama, N. and Kurihara, K. 1972. Mechanism of bitter taste reception: interaction of bitter compounds with monolayer of lipids from bovine circumvallate papillae. *Biochim. Biophys. Acta, 288,* 22.

Koziol, M. 1991. Afrosimetric estimation of threshold saponin concentration for bitterness in quinoa (*Chenopodium quinoa* Willd). *J. Sci. Food Agric.*, *42*(2), 211–219.

Kuang-chih, T. and Hua-zhong, H. 1987. (Neodiosmin is incorrectly identified as Rhoifolin in this reference.) Structural theories applied to taste chemistry. *J. Chem. Ed.*, *64*(12), 1003.

Kuchiba-Manabe, M., Matoba, T. and Hasegawa, K. 1991. Sensory changes in umami taste of inosine-5′-monophosphate solution after heating. *J. Food Sci.*, *56*(5), 1429–32.

Kuepcue, S., Mader, C. and Sara, M. 1995. The crystalline cell surface layer . . . as an immobilization matrix. *Biotechnol. Appl. Biochem.*, *21*(3), 275–86.

Kumai, S. 1995. Manufacture of health beverages and jellies, JP 07,115,942.

Kurihara, Y., Shimada, T., Saitoh, M., Ikeda, K., Sugiyama, H. and Kohno, K. 1990. Patent Cooperation Treaty WO 91/17671.

Kurtz, R. J. and Fuller, W. D. 1993. Ingestibles containing substantially tasteless sweetness inhibitors as bitter taste reducers or substantially tasteless bitter inhibitors as sweet taste reducers, U.S. Pat. No. 5,232,735 to Bioresearch Inc.

Kurz, O. 1986. Process and device for thermal treatment of soybeans, European Patent Appl. EP 0193633.

Lai, K. G., Pokomy, P., Davidek, J. and Valentova, H. 1987. Effect of caffeine content on the bitterness of tea infusions. *Sb. Vys. Sk. Chem.-Technol. Praze, Potraviny, E61,* 95–144.

Lee, K.-P. D. and Warthesen, J. J. 1996. Preparative methods of isolating bitter peptides from cheddar cheese. *J. Agric. Food Chem., 44,* 1058–1063.

Levy, E. 1992. Removal of chlorine from beverages with thiosulfates, U.S. Pat. No. 5,096,721 to Selecto, Inc.

Li, C. 1992. Mechanism and removal of bitterness of citrus juices. *Shipin Kexue (Beijing), 145,* 12–14, in Chinese; *Chem. Abstr.* 117: 129991n.

Lindley, M. G. and Rathbone, E. B. 1985. Lactisole. UK Patent Application GB2157148A to Tate and Lyle.

Lowrie, R. J. and Lawrence, R. C. 1992. Cheddar cheese flavor. IV. A new hypothesis to account for the development of bitterness. *N. J. Dairy Sci. Tech., 7,* 51.

Maack, E., Schneeweiss, V. and Schramm, G. 1984. Process for manufacture of partially hydrolysed pea milling products, GDR Patent DD 212646.

Magnolato, D. 1983. Nestle SA, Swiss Patent CH 635732 A5.

Manjon, A., Bastida, J., Romero, C., Jimeno, A. and Iborra, J. L. 1985. Immobilization of naringinase on glycophase-coated porous glass. *Biotech. Lett., 7*(7), 477.

Manlan, M., Matthews, R. F., Rouseff, R. L., Littell, R. C., Marshall, M. R., Moye, H. A. and Teixeira, A. A. 1990. Evaluation of the properties of polystyrene-divinylbenzene adsorbents for debittering grapefruit juices. *J. Food Sci., 55*(2), 440–5, 449.

Margolis, G., Chiovini, J. and Pagiaro, F. A. 1989. U.S. Patent 4,861,607.

Marwaha, S. S., Puri, M., Bhullar, M. K. and Kothari, R. M. 1994. Optimization of parameters for hydrolysis of limonin for debittering of kinnow mandarin juice by *Rhodococcus fascians. Enzyme Microb. Technol., 16*(8), 723–5; *Chem. Abstr.* 121: 106986h.

Masanobu, H. and Keiji, W. 1994. Central nervous system activator and taste enhancing food additive, U.S. Pat. No. 5,344,648 to Latron Laboratories, Inc.

Matsumoto, R. 1995. Studies on genetics of bitterness of citrus fruit and aplication to breeding of bitterless citrus cultivars with special reference to the bitterness caused by flavonone glycosides. *Kaju Shikenjo Hokoku, Extra 6,* 1–74, in Japanese/English; *Chem. Abstr.* 1996: 647890.

Matsushita, I. and Ozaki, S. 1995. Purification and sequence determination of tasty tetrapeptide (Asp-Asp-Asp-Asp) from beer yeast seasoning and its enzymic synthesis. *Pept. Chem. 1994 32nd* (Pub. 1995), 249–52.

McGillivray, T. D., Groom, D. R., Brown, D., Fergle, R. R. and Haakenson, G. 1993. Method of preparation of sugar beet fiber material in U.S. Pat. No. 5,213,836 to American Crystal Sugar Co.

Mega, A. 1994. Studies on peptide components in soup stock. *Monatsh. Chem., 125*(8–9), 99–103, in Japanese; *Chem. Abstr.* 121: 229271d.

Merck Encyclopedia of Chemical Drugs and Biologicals, The Merck Index, 10th ed. 1983. No. 9116; No. 622; No. 9107; No. 6948; No. 5471; Nos. 7970, 7973, 7981, 7983; Denatonium benzoate, No. 2863.

Meyer, B. N., Heinstein, P. F., Burnouf-Radosevich, M., Delfel, N. E. and McLaughlin, J. L. 1990. *J. Agric. Food Chem., 38,* 205–208.

Michikawa, K. and Konosu, S. 1995. Sensory identification of effective components for masking bitterness of arginine in synthetic extract of scallop. *Nippon Shokuhin Kagaku Kogaku Kaishi, 42*(12), 982–8, in Japanese.

Miller, H. W., Jr. 1981. Method and apparatus for debittering soybeans, U.S. Patent No. 4,248,141.

Minamiura, N., Matsumura, Y. and Yamamoto, T. 1972. Bitter peptides in the casein-digests with bacterial proteinase. II. A bitter peptide consisting of tryptophan and leucine. *J. Biochem. (Tokyo), 72*(4), 841.

Mogenson, L. and Adler-Nissen, J. 1988. Bitterness intensity of protein hydrolysates—chemical and organoleptic characterization. *Proc. 5th Int. Flavor Conf.*, July 1–3, 1987 in *Frontiers of Flavor,* Charalambous, G., Ed., Elsevier, Amsterdam, 63.

Molimard, P., Lesschaeve, I., Bouvier, I., Vassal, L., Schlich, P., Issanchou, S. and Spinnler, H. E. 1994. Bitterness and nitrogen fractions of mold ripened cheese of Camembert type: impact of the association of *Penicillium camemberti* with *Geotrichum candidum. Lait, 74*(5), 361–74, in French.

Molimard, P., Bouvier, I., Issanchou, S., Lesschaeve, I., Vassal, L. and Spinnler, H. E. 1995. Cooperation between *Penicillium camemberti* and *Geotrichum candidum:* effect on taste and flavor qualities of Camembert type cheese. *Colloq.-Inst. Natl. Rech. Agron., 75*(Bioflavour 95), 173–5.

Moll, D. 1990. Manufacturing protein hydrolysates without giving rise to bitter taste. Röhm GmbH, *Conference Proceedings of the F.I.E.*, Dusseldorf, Germany.

Moncreiff, R. W. 1967. *The Chemical Senses,* 2nd ed., Leonard Hill Books, London.

Moncrieff, R. W. 1970. *Flavor Ind., 1,* 583.

Moore, D. E. and McAnalley, W. H. 1995. Drink containing mucilaginous polysaccharides and its preparation, U.S. Pat. No. 5,443,830 to Carrington Labratories, Inc.

Moriguchi, T. 1990 (Pub. 1991). Thermostability and taste of soy sauce with addition of taurine. *Kagawa-ken Hakko Shokuhin Shikenjo Hokoku, 83,* 38–9, in Japanese; *Chem. Abstr.* 117: 68703w.

Morishita, T. 1993 Pub. 1994). Changes in bitter substances of the grapefruits (*Citrus paradici* MACF) juice during preservation. *Mukogawa Joshi Daigaku Kiyo, Shizen Kagakau-hen, 41,* 35–8, in Japanese.

Morishita, T. 1994. Effect on the taste and flavor of marmalade by combining citrus fruits. *Nippon Kasei Gakkaishi, 45*(8), 709–12.

Murakami, F. and Minami, M. 1992. Soyasaponin preparations containing γ-cyclodextrin and beverages containing them, JP 04,364,130 to Maruzen Chemical Co. Ltd.; *Chem. Abstr.* 118: 146624h.

Murray, D. G. and Shackelford, J. R. 1991. U.S. Patent 5,064,663.

Nagabori, T., Hashimoto, H. and Takeishi, S. 1993. Bitterness removal from herring eggs with chelation agents, JP 05,161,478 to Kikkoman Corp.; *Chem. Abstr.* 119: 115972m.

Nagaoka, T., Hane, H., Yamashita, H. and Kensho, I. 1990. *Seito Gijutsu Kenkyukaishi 38,* 61–70.

Nair, M., Burke, B. and Mudd, B. 1990. *Phytochem. Anal., 1*(1), 31–35.

Nakagawa, T. 1992a. Removal of bitterness of potassium chloride with carrageenan, JP 04,262,758 to San Ei Chemical Industries, Ltd.; *Chem. Abstr.* 118: 21392z.

Nakagawa, T. 1992b. Bitter taste attributable to potassium chloride and its removing, JP 04,108,358 to San-Ei Kagaku Kogyo K.K.; *Chem. Abstr.* 117: 110495g.

Nakajima, N. 1984. Sweetener composition and sweetening method, Eur. Pat. Appl. 0122400 to Takeda Chem. Ind. Ltd.

Nakamura, T., Syukunobe, Y., Doki, R., Hirano, K. and Itoh, H. 1991. *Nippon Shokuhin Kogyo Gakkaishi 38*(5), 377–383.

Nakamura, T., Sado, H. and Syukunobe, Y. 1992. Antigenicity of whey protein hydrolyzates prepared with various proteases against rabbit antiserum to bovine β-lactoglobulin (Snow Brand Milk Product Co. Ltd.). *Animal Science Technology, 63*(8), 814–17.

Nakamura, T., Syukunobe, Y., Sakuari, T. and Idota, T. 1993. Enzymatic production of hypoallergenic peptides from casein (Snow Brand Milk Product Co. Ltd.). *Milchwissenschaft, 48*(1), 11–14.

Nakatani, M., Nakata, T., Kouge, K. and Okai, H. 1994. Studies on bitter peptides from casein hydrolysate. XIV. Bitter taste of synthetic analogs of octapeptide, Arg-Gly-Pro-Phe-Pro-Ile-Ile-Val, corresponding to the C-terminal portion of β-casein. *Bull. Chem. Soc. Jpn., 67*(2), 438–44; *Chem. Abstr.* 121: 57946j.

Narubayashi, I. 1991. Taste improvement of glycerin by calcium salts, JP 03,151,849 to Sakamoto Yakuhin Kogyo Co., Ltd.; *Chem. Abstr.* 115: 254747x.

Ney, K. H. 1986. Cocoa flavor-bitter compounds as its essential taste components. *Gordian, 86*(5), 84; *Chem. Abstr.* 105: 112769a.

Nielsen, P. M. and Jimenez, M. J. 1995. Enzyme technology for the manufacturing of protein-based flavorings. *Aliment., Equipos Tecnol., 14*(6), 91–5, in Spanish.

Noguchi, M., Yamashita, M., Arai, S. and Fujimaki, M. 1975. *J. Food Sci., 40,* 367.

Noomhorm, A. and Kasemsuksakul, N. 1992. Effect of maturity and processing on bitter compounds in Thai tangerine juice. *Int. J. Food Sci. Technol., 27*(1), 65–72.

Nout, M. J. R., Tuncel, G. and Brimer, L. 1995. Microbial degradation of amygdalin of bitter apricot seeds *(Prunis armeniaca). Int. J. Food Microbiol., 24*(3), 407–12.

Oberdieck, R. 1987. Geschmacksverstarker, Alkohol-Industrie 80/7, p. 156 in van Eijk, T., *Umami substances—flavor enhancers or modifiers?* Dragoco Report Flavoring Information Service, *3,* 64.

Okada, G., Totsuka, A., Nakakuki, T. and Unno, T. 1992. Removal of bitterness from β-glucooligosaccharides, JP 04,148,661 to Nippon Shokuhin Kako K.K.; *Chem. Abstr.* 117: 21188x.

Okazawa, S., Hashimoto, M. and Hatsutori, Y. 1993. Treatment of adzuki bean with enzymes, JP 05,244,889 to Morinaga & Co.; *Chem. Abstr.* 120: 7310f.

Okubo, K. and Otomo, K. 1991. Off-flavor free soy sauce lees for use in food, Japanese Patent JP 03 76,552 to Miyagiken Shoyu Jozo Kyodo Kumiai; Life Engineering Co., Ltd.; *Chem. Abstr.* 115: 134672x.

Omari, Y. J. 1992. Process for removing the bitterness from potassium chloride, U.S. Pat. No. 5,173,323.

Oonishi, T., Koiso, H., Tamiya, T. and Ishii, T. 1995a. PCT Int. Appl. WO 95 04,477, WO 95 04,478 and WO 95 04,478 to San-ei Gen F.F.I., Inc.

Oonishi, T., Koiso, H., Tamya, T. and Ishii, T. 1995b. Food additives containing thaumatins and phenoxyalkanoic acids and salts thereof, JP 07 79,730 and 07 79,731 to San-ei Gen Efu Efu Ai Kk.

O'Rourke, T. 1980. Efficient handling of brewery byproducts. *Brew. Guardian, 109*(9), 39.

Osaka, Y. 1990. Japanese Patent JP 02,255,067.

Ozawa, O. and Hino, S. 1993. Sweetener compositions containing cellobiulose, JP 05,207,861 to Nissan Sugar Mfg.; *Chem. Abstr.* 119: 224813f.

Pawlett, D. and Fullbrook, P. 1988. *Proc. Food Ingredients Europe, 3rd Int. Conf.*, London, November 15–17; Debitrase™, Imperial Biotechnology Inc.

Peterson, R. 1992. Sports beverage . . . growth, in *Food Processing*, Labell, F., Ed., December, pp. 36–40.

Pickenhagen, W., Dietrich, P., Keil, B., Polonsky, J., Nouaille, F. and Lederer, E. 1975. Identification of the bitter principle of cocoa. *Helv. Chim. Acta, 58,* 1078.

Premi, B. R., Lal, B. B. and Joshi, V. K. 1995. Debittering of kinnow juice with Amberlite XAD-16 resin. *Indian Food Packer, 49*(1), 9–17.

Price, K. R., Griffiths, N. M., Curl, C. L. and Fenwick, G. R. 1985. Undesirable sensory properties of the dried pea *(Pisum sativum)*. The role of saponins. *Food Chem., 17*(2), 105.

Puri, M., Marwaha, S. S. and Kothari, R. M. 1996. Studies on the applicability of alginate-entrapped naringinase for the debittering of kinnow juice. *Enzyme Microb. Technol., 18*(4), 281–5.

Rahma, E. H. and Rao, M. S. N. 1984. Effect of debittering treatment on the composition and protein components of lupin seed *(Lupinus termis)* flour, *J. Agric. Food Chem., 32*(5), 1026.

Rathbone, E. B., Butters, R. W., Cookson, D. and Robinson, J. L. 1989a. Chirality of 2-(4-methoxyphenoxy)propanoic acid in roasted coffee beans: analysis of the methyl esters by chiral HPLC. *J. Agric. Food Chem., 37,* 58.

Rathbone, E. B., Patel, G. D., Butters, R. W., Cookson, D. and Robinson, J. L. 1989b. Occurrence of 2-(4-methoxyphenoxy)propanoic acid in roasted coffee beans: analysis by GLC and HPLC. *J. Agric. Food Chem., 37,* 54.

Riemer, J. 1994. Bitterness inhibitors, U.S. Pat No. 5,336,513 to Kraft General Foods, Inc.

Roozen, J. P. 1989. In *Food Science: Basic Research for Technological Progress, Proceedings of the Symposium in Honour of Prof. W. Pilnik,* Wageningen, Nov. 25, 1988, Roozen, J. P., Rombouts, F. M. and Voragen, A. G. J., Eds., pp. 171–177, Pudoc Publishing, Wageningen, The Netherlands.

Roudot-Algaron, F., LeBars, D., Einhorn, J., Adda, J. and Gripon, J. C. 1993. Flavor constituents of aqueous fraction extracted from Comte cheese by liquid carbon dioxide. *J. Food Sci., 58*(5), 1005–9.

Roy, G. M. 1990. The applications . . . of bitterness reduction and inhibition in food products. *Crit. Rev. Food Sci. Nutr., 29*(2), 59–71.

Roy, G. M. 1992. Bitterness: reduction and inhibition. *Trends in Food Science and Nutrition, 3*(4), 85–91.

Roy, G. M. 1994. *Activated Carbon Applications in the Food and Pharmaceutical Industries,* Technomic Publishing Co., Inc., Lancaster, PA.

Saito, S. and Misawa, K. 1990. Taste improvement of casein hydrolyzates with cyclodextrin, Japanese Patent JP 02,283,246.

Sakanaka, S., Higuchi, T., Kuwano, K., Nishimoto, K., Kanetake, M. and Yamazaki, N. 1992. Tea extract free of bitterness for health food, JP 04,104,773 to Taiyo Kagaku Co., Ltd.; *Chem. Abstr.* 117: 68889m.

Shaw, P. E. and Wilson, C. W., III. 1983. Debittering citrus juices with β-cyclodextrin polymer. *J. Food Sci., 48*(2), 646.

Shaw, P. E. and Wilson, C. W., III. 1985. Reduction of bitterness in grapefruit juice with β-cyclodextrin polymer in a continuous-flow process. *J. Food Sci.*, *50*(4), 1205.

Shaw, P. E., Calkins, C. O., McDonald, R. E., Greany, P. D., Webb, J. C., Nisperos-Carriedo, M. O. and Barros, S. M. 1991. Changes in limonin and naringin levels in grapefruit albedo with maturity and the effects of gibberellic acid on these changes. *Phytochemistry, 30*(10), 3215–19.

Shiba, T. and Nunami, K.-ichi. 1974. Structure of a bitter peptide in casein hydrolysate by bacterial proteinase. *Tetrahedron Lett.*, 509.

Shidehara, N., Kuramoto, T., Yumoto, T. and Takaya, I. 1990. Japanese Patent JP 02 207,768.

Shimazu, Y., Hashimoto, H., Nakajima, Y. and Watabe, Y. 1992. Manufacture of tannin- and sourness-free cranberry juice, JP 04,316,468 to Kikkoman Corp. and Nippon Delmonte K.K.; *Chem. Abstr.* 118: 100833x.

Shin, W., Kim, S. J., Shin, J. M. and Kim, S.-H. 1995. Structure-taste correlations in sweet dihydrochalcone, sweet dihydroisocoumarin, and bitter flavone compounds. *J. Med. Chem.*, *38*(21), 4325–31.

Shinoda, I., Nosho, Y., Kouge, K., Ishibashi, N., Okai, H., Tatsumi, K. and Kikuchi, E. 1987. Variation in bitterness potency when introducing Gly-Gly residue into bitter peptides. *Agric. Biol. Chem.*, *51*(8), 2103.

So, R., Suguira, Y. and Ootsuji, K. 1995. Fatty acid glycerol diesters to control bitter taste of fruit and vegetable juice, JP 07 51,034, to Kao Corp.

Sproessler, B. and Plainer, H. 1990. Preparation of non-bitter protein hydrolyzates using proteinases and a heat-treated peptidase, FRG Patent DE 3,905,194 to Roehm GmbH.

Steinkraus, K. H. 1984. Process for producing defatted and debittered soybean meal (Cornell Res. Found., 1984), UK Patent Appl. GB 2136667 A.

Sugakawa, M. and Masuda, T. 1993. Taste improvers containing protamines, JP 05,328,935 to S&B Shokuhin Co., Ltd.; *Chem. Abstr.* 120: 190208r.

Sugakawa, M. and Masuda, T. 1994. Removal of bitter taste from beverages, JP 06,153,875 to S&B Shokuhin Co., Ltd.; *Chem. Abstr.* 121: 156269x.

Sugiyama, K., Egawa, M., Onzuka, H. and Oba, K. 1991. Characteristics of sardine muscle hydrolyzates prepared by various enzymic treatments. *Nippon Suisan Gakkaishi, 57*(3), 475–9; *Chem. Abstr.* 116: 19929v.

Swientek, R. J., Ed. 1988. Cyclodextrins debitter citrus juices, in *Food Processing,* p. 54.

Takahashi, M., Nakata, T., Nakatani, M., Kataoka, S., Nakamura, K. and Okai, H. 1995. Conversion of bitterness of C-terminal octapeptide of bovine β-casein (Arg-Gly-Pro-Phe-Pro-Ile-Ile-Val) into sweetness. *Pept. Chem.*, *1994* (Pub. 1995), *32nd,* 281–4.

Takemori. T., Ito, Y., Ito, M. and Yoshama, M. 1992. Flavor and taste improvement of cacao nib by enzymic treatment, JP 04,126,037 to Lotte Co., Ltd.

Tamura, M., Miyoshi, T., Mori, N., Kinomura, K., Kawaguchi, M., Ishibashi, N. and Okai, H. 1990a. Mechanism for the bitter tasting potency of peptides using *O*-aminoacyl sugars as model compounds. *Agric. Biol. Chem.*, *54*(6), 1401–1409.

Tamura, M., Mori, N., Miyoshi, T., Koyama, S., Kohri, H. and Okai, H. 1990b. Practical debittering using model peptides and related compounds. *Agric. Biol. Chem.*, *54,* 41–51.

Tanaka, O., Akano, H., Kako, N., Nakagawa, H., Okumura, H. and Kawamura, K. 1991. Japanese Patent JP 03 34,990.

Tanimoto, S., Tanabe, S., Watanabe, M. and Arai, S. 1991. *Agric. Biol. Chem.*, *55*(4), 1119–1123.

Tateo, F. and Caimi, P. 1993. Debittering of citrus juices: pilot experience with adsorbent resins. *Mitt. Geb. Lebensmittelunters. Hyg.*, *84*(4), 498–508.

Terayama, H. 1991. Regeneration of synthetic adsorbents used in removal of bitterness from citrus juice, JP 03,266,962 to Nippon Rensui Co.; *Chem. Abstr.* 116: 104844h.

Triani, R. J. and Meczkowski, F. J. 1987. Debittering bran flakes using citrus peel, U.S. Patent No. 4,661,362 to General Foods Corp.

Tsen, H. Y. 1984. Rhamnosidase is markedly inhibited by citric acid. Factors affecting the inactivation of naringinase immobilized on chitin during debittering of fruit juice. *J. Ferment. Technol. (Hakko Kogaku Zasshi)*, *62*(3), 263.

Tsen, H. Y. 1990. U.S. Patent 4,971,812.

Tsen, H. Y. and Yu, G. K. 1991. Limonin and naringin removal from grapefruit juice entrapped in cellulose triacetate fibers. *J. Food Sci.*, *56*(1), 31–4.

Tuncel, G., Nout, M. J. R. and Brimer, L. 1995. The effects of grinding, soaking and cooking on the degradation of amygdalin of bitter apricot seeds. *Food Chem.*, *53*(4), 447–51.

Turos, S. 1982. Process for preparing crumb products, U.S. Patent No. 4,346,121.

Uesugi, S., Matsuzaki, K. and Hashimoto, Y. 1994. Bitterness- and pungency-free protein hydrolyzates preparation with protease, JP 06,197,788 to Fuji Oil Co., Ltd.; *Chem. Abstr.* 121: 203553x.

Umetsu, H., Matsuoka, H. and Ichishima, E. 1983. Debittering mechanism of bitter peptides from milk casein by wheat carboxypeptidase. *J. Agric. Food Chem.*, *31*(1), 50.

University of Giessen authors. 1983. *Plant Foods Hum. Nutr.*, *32*(2), 125.

van Eikeren, P. and Brose, D. J. 1993. Membrane extraction of citrus bittering agents for producing citrus juice of reduced bitterness, U.S. Pat. No. 5,263,409 to Bend Research Inc.

Velisek, J. and Dolezal, M. 1994. Flavor changes during alkaline treatment of protein hydrolyzates. *Dev. Food Sci.*, *35 (Trends in Flavor Research)*, 367–71.

Vellucci, D. J., Barnett, R. E. and Zanno, P. R. 1987. 4-Alkoxy benzoyloxyacetic acids, UK Patent Application GB 2180534A.

Vieira, G. H. F., Martin, A. M., Saker-Sampaiao, S., Sobreira-Rocha, C. A. and Goncalves, R. C. F. 1995. Production of protein hydrolyzate from lobster (*Panulirus* spp.). *Dev. Food Sci.*, *37B*, 1405–15.

Visser, S. 1993. Proteolytic enzymes and their relation to cheese ripening and flavor; an overview. *J. Dairy Sci.*, *76*(1), 328–50.

Voigt, J., Voigt, G., Henrichs, H., Wrann, D. and Biehl, B. 1994. In vitro studies on the proteolytic formation of the characteristic aroma precursors of fermented cocoa seeds: the significance of endoprotease specificity. *Food Chem.*, *51*(1), 7–14.

Vreeman, H. J., Both, P. and Slangen, C. J. 1994. Rapid procedure for isolating the bitter carboxyl-terminal fragment 193–209 of β-casein on a preparative scale. *Neth. Milk Dairy J.*, *48*(2), 63–70.

Wakayama, T. 1992. Masking of bitterness of kojic acid by amino acids, JP 04 40,881 to Kongo Yakuhin K.K.; *Chem. Abstr.* 117: 25094f.

Warmke, R. and Belitz, H.-D. 1993. Influence of glutamic acid on the bitter taste of various compounds. *Z. Lebensm.-Untersuch. Forsch.* *197*(2), 132–3.

Watanabe, M., Shimizu, J. and Arai, S. 1990. *Agric. Biol. Chem.*, *54*(12), 3351–3353.

Watanabe, M., Arai, S., Tanimoto, S. and Seguro, K. 1992. Manufacture of food peptides without undesirable flavors, JP 04,126,039 to Ajinomoto Co., Inc.; *Chem. Abstr.* 118: 79809g.

We, Z., Zhou, G., Zhang, Y. and Li, X. 1992. Bitter essence in citrus products-limonin. *Shipin Yu Fajiao Gongye* (3), 72–6, 16, in Chinese; *Chem, Abstr.* 117: 232255z.

Weir, G. S. D. 1992. Proteins as a source of flavor. *Biochem. Food Proteins,* 363–408, Hudson, B. J. F., Ed., Elsevier, London, UK.

Wu, H., Calvarano, M. and DiGiacomo, A. 1991. Improvements of extracting naringin from grapefruit peel. *Essenze Deriv. Agrum.*, *61*(3), 187–91.

Xu, Z. and Liu, X. 1992. Bitterness reduction of citrus juice by three methods. *Shipin Yu Fajiao Gongye* (4), 60, 16, in Chinese.

Yano, H., Shimizu, H. and Araki, Y. 1995. Control of bitter taste of vitamin B1 derivatives in beverages, JP 07 95,867 to Takeda Chemical Ind. Ltd.

Yasumoto, R. 1994. Preparation of angiotensin converting enzyme inhibiting proteins without bitterness, JP 06,298,794 to Nippon Synthetic Chem. Ltd.

Yeom, H. W., Kim, K. S. and Rhee, J. S. 1994. Soy protein hydrolyzate debittering by lysine acetylation. *J. Food Sci.*, *59*(5), 1123–6.

Yoshida, T., Matsudaira, M., Aochi, M. and Uejima, M. 1990. JP 02,163,101.

Yoshimura, M., Tsuruya, R., Kibune, K., Kin, B., Hatsuta, H., Ogasawara, Y. and Nagato, Y. 1993. Tea extracts containing therapeutic polyphenols, JP 05,279,264 to Unitika Ltd., Taiyo Kagaku Kk; *Chem. Abstr.* 120: 86425y.

Zemanek, E. C. and Wasserman, B. P. 1995. Issues and advances in the use of transgenic organisms for the production of thaumatin, the intensely sweet protein from *Thaumatoccus danielli. Crit. Rev. Food Sci. Nutr.*, *35*(5), 455–66.

Zhao, J. and Whistler, R. L. 1994. Spherical aggregates of starch granules as flavor carriers. Discovery of a novel property of starch granules leads to a method for controlled release of flavor in food products. *Food Technology, 48*(7), 104–5.

Zhou, J. and Wen, Q. 1994. New technology for debittering of steviosides. *Zhongguo Tiaoweipin* (8), 15–16, in Chinese.

Zhu, B., Sun, H., Zhao, M. and Jiao, P. 1993. Influence of soaking solution with different pH value on removing bitterness and toxins of bitter almond. *Shipin Gongye Keji* (2), 3–6, in Chinese.

Ziajka, S., Dzwolak, W. and Zubel, J. 1994. The effect of processing variable on some properties of whey protein hydrolyzates. *Milchwissenschaft, 49*(7), 382–5.

SECTION II

A SYMPOSIUM

CHAPTER 6

Interactions between Sweet and Bitter Tastes

D. ERIC WALTERS[1]

INTRODUCTION

THE old adage "a spoonful of sugar helps the medicine go down" reminds us that sweet taste has been used to mask bitter taste for a very long time. Sweet taste is usually considered desirable whereas bitter taste is usually (but not always) considered undesirable, so sweetness is frequently used to cover or balance bitter taste. Sweet taste and bitter taste may interact in several ways. There can be a *masking* effect, in which sweet taste is used to cover up, balance, or overwhelm bitter taste. There may also be ways in which some sweeteners *inhibit* perception of some bitter substances.

Taste receptors are proteins on the surface of taste bud cells that recognize certain chemical structures and initiate the signals that the brain recognizes to mean *sweet, bitter, salty,* or *sour.* We often think of receptors using a lock-and-key analogy. For instance, sweet-tasting compounds are like *keys* that can fit into the sweet-receptor *locks.*

It is likely that there are several different (but related) kinds of sweet receptors and several kinds of bitter receptors. In the case of sweeteners, at least three lines of evidence point to multiple receptor types:

(1) Some sweeteners are perceived nearer the front of the tongue, while others appear to be more intensely sweet near the center of the tongue.
(2) The incredible structural diversity of sweet-tasting compounds argues strongly against a single receptor site, which could recognize all substances known to taste sweet.

[1]Associate Professor of Biological Chemistry, Finch University of Health Sciences, the Chicago Medical School, 3333 Green Bay Road, North Chicago, IL 60064, U.S.A.

(3) Some mixtures of two different sweeteners (e.g., aspartame + acesulfame) produce synergy, a higher level of sweetness for the mixture than would be predicted from the sweetness of the individual compounds; other mixtures (e.g., saccharin + acesulfame) are strictly additive and probably compete for a common receptor site (Carr et al., 1993).

All of these phenomena probably apply to bitter compounds as well, although there is far less sensory research on bitter taste. Despite numerous efforts, no one has yet been successful in isolating and characterizing sweet or bitter taste receptors at the molecular level.

In this chapter, we summarize a number of experiments that point to interactions between sweet and bitter taste, propose a likely mechanism by which sweet and bitter tastes may interact, and finally look at the practical implications of such a mechanism for masking or inhibiting bitter taste.

EXPERIMENTAL OBSERVATIONS

Numerous experiments suggest the existence of several related receptor types for sweet and bitter tastes. Also, important experiments indicate a relationship between sweet and bitter taste receptors. And there are experiments that give us some clues about the nature of these taste receptors. The following section summarizes some of the key experiments.

The chemical literature is filled with instances of the relationships between molecular structure and sweet and bitter taste. For example, Janusz has reviewed the more than 1000 analogs of aspartame that have been synthesized and tested in laboratories all over the world (Janusz, 1989). Besides hundreds of sweet-tasting analogs, there are many compounds that are bitter, some are both bitter and sweet, and a few are tasteless. Often, very subtle changes in structure can convert potently sweet compounds into potently bitter compounds. Figure 1 illustrates some examples of this phenomenon.

Aspartame [Figure 1(a)] is formed from the natural amino acids L-aspartate and L-phenylalanine; if D-phenylalanine is substituted [Figure 1(b)], a bitter compound is formed (Mazur et al., 1969). A sweet-tasting oxime [Figure 1(c)] can be converted to a bitter one [Figure 1(d)] just by altering the pattern of double bonds (Acton et al., 1960). Acesulfame [Figure 1(e)] tastes sweet, and to some people, has a bitter aftertaste; addition of an ethoxy group [Figure 1(f)] produces a substance that is only bitter (Clauss and Jensen, 1973; Clauss, 1980). In at least one instance, modification of a bitter substance has produced a sweet compound. Neohesperidin [Figure 1(h)] is a bitter component of citrus peel; breaking a

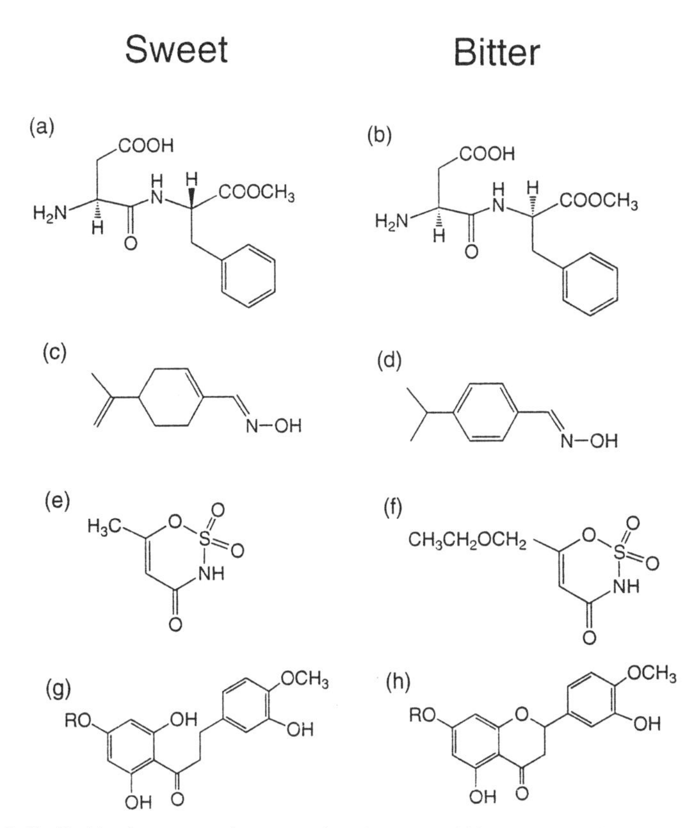

Figure 1 Similarities in structure between selected sweet and bitter compounds: (a) aspartame, L-aspartyl-L-phenylalanine methyl ester; (b) L-aspartyl-D-phenylalanine methyl ester; (c) a sweet oxime; (d) a bitter oxime; (e) acesulfame; (f) bitter ethoxy-substituted analog of acesulfame; (g) neohesperidin dihydrochalcone, R = b-neohesperidosyl; (h) neohesperidin, R = b-neohesperidosyl.

single bond in the molecule produces neohesperidin dihydrochalcone [Figure 1(g)], which is sweet (Horowitz and Gentilli, 1969). In the lock-and-key analogy, these results are an indication that some sweet "locks" are very similar to some bitter "locks."

Extending the lock-and-key analogy, taste inhibitors are compounds that "fit into the lock" but cannot open it. Bitter taste inhibition has been thoroughly reviewed by Roy (1992, 1994). There seem to be no bitterness competitive inhibitors with broad activity against many bitter compounds; instead, each inhibitor blocks some small set of bitter substances. Conversely, the known sweetness inhibitors appear to act against all sweeteners. These include an arylurea that blocks all 10 structural classes of sweeteners against which it was tested (Muller et al., 1992); lactisole, which blocks at least five classes of sweeteners (Lindley, 1991); and gymnemic acid, which blocks 11 different sweeteners (Hellekant and Ninomiya, 1991). Curiously, the arylurea also blocked some (but not all) bitter compounds, with no effect on salty or sour tastes, again indicating a link between sweet and bitter mechanisms.

It is well known that there is mixture suppression when bitter and sweet tastes are combined. That is, in a mixture containing both sweet and bitter substances, the perception of both sweet and bitter tastes is decreased. There is not agreement, however, on whether this suppression occurs at the level of the taste bud (Lawless, 1982) or in the brain (Kroeze and Bartoshuk, 1985).

One very curious aspect of sweetener perception is the phenomenon of temporal profile (Figure 2a)—the time of onset of taste perception and the duration of that perception (Carr et al., 1993). Sucrose, which is generally considered ideal, has a fairly rapid onset and clears fairly quickly. Sweeteners such as saccharin and acesulfame have a more rapid onset than sucrose and have a very short duration. At the other extreme, glycyrrhizin has a very slow onset and lingers for a very long time. Is this an indication that different sweeteners may act by different mechanisms? Again, bitter taste is not nearly so well studied, although qualitatively bitterness seems to be more of a slow onset/lingering taste phenomenon.

A common phenomenon in the study of pharmaceutical receptors is the occurrence of partial agonists. These compounds interact with their receptor and can cause a partial response. They are of interest because they are not able, even at high concentration, to produce a maximal response, and they can block the ability of other drugs to bind to the receptor. The classical example of this is nalorphine. This is an analgesic drug related to morphine. It can be used to treat severe pain, but it is also useful in treating an acute overdose of other narcotic analgesics, since it only partially triggers the receptors and prevents other drugs from triggering a full response.

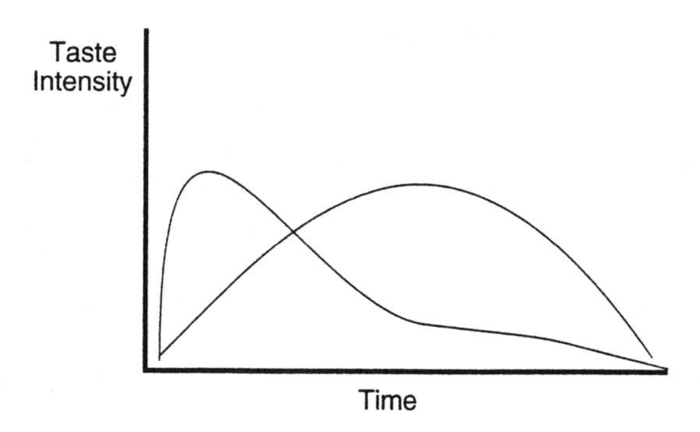

Figure 2a Schematic representation of rapid and slow temporal profiles.

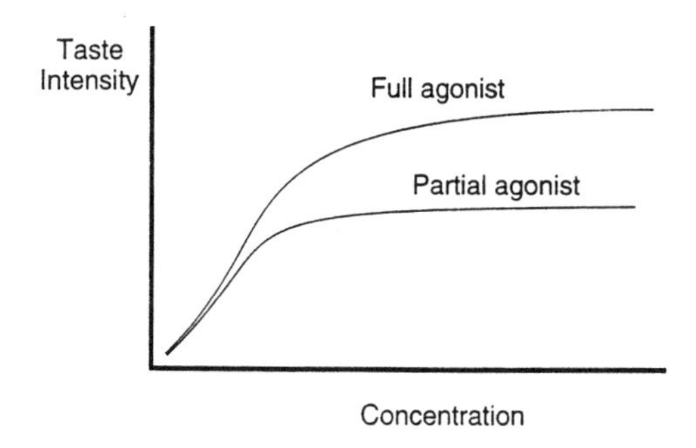

Figure 2b Illustration of the "partial agonist" phenomenon.

In the lock-and-key analogy, these keys open the lock "part of the way," and they prevent other, more active keys from being inserted into the lock. Partial agonists (Figure 2b) have been found among sweeteners (DuBois et al., 1991). These are compounds that have a high potency relative to sucrose at threshold levels, but cannot match the sweetness of a 10% sucrose solution no matter how high their concentration. The phenomenon of partial agonism has not been documented for bitter compounds, but this is very likely because no one has studied the sensory properties of bitter compounds in sufficient detail.

In recent years, it has become clear that sweet and bitter taste receptors are part of a large family of receptors called the G protein-coupled receptors (GPCR) (Akabas, 1990). Members of the GPCR family are utilized in olfaction, vision, neurotransmitters and peptide hormones. The GPCR is a very old family of receptor proteins, since some GPCRs serve as chemosensory receptors in single-celled organisms. As shown in Figure 2c, the receptor is a protein embedded in the cell membrane. The signal molecule on the outside of the cell is recognized by the receptor, which triggers a series of events inside the cell. First, the activated receptor binds to a GTP-binding protein (the G protein). This activates the G protein, which may cause initiation of a second messenger signal inside the cell. Ultimately, an ion channel may be opened, depolarizing the cell and triggering an electrical signal along a nerve fiber. Second messengers commonly used by GPCR include cyclic adenosine monophosphate (cAMP), inositol phosphates (IP3) and calcium ions (Ca^{2+}).

It has been shown in Naim's laboratory (Striem et al., 1989) that some sweeteners produce increased levels of cAMP in cell membranes from rat tongue. Further, this group has demonstrated (Striem et al., 1990) that an inhibitor of sweet taste is able to block the sucrose-stimulated production of cAMP in these membranes. These findings clearly implicate GPCR in sweet taste transduction.

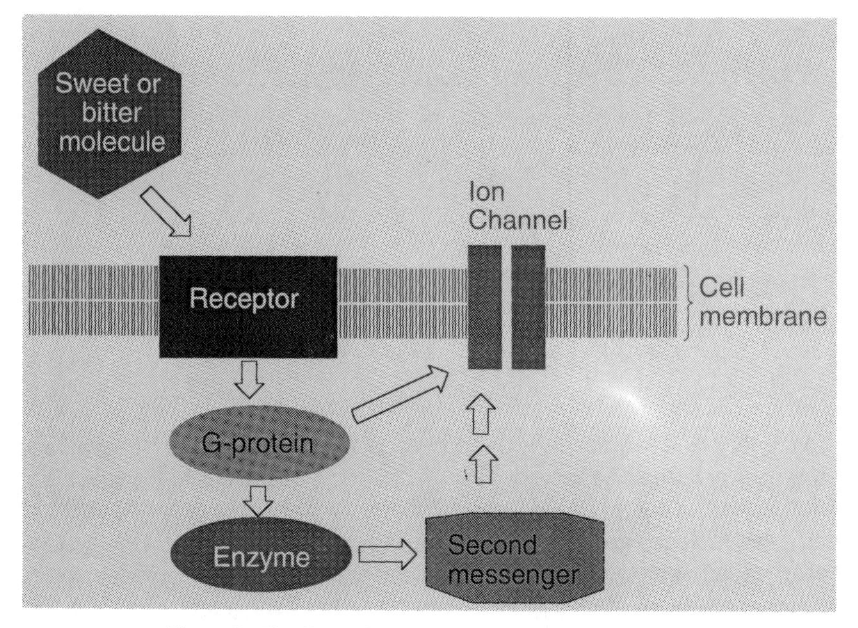

Figure 2c The G protein-coupled receptor (GPCR) system.

The transduction of bitter taste is even more clearly linked to GPCR. The bitter-tasting substance denatonium was shown (Akabas et al., 1988) to stimulate intracellular calcium ion release in rat taste cells; calcium release by the second messenger phosphatidyl inositol is a common pathway for GPCR in other tissues. Phosphatidyl inositol was demonstrated to be involved in bitter taste transduction (Hwang et al., 1990). Finally, McLaughlin et al. (1992) found a G protein in taste cells, which they named gustducin. This protein is closely related to the retinal G protein transducin, which is involved in vision, and it is likely that gustducin is involved in bitter taste transduction.

MECHANISTIC IMPLICATIONS

The GPCR is a vast family of related receptors that detect a broad range of chemical signals. The likely relationship between sweet and bitter taste receptors may be inferred by looking at the relationships among some other, more extensively studied families of GPCR. For example, the neurotransmitters norepinephrine, dopamine, and serotonin are structurally related (Figure 3). Their GPCRs are also structurally related, as determined by sequencing the genes for these receptors. In fact, there are multi-

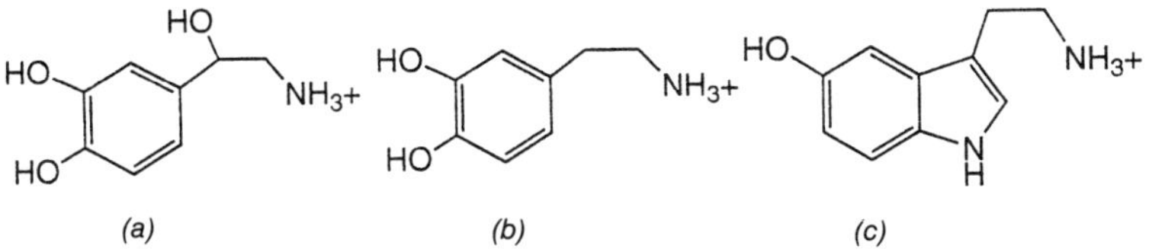

Figure 3 Structures of the neurotransmitters: (a) norepinephrine, (b) dopamine and (c) serotonin.

ple receptors for each of these neurotransmitters. The adrenergic (norepinephrine) receptors have been classified into α and β types, and these have been further classified into alpha-1, alpha-2, beta-1, and beta-2 subtypes.

Recent evidence points to subtypes of these subtypes. Norepinephrine is able to activate all of these subtypes, but some drugs act with varying degrees of selectivity: some drugs act at all adrenergic receptors, others are selective for α and β receptors, and yet others are selective for only α-2, for example. Similarly, dopamine and serotonin receptors have subtypes, and there are drugs with varying degrees of selectivity for these subtypes. Most interesting is the observation that many drugs are even less selective, with affinities for all three major classes of receptors. Figure 4 shows schematically the relationship among adrenergic receptor types. Comparison of gene sequences for many neurotransmitter GPCR (Harrison et al., 1991) indicates that all of these receptor types evolved from some common ancestral neurotransmitter receptor.

I propose that the sweet and bitter receptors similarly evolved from a common ancestral chemoreceptor. These receptors have differentiated

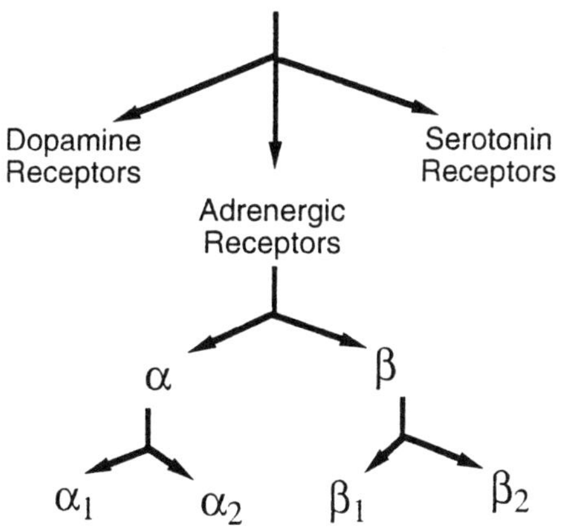

Figure 4 The adrenergic (norepinephrine) receptor family tree.

broadly into sweetness types and bitterness types, and there are probably many subtypes of each. As is the case for adrenergic receptors, there are probably some sweet compounds that act upon many sweet receptor types, and others that are selective for one or a few types; there are probably some bitter compounds that act upon many bitter receptor types, and others that are selective for one or a few types; and there may well be some compounds that act upon both sweet receptors and bitter receptors. Figure 5 shows the proposed evolutionary relationship of these receptor types. Such a scheme could account for the following observations:

- the structural diversity among sweet-tasting compounds
- the structural diversity among bitter-tasting compounds
- the variations in taste quality among sweeteners
- the variations in location of perceived tastes on the tongue
- the ability of some sweetness inhibitors to block many different sweeteners
- the structural similarities among some sweet and bitter compounds
- the observation that some compounds taste both bitter and sweet
- the observation that sweeteners partially block bitter taste and bitter compounds partially mask sweet taste

What about the observation that different sweeteners have different temporal profiles? A closer look at Figure 2c shows that there are several different places where a chemical substance may affect the signal transduction pathway. The receptor site is the obvious one, but compounds that affect the ultimate ion channel could also activate or block the signal to the nervous system. Similarly, compounds that enter the cell and modulate the G protein or the second messenger activity would also modify taste perception. Likely candidates in this category include the sweetness inhibitor gymnemic acid (which requires pretreatment of the taste bud in order to be effective) and caffeine (which is an inhibitor of the enzyme phosphodiesterase, responsible for breaking down cAMP). Different sites of action could account for different temporal profiles. For example, a compound

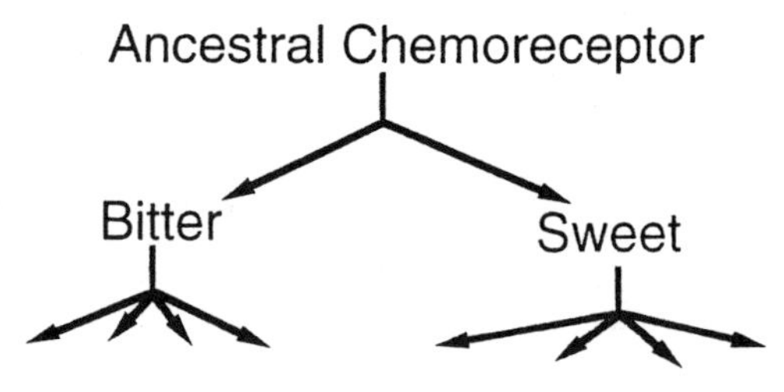

Figure 5 Proposed evolutionary relationship among sweet and bitter taste receptors.

that blocks ion channels could be a sweetness inhibitor with a broad range of activity. Finally, compounds that perturb the lipid membrane in which the receptor is embedded could *indirectly* activate the receptor. This might be the mechanism of action for sugars, which are not stereoselective in their activity (L-sugars and D-sugars are equally sweet) and which taste sweet only at very high concentration. Most GPCRs bind to their ligands at concentrations of 0.000001 molar or less; we taste sugars only in the 0.1 molar concentration range.

PRACTICAL IMPLICATIONS

If this proposed rationale is indeed the case, what lessons can we derive for controlling the perception of bitter taste?

First, we may look for and exploit specific bitter taste inhibitors. There are no bitter taste inhibitors that are universally effective, but many inhibitors that block specific bitter tastes are known. These have been thoroughly reviewed by Roy (1992, 1994).

Further, we can look in specific places for taste inhibitors. When we find a tasteless compound that is a close structural analog of known sweet or bitter compounds, we should evaluate it as a taste inhibitor. This approach has been successful in at least two instances. Lindley discovered lactisole while working on sweeteners in the Tate & Lyle research laboratories (Lindley, 1991). Similarly, a study of analogs of the sweet compound suosan carried out by the NutraSweet Company resulted in the discovery of a compound that inhibits sweet taste and some bitter tastes as well (Muller et al., 1992).

Cross-reactivity among receptors is a possible mechanism for the observed mixture suppression that occurs when bitter and sweet compounds are mixed. Receptor cross-reactivity is known to occur among compounds that act upon neurotransmitters, and this could happen with related sweet and bitter receptors as well. If this is the case, then it is a good idea to try as many different sweeteners as possible when trying to mask a bitter taste with sweetness. Combinations of sweeteners might also be expected to increase the possibility of inhibiting appropriate classes of receptors.

Since most bitter tastes seem to have a long temporal profile, it will clearly be beneficial to use sweeteners with longer temporal profiles in order to mask bitterness. Saccharin and acesulfame are of very short duration, so they may prove to be less useful than, for example, aspartame.

Finally, although it may seem counterintuitive, really intense bitter taste might benefit by addition of other, less intense bitter-tasting compounds. This would be the case if moderately bitter compounds act as partial agonists, occupying the receptor, inducing a less-than-maximal bitter taste, and preventing other bitter compounds from accessing the receptor.

CONCLUSION

As we learn more about the mechanisms of taste perception at the molecular level, we will be better able to control perception of sweet and bitter taste. Particularly, for blocking bitter taste, it would be useful to actively look for specific bitter taste inhibitors, and to try several different sweeteners or combinations of sweeteners.

REFERENCES

Acton, E. M., Leaffer, M. A., Oliver, S. M. and Stone, H. 1970. Structure-taste relationships in oximes related to perillartine, *J. Agric. Food Chem.*, 18: 1061–1068.

Akabas, M. H., Dodd, J. and Al-Awqati, Q. 1988. A bitter substance induces a rise in intracellular calcium in a subpopulation of rat taste cells, *Science,* 242: 1047–1050.

Akabas, M. H. 1990. Mechanisms of chemosensory transduction in taste cells, *Int. Rev. Neurobiol.*, 32: 241–279.

Carr, B. T., Pecore, S. D., Gibes, K. M. and DuBois, G. E. 1993. Sensory methods for sweetener evaluation, in *Flavor Measurement,* Ho, C.-T. and Manley, C. H., eds., New York: Marcel Dekker, pp. 219–237.

Clauss, K. and Jensen, H. 1973. Oxathiazinone dioxides—A new group of sweetening agents, *Angew. Chem. Internatl. Ed. Engl.*, 12: 869–876.

Clauss, K. 1980. Neue, süss schmeckende oxathiazinon-dioxide, *Liebigs Ann. Chem.*, 494–502.

DuBois, G. E., Walters, D. E., Schiffman, S. S., Warwick, Z. S., Booth, B. J., Pecore, S. D., Gibes, K., Carr, B. T. and Brands, L. M. 1991. Concentration-response relationships of sweeteners: A systematic study, in *Sweeteners: Discovery, Molecular Design, and Chemoreception.* Walters, D. E., Orthoefer, F. T. and DuBois, G. E., eds., Washington, DC: American Chemical Society, pp. 261–276.

Harrison, J. K., Pearson, W. R. and Lynch, K. R. 1991. Molecular characterization of alpha-1- and alpha-2-adrenoceptors, *Trends Pharmacol. Sci.*, 12: 62–67.

Hellekant, G. and Ninomiya, Y. 1991. On the taste of umami in chimpanzee, *Physiol. Behav.*, 49: 927–934.

Horowitz, R. M. and Gentili, B. 1969. Taste and structure in phenolic glycosides, *J. Agric. Food Chem.*, 17: 696–700.

Hwang, P. M., Verma, A., Bredt, D. S. and Snyder, S. H. 1990. Localization of phosphatidylinositol signaling components in rat taste cells: role in bitter taste transduction, *Proc. Natl. Acad. Sci. USA,* 87: 7395–7399.

Janusz, J. M. 1989. Peptide sweeteners beyond aspartame, in *Progress in Sweeteners,* Grenby, T. H., ed., Essex: Elsevier Applied Science Publishers, pp. 1–46.

Kroeze, J. H. A. and Bartoshuk, L. M. 1985. Bitterness suppression as revealed by split-tongue taste stimulation in humans, *Physiol. Behav.*, 35: 779–783.

Lawless, H. 1982. Paradoxical adaptation to taste mixtures, *Physiol. Behav.*, 25: 149–152.

Lindley, M. G. 1991. Phenoxyalkanoic acid sweetness inhibitors, in *Sweeteners: Discovery, Molecular Design, and Chemoreception*, Walters, D. E., Orthoefer, F. T. and DuBois, G. E., eds., Washington, DC: American Chemical Society, pp. 251–260.

Mazur, R. H., Schlatter, J. M. and Goldkamp, A. H. 1969. Structure-taste relationships of some dipeptides, *J. Amer. Chem. Soc.*, 91: 2684–2691.

McLaughlin, S. K., McKinnon, P. J. and Margolskee, R. F. 1992. Gustducin is a taste-cell-specific G protein closely related to the transducins," *Nature,* 357: 563–569.

Muller, G. W., Culberson, J. C., Roy, G., Ziegler, J., Walters, D. E., Kellogg, M. S., Schiffman, S. S. and Warwick, Z. S. 1992. Carboxylic acid replacement structure-activity relationships in suosan type sweeteners. A sweet taste antagonist, *J. Med. Chem.*, 35: 1747–1751.

Roy, G. M. 1992. Bitterness: reduction and inhibition, *Trends Food Sci. Technol.*, 3: 85–91.

Roy, G. 1994. *Pharmaceutical Technology,* April, pp. 84–99; May, p. 36.

Striem, B. J., Pace, U., Zehavi, U., Naim, M. and Lancet, D. 1989. Sweet tastants stimulate adenylate cyclase coupled to GTP-binding protein in rat tongue membranes, *Biochem. J.*, 260: 121–126.

Striem, B. J., Yamamoto, T., Naim, M., Lancet, D., Jakinovich, W., Jr. and Zehavi, U. 1990. The sweet taste inhibitor methyl 4,6-dichloro-4,6-dideoxy-α-D-galactopyranoside inhibits sucrose stimulation of the chorda tympani nerve and of the adenylate cyclase in anterior lingual membranes of rats, *Chem. Senses,* 15: 529–536.

CHAPTER 7

Factors Affecting the Perception of Bitterness: A Review

J. H. THORNGATE III[1]

INTRODUCTION

MITIGATING bitter taste remains a primary goal for many in the food and beverage industry. Bitterness is associated with a wide array of foodstuffs, including fruits (e.g., grapefruit, oranges), vegetables (e.g., cucumbers, avocados), beverages (eg., beer, coffee) and protein commodities (e.g., dairy and soy products). Although bitterness is desirable in a few select commodities (notably coffee), evidence suggests that bitterness acceptance is a learned phenomenon. The aversive response of newborns to bitter stimuli is perhaps indicative of an innate toxicity rejection mechanism (Bartoshuk and Beauchamp, 1994; cf. Glendinning, 1994).

Bitterness is most commonly modified in palatables through debittering processes, such as those used in the citrus industry, or through use of masking agents, such as sugar or dressings. Debittering works through direct removal of the putative bitter agents, while masking works by overriding the perceived bitterness signal with a competing sensory signal (e.g., sweetness). To truly inhibit bitter perception, a compound must compete with bitter sapophores for receptor sites or actually block signal transduction at the receptor site.

Modification of bitterness has tremendous financial repercussions for the food industry. Roy (1990) has estimated that the market value of bitter consumables made palatable by the addition of sugar alone is on the order of $150 billion. Bitterness abatement can be achieved by increasing the efficacy of existing masking agents or creating new ones, by modifying

[1]Department of Food Science & Toxicology and the School of Family and Consumer Sciences, University of Idaho, Moscow, ID 83844-1053, U.S.A.

systems to slow the release of bitterants, or by designing a true bitterness inhibitor. Real progress in this field demands an understanding of the mechanisms by which bitterness is perceived and can only be achieved by integrating psychophysical studies with biochemical and physiological research.

BITTER TASTE

The study of bitter taste is especially problematic; the number of compounds eliciting bitterness exceeds those eliciting sweetness (Bartoshuk and Beauchamp, 1994) and the structure-activity relationships are for the most part poorly understood (Shallenberger and Acree, 1971). The compounds that evoke bitterness are structurally diverse, and include ions, amino acids and peptides, alkaloids, acylated sugars, glycosides, nitrogenous compounds and thiocarbamides.

RECEPTOR EVENTS

Taste receptor cells (TRC) are the locus of taste transduction in humans; they are primarily associated with the lingual papillae, although the mucosal surfaces of the esophagus, trachea and oral cavity may also contain taste receptor cells (Henkin and Christiansen, 1967). It has been postulated that bitter taste is transduced through a proteinaceous membrane-bound receptor system analogous to that employed in the olfactory system (Bartoshuk and Beauchamp, 1994). Akabas et al. (1988), using denatonium chloride as the bitter stimulus, obtained results indicative of an inositol trisphosphate second-messenger system in rat circumvallate TRC patch-clamp studies. Denatonium is membrane impermeable and is thus postulated to bind to an apical membrane receptor, although the putative receptors have yet to be isolated (Naim et al., 1994).

Membrane-bound receptors, however, are not requisite for signal transduction. At least one signal pathway appears to be receptor-independent and involve direct blockage of potassium channels (Shepherd, 1991). In addition, nonionic bitter sapophores are capable of adsorbing directly onto, and thus depolarizing, the lipid membrane (Kurihara et al., 1978). Quinine, for example, is membrane permeable and may block ion channels from within the cell (Cummings and Kinnamon, 1992); recently, Naim et al. (1994) demonstrated that quinine is a direct activator of the G protein cascade. The diversity of taste mechanisms is nicely reviewed (Kinnamon, 1988).

Psychophysically, the nonsingularity of the bitterness receptor system, was first demonstrated over sixty years ago; in 1931 Fox reported the existence of a bimodal distribution for the ability to taste bitter compounds containing thiocarbamide functional groups, notably 1-phenyl-2-thiourea (PTC) and its *p*-ethoxy counterpart, the thiocarbamide analog to dulcin (Fox, 1931, 1932). The inability to taste PTC is a recessive Mendelian characteristic (Snyder, 1931; Blakeslee, 1932). Because of the toxicity of PTC, an odorless thiocarbamide compound, 6-propyl-2-thiouracil (PROP) has been used for recent studies (Lawless, 1980). Research by Miller and coworkers indicates that sensitivity to PROP may be directly related to taste bud density (Miller and Whitney, 1989; Miller and Reedy, 1990).

Consistent with the evidence that PROP tasters can be either heterozygous or homozygous to the dominant allele (Anliker et al., 1991), Bartoshuk (1992) characterized tasters as either medium tasters or supertasters. "Supertasters" perceive bitterness as more intense, and show a gender differentiation, with more females belonging to the homozygous population (Bartoshuk et al., 1994).

INNERVATION AND CODING

The TRC and the fungiform papillae are innervated by the chorda tympani branch of cranial nerve VII, the facial, whereas the TRC of the vallate papillae are served by the ninth cranial nerve, the glossopharyngeal. Cranial nerve X, the vagus, primarily innervates the palatal and laryngeal regions. However, none of these nerves are exclusively gustatory afferents; the X cranial nerve response in particular appears to respond to mechanical stimulation (Norgren, 1984). Recently it has been demonstrated that the chorda tympani nerve inhibits sensation in the contralateral glossopharyngeal nerve (Bartoshuk, 1992), indicating that the taste system is far more complex than originally thought.

Although it appears that the anterior portion of the tongue (especially the tip) is more sensitive to sweet taste, with bitter tastants having lower thresholds in the posterior vallate taste buds (Beidler, 1961), more recent investigations have revealed that the picture is not that clear. Collings (1974) found that the lowest threshold for bitterness was actually in the receptor cells of the soft palate. With regard to innervation, however, electrophysiological evidence strongly suggests that bitter substances elicit stronger responses in the glossopharyngeal nerve than in the chorda tympani (Ninomiya and Funakoshi, 1989), although the glossopharyngeal also conveys important information on nutritionally important nontoxic substances (Ninomiya et al., 1994).

Currently, the two prevailing schools of taste encoding theory are the labeled line (LL) theory (in which only primary taste-specific neurons carry gustatory information) and the across-fiber pattern (AFP) theory (in which all responding neurons encode the quality and intensity information, regardless of "best-stimulus" specificity) (Maes, 1984). The LL theory is consistent with the concept of taste primaries, whereas the AFP theory is supportive of a taste continuum. Many arguments using electrophysiological and psychophysical data have been presented to support LL or AFP (McBurney, 1974; Pfaffmann et al., 1979; Schiffman and Erickson, 1980; Schiffman et al., 1981; Faurion, 1987, Kurihara et al., 1989) and as Faurion and Vayssettes-Courchay note (1990), often with the same data. It is clear that this issue has yet to be fully resolved (Erickson, 1984; Frank, 1984), although recent research in the rat (Scott and Giza, 1990) and the chimpanzee (Hellekant and Ninomiya, 1994) strongly supports the labeled line theory.

BITTERNESS MEASUREMENT

Bitter intensity, as with the other primary tastes, has traditionally been measured using scalar techniques. However, for bitterness and a number of oral sensations (e.g., astringency, chemical burn), temporal information is necessary for complete characterization of the perceived sensation, due to lingering sensations (Leach and Noble, 1986). As Cliff and Heymann (1993) have noted, intensity quantification for these sensations is not limited by individual judge reaction time, thus making these sensations particularly well suited to temporal analysis.

SCALAR STUDIES

The literature is replete with highly variable threshold data for bitterants and other sapophores, as Amerine et al. (1965) noted. While some of the variability may be attributed to testing logistics (type of test, chemical purity, lack of statistical power), much of the variability is inherent in the judges themselves. As Peryam (1963) found in suprathreshold testing, variation in sensitivity and scaling is related to both stimulus familiarity and learning effects.

TEMPORAL STUDIES

Temporal analysis methods (time-intensity, *TI*) have been reviewed extensively elsewhere (Lee and Pangborn, 1986; Cliff and Heymann, 1993) and will not be detailed here. Three parameters are commonly extracted from the two-dimensional intensity vs. time curves: time to maximum intensity, intensity at maximum and total duration (Noble, 1995). These parameters generally allow for adequate intercomparison of stimuli temporal properties. Lewis et al. (1980) compared the time-intensity methodology to a scalar category-scaling methodology for perception of iso-α-acid bitterness in water and beer, and demonstrated that the data derived from both methods were correlated.

FACTORS AFFECTING BITTERNESS PERCEPTION: COMPOUND EFFECTS

Quinine has been demonstrated to have a threshold value in water on the order of two magnitudes less than caffeine in water (Pfaffman et al., 1971); these data are in good agreement with those presented by Amerine et al. (1965). At suprathreshold levels, equi-bitter solutions of caffeine and quinine solutions also varied by a hundredfold differential in concentration (Leach and Noble, 1986). Further, for these equi-bitter solutions, caffeine had a longer total duration of sensation, T_{tot}, than quinine.

In wines, bitterness is elicited by the monomers and polymers of the flavan-3-ols (+)-catechin and (−)-epicatechin (Noble, 1994). The sensory properties of this class of compounds have been extensively researched in both scalar (Rossi and Singleton, 1966; Arnold and Noble, 1978; Lea and Arnold, 1978, 1983; Arnold et al., 1980; Delcour et al., 1984) and temporal studies (Robichaud and Noble, 1990; Fischer and Noble, 1994; Thorngate and Noble, 1995). By both approaches, bitterness has been found to be dependent on both molecular weight and molecular configuration.

The bitterness of condensed tannin fractions increases from monomer to tetramer (Arnold et al., 1980; Lea and Arnold, 1983); bitterness of higher oligomers is masked in part by a concomitant increase in astringency, although Arnold et al. (1980) found that on a constant weight basis the highest molecular weight fractions were the most bitter. The bitterness of the monomeric flavan-3-ols (+)-catechin and (−)-epicatechin, which are optical isomers about the heterocyclic hydroxyl group, has been shown to differ, with the more lipophilic epicatechin having both a greater maximum intensity (I_{max}) and T_{tot} for bitterness (Thorngate and Noble, 1995).

Benzoic acid derivatives, also found in wine, have been studied by Noble and coworkers (Robichaud and Noble, 1990; Peleg and Noble, 1995). Gallic acid (3,4,5-trihydroxybenzoic acid) was found to be more bitter than astringent (Robichaud and Noble, 1990; Peleg and Noble, 1995). Gentisic acid (2,5-dihydroxybenzoic acid) was the most bitter of a series of hydroxy-substituted benzoic acids (Peleg and Noble, 1995), although not significantly more so than benzoic acid or protocatechuic acid (3,4-dihydroxybenzoic acid). *m*-Hydroxybenzoic acid (3-hydroxybenzoic acid) and salicylic acid (2-hydroxybenzoic acid) were the least bitter. The underlying chemical reasons for these changes in bitterness intensity are not clear. However, the di-substituted derivatives were found to be less sweet (Peleg and Noble, 1995), which may have exaggerated their bitterness, whereas benzoic acid itself would be expected to be more lipophilic, perhaps increasing its bitterness.

Guinard et al. (1994) have recently summarized the perceptual data for a number of bitterants; Beets (1978) and Belitz and Weiser (1985) have exhaustively reviewed the structure-activity relationships underlying bitterness.

TASTE INTERACTIONS

As Lawless (1986) noted, a common occurrence in taste psychophysics is masking of intensity in mixtures containing two or more components. The individual components retain their specific sensory qualities, but are perceived as being less intense than when assessed singularly. Enhancement is occasionally seen, although far less commonly and usually in association with low concentration components (Pfaffman et al., 1971). Early studies of taste interrelationships were summarized by Pangborn (1960).

As noted previously, sugar is a commonly used masking agent for bitterants. Sweetness suppression of bitterness has been repeatedly demonstrated for sucrose and quinine mixtures (Bartoshuk, 1975; Lawless, 1979), sucrose and caffeine mixtures (Pangborn, 1960; Kamen et al., 1961; Calviño et al., 1990) and in more complex systems (Guadagni et al., 1974; Burns and Noble, 1985). The magnitude of the bitterness suppression is directly related to the sweetness intensity (in this case, sucrose concentration).

Saltiness has also been found to mask bitterness; Pangborn (1960) reported that a subthreshold level of sodium chloride decreased caffeine bitterness, although Kamen et al. (1961) found no effect. Bartoshuk (1975) and Frijters and Schifferstein (1994) have observed that sodium chloride reduces quinine bitterness. This masking is markedly asymmetric; sodium chloride is only weakly masked by addition of quinine (Schifferstein and Frijters, 1992).

The effect of sour sapophores on bitterness has received little attention. Early work by Pangborn (1960) found that subthreshold levels of citric acid decreased caffeine bitterness. Kamen et al. (1961), however, observed pronounced caffeine bitterness enhancement by citric acid, although their study is believed to have suffered from bitter/sour taste confusion (Pfaffman et al., 1971).

In white wine, pH changes produced nonlinear effects on intensity; bitterness was higher at pH 3.2, and lower at pH 2.9 or pH 3.8 (Fischer and Noble, 1994). A similar inconsistent effect of sourness was previously observed for caffeine bitterness; addition of tartaric acid had minimal and unpredictable effects on bitterness perception (Pyle, 1986).

It should be noted that the observation of suppression provides little insight into the physical or psychophysical nature of the phenomenon. Sweetness enhancement due to threshold concentrations of sodium chloride has been attributed to the sweet taste NaCl possesses at low concentration, but this is speculative. As Lawless and Stevens (1984) state, the nature of suppression may be due to such diverse mechanisms as salivary flow changes, neural inhibitory effects, chemical interactions or recruitment of gustatory channels to carry competing signals. Additionally, the stimulus set employed in such experiments may have a great effect on the levels of suppression observed (Schifferstein, 1994).

MEDIUM OF PRESENTATION

Perception of taste is not only dependent upon the compound(s) present, but also upon the physical and chemical properties of the medium of presentation. For example, Mackey (1985) reported that the bitterness of caffeine and quinine was reduced when these compounds were presented in refined peanut oil as opposed to water. The hydrophobic nature of the solvent may have altered the distribution ratio, lowering salivary concentration of the tastants, or the viscosity of the oil may have influenced perceptual response.

VISCOSITY

Mackey (1958) investigated the role of viscosity upon taste discrimination. Minimal suppression of bitterness was found in solutions thickened with methylcellulose to the viscosity of the peanut oil (vide supra) and greater reduction of bitterness was found in samples with the oil.

Moskowitz and Arabie (1970) also observed that increasing solution viscosity (using carboxymethylcellulose) decreased quinine bitterness intensity; Pangborn and coworkers found that caffeine bitterness was suppressed by a variety of hydrocolloids both in model systems (Pangborn et al., 1973) and in coffee (Pangborn et al., 1978). Burns and and Noble (1985) found similar results in their study of vermouth. At constant sucrose levels, perceived bitterness was reduced with increasing viscosity due to addition of Polycose (a modified cornstarch product).

Calviño et al. (1993) studied the masking effects on bitterness due to presentation in carboxymethylcellulose (CMC) or gelatin solutions versus water. As the viscosity due to vehicle increased, perceived bitterness decreased, consistent with earlier findings. The suppression was only statistically significant for the water-gelatin comparison, however. More recent *T-I* studies with grape seed tannin found no reduction in bitterness as viscosity was increased with CMC (Smith et al., 1995).

TEMPERATURE

Temperature effects typically exhibit an inverted U-shaped curve, with optimal response occurring at some mid-range temperature (Pfaffman et al., 1971). Although human studies have proven to inconsistent, there are data to suggest that the optimal temperature range for taste discrimination in humans falls just below body temperature, in the range of 30 to 35°C (Pfaffman et al., 1971). However, as Green and Frankmann (1987) state, the role of temperature in taste perception remains poorly understood.

Hahn and Günther (1932, cited in Pfaffman et al., 1971) reported that quinine sensitivity decreased with increasing temperature, although Maurizi and Cimino (1961, cited in Amerine et al., 1965) found that bitter thresholds decreased with increasing temperature. McBurney et al. (1973) reported a temperature optimum for perception of suprathreshold levels of quinine of 22°C, with sensitivity decreasing at lower or higher temperatures.

In agreement with Maurizi and Cimino and McBurney et al., Green and Frankmann (1987) found that cooling both the tongue and the taste solution led to suppression of bitter response, although this reduction was not statistically significant. When tongue and solution cooling was studied factorially, the authors found that cooling the tongue had the greater suppressive effect. Larson-Powers and Pangborn (1978), in a time-intensity study of four sweeteners, did not observe any suppression of bitterness for cyclamate or saccharin when the presentation temperature was lowered to 3°C from 22°C. The authors attributed the lack of suppression to mouth-warming of the sample before it reached the vallate receptors at the base of the tongue.

For urea, persistence of bitter sensation was greatest at 50°C, the highest temperature studied; however, there was no clear pattern of temperature influence on persistence of taste sensation (Calviño, 1984).

ORAL CHEMICAL IRRITANTS

In addition to taste compounds, oral chemical irritants have been proposed as perceptual masking agents; Lawless and Stevens (1984) found that intensity ratings of quinine following a capsicum oleoresin or piperine rinse diminished the bitter response. A follow-up study, using capsaicin, yielded similar results, with the perceived bitterness of quinine moderately suppressed by the oral irritant prerinse (Lawless et al., 1985). Cowart (1987), though, found no suppressive effects of oral irritants when they were presented simultaneously with taste compounds, and concluded that taste perception was independent of oral chemical irritation. However, as Prescott et al. (1993) note, the level of capsaicin Cowart used, 2 mg/L, may have precluded her observing suppressive effects; the authors conclude that, far from being independent, the interactions between the trigeminal system (which responds to irritation) and the gustatory system are highly complex, and poorly understood.

ETHANOL

Although Mackey (1958) concluded that bitter taste perception was dependent on an aqueous (as opposed to lipid) delivery medium, later studies by Martin and Pangborn (1970) indicated that ethanol, a solvent more lipophilic than water, potentiated bitter response to quinine. Lea and Arnold (1978) found that ethanol increased the bitterness of cider tannin extract; they speculated that the ethanol potentiates bitterness through a cosolubility effect. In a study of vermouth, bitterness was enhanced by increasing ethanol concentration (Burns and Noble, unpublished data, reported in Noble, 1994). Similarly, increasing ethanol to 8% enhanced bitterness of caffeine in model solutions (Pyle, 1986).

Fischer and Noble (1994) found that relatively small increases in ethanol concentration (3%) increased the bitterness of a white wine system more than the addition of 1400 mg/L (+)-catechin. Pangborn et al. (1983) had observed a similar potentiation of bitter iso-α-acids in beer upon addition of 2.6% ethanol. Potentiation of the bitter taste of the flavan-3-ols (+)-catechin and (−)-epicatechin by 5% ethanol was also observed by Thorngate (1992).

In a series of time-intensity experiments, Fischer (1990) found that ethanol increased the maximum bitterness intensity and the total duration of bitter aftertaste for both (+)-catechin and tannic acid in white wine. Pangborn et al. (1983) also observed an increase in bitterness intensity due to ethanol; however, the total duration of bitter aftertaste remained constant.

MODE OF PRESENTATION

Little research has specifically studied the effect of mode of presentation on bitterness perception. Noble (1995) cites Hummer's work showing no difference in perception of bitterness in solutions that were swallowed versus expectorated. Effects of sample size were demonstrated by Guinard et al. (1986), with ingestion of 20 mL samples increasing I_{max} over ingestion of 10 mL samples. Indirect evidence for presentation variation is found in the idiosyncratic nature of the time-intensity curve shape; in a study of bitterness of iso-α-acids, Pangborn et al. (1983) note that individual *T-I* curve shape may be dependent on the manner in which the samples were ingested. However, no formal work has been done in this area.

Guinard et al. (1986) elegantly demonstrated the effect of repeated ingestion on bitterness perception in beer using a time-intensity methodology. Repeated ingestion at 20-sec intervals yielded increased I_{max}, whereas 40-sec intervals failed to show the same increase. As Noble (1995) noted, this paradigm is far closer to that of normal consumption.

TASTE MODIFIERS

A number of compounds modify taste, apparently through interaction with the taste receptor cells (Kurihara, 1971). Taste modifiers include anesthetics, which uniformly depress all tastes (Kurihara, 1971), as well as the familiar extracts derived from *Gymnema sylvestre, Richadella dulcifica* (*Synsepalum dulcificum,* miracle fruit) and *Ziziphus jujuba,* which specifically inhibit sweet taste (Kurihara, 1992).

Bartoshuk et al. (1969) reported that purified *Gymnema sylvestre* extracts showed no effect on bitter taste perception of quinine; earlier reports to that effect were attributed to cross-adaptation to the bitter taste of crude *Gymnema sylvestre* extracts. Kurihara (1971) also noted that a purified gymnemic acid fraction had no effect on other tastes; however, higher concentrations of the same fraction reportedly did affect other unspecified tastes. The specificity of *Gymnema sylvestre* for sweet taste has also been demonstrated in paired psychophysical and electrophysiological studies (Diamant et al., 1965); bitter perception of quinine was not affected.

Miracle fruit extracts have also been shown to have no effect on bitter taste perception (Bartoshuk et al., 1969; Kurihara et al., 1969); similar negative results have been found for gurmarin (an extract of *Gymnema sylvestre* leaves), ziziphin (Kurihara, 1992) and hodulcin (an extract of *Hovenia dulcis* leaves; Kennedy et al., 1988).

Sodium dodecyl sulfate, SDS, a common surfactant, was found to depress perceptual response to quinine (DeSimone et al., 1980). However, the results were not consistent, and further confounded by the bitter taste of SDS. More recently, two specific inhibitors for bitter taste have been reported, the sodium salt of 2-(4-methoxyphenoxy)propanoic acid (Na-PMP) (Fuller and Kurtz, 1991, and cited in Johnson et al., 1994) and a lipoprotein supermolecule for taste as prepared from phosphatidic acid and β-lactoglobulin (PA-LG) (Katsuragi and Kurihara, 1993). While PA-LG looks promising as a true bitterness inhibitor, Johnson et al. (1994) demonstrated that Na-PMP had no bitterness inhibitory effect at the concentrations allowed by the Fragrance and Extract Manufacturer's Association. This is in agreement with the data reported by Lindley (1991).

PROP STATUS

The perceptual relationship of bitter tastants to PTC or PROP taster status is inconclusive; quinine and caffeine have been shown to be both correlated and uncorrelated to PROP taster status (Schifferstein and Frijters, 1991). As Lawless (1987) cautions, the correlations observed, though statistically significant, are frequently small and merit critical examination. Thus, the only compounds that can be said to unambiguously demonstrate PTC or PROP bimodal effecst are those containing the N$-$C$=$S group (Kalmus, 1971).

Jefferson and Erdman (1970) found a significant correlation between PTC sensitivity and quinine sensitivity. Leach and Noble (1986) reported that time-intensity measures of maximum bitterness intensity for quinine exhibit significant PTC status effects; Gent and Bartoshuk (1983) also observed evidence for PROP status effects on quinine bitterness perception. The majority of studies, however, have shown no such effects for quinine (Blakeslee, 1932; Hall et al., 1975; Bartoshuk, 1979; Bartoshuk et al., 1988; Mela, 1989; Schifferstein and Frijters, 1991; Yokomukai et al., 1993).

With regard to caffeine bitterness perception, Hall et al. (1975) found a significant PTC status effect, whereas Leach and Noble (1986) and Mela (1989) did not. Potassium chloride, sodium benzoate (Bartoshuk et al., 1988) and urea and sucrose octaacetate (Mela, 1989) have all been reported to exhibit significant PTC or PROP status effects. Schifferstein and Frijters (1991), however, found no effects for potassium chloride; Yokomukai et al. (1993) reported no effects for urea, whereas Hall et al. (1975) found effects for urea only at low concentrations. Bartoshuk (1979) reported that saccharin demonstrated PROP status effects, although in a study of context effects Rankin and Marks (1992) found no evidence for PROP effects on saccharin bitterness perception.

Much of the interest in PROP status effects and bitterness perception has been related to prediction of dietary intake, as it is assumed that persons who are acutely sensitive to bitterness will be more discriminating in their food selections. Fischer et al. (1961) reported that food dislikes were related to sensitivity to quinine and PROP; a similar study by Glanville and Kaplan (1965) using a larger population verified these findings. Jefferson and Erdman (1970) observed that low thresholds for PTC were significantly correlated to high percentages of food dislikes. Anliker et al. (1991), in a study of children's food preferences, found that nontasters rated cheese higher than did tasters. These results were in accord with unpublished studies of the authors (reported in Anliker et al., 1991) showing adult tasters perceived stronger bitter/sour tastes in cheeses.

However, Mattes and Labov (1989) found that PROP taster status was not related to goitrogen intake, although it had earlier been reported that nontasters ingested greater quantities of the palatables containing these bitter compounds. As Mattes (1994) has note, the PROP sensitivity/food dislikes relationships reported for particular foods have tended to be inconclusive, symptomatic of the complexity of assessing food acceptability.

SALIVARY STATUS

In order for a sapophore to access a taste receptor cell it must be solubilized in an aqueous medium; in human taste perception, this medium is provided by saliva (Amerine et al., 1965; cf. Weiffenback et al., 1986). The interrelationships of saliva and taste have been reviewed elsewhere (Christiansen, 1986; Spielman, 1990); this section will focus specifically on bitterant-related effects.

Saliva flow is promoted by a variety of chemicals, especially those causing oral irritation (Christiansen, 1986). Thus, among the primary tastants, acids are the most effective at stimulating salivary flow. Bitter compounds, however, can increase salivary flow, albeit weakly. Funakoshi and Kawamura (1967) found that 0.02% quinine yielded 0.57 mL of parotid saliva over three minutes of collection; 18.8% sucrose or 3% sodium chloride stimuli stimulated similar volumes of saliva. Tartaric acid (0.28%), however, yielded 1.83 mL of saliva over three minutes, indicative of the efficacy of acids on salivary stimulation. Guinard et al. (1994) found no increase in salivary flow rates when subjects were given isohumulones added to beer. However, the presentation medium itself, a lager beer, is bitter to begin with, and a strong sialagog. It could well be that isohumulones stimulate salivary flow when compared to distilled water. Regardless, the evidence suggests that dilutatory effects due to increased saliva flow should be minimal, and not greatly affect bitterness perception.

Salivary composition is also known to be influenced by gustatory stimuli; quinine increases Ca^{2+} in saliva, although not as markedly as sodium chloride or citric acid; quinine had no effect on increasing salivary protein content, or the relative proportions of proteins present (Dawes, 1984). Spielman (1988, cited in Spielman, 1990) obtained similar results in studies of proline-rich proteins (PRP); the relative proportions of the PRPs were unaffected by stimulation with quinine or the other primary tastants.

Proline-rich proteins constitute the majority of proteins found in parotid saliva and 70% of the salivary proteins (Spielman, 1990). Azen and coworkers (1986) found that the distribution pattern for the genes encoding PRPs and ablity to taste bitterants in mice were identical. Speculation as to a "taste-carrier" role for PRPs is premature, however, experiments by Spielman (1988, cited in Spielman, 1990) investigating PRP binding to quinine sulfate were inconclusive.

A similar "taste-carrier" role has been proposed for proteins found in von Ebner gland (VEG) secretions (Schmale et al., 1990). The VEG proteins showed a high degree of homology to known hydrophobic carrier proteins; the authors speculated that the VEG proteins either serve as lipophilic transporters to taste receptor cells, or serve to clear the TRC of lipophilic tastants (notably bitterants). Gurkan and Bradley (1988) found that the von Ebner secretions had a direct effect on gustatory afferent response, which in turn affected efferent activity to the glands. Gurkan and Bradley postulated that this was necessary to ensure flushing of stimuli from the clefts of the foliate and vallate papilla, a mechanism in agreement with Schmale et al.'s (1990) role for PRPs.

Compositionally, saliva also contains components that could elicit or affect bitterness: urea, thiocyanate and potassium chloride (Spielman, 1990). High levels of urea could affect bitter perception, although urea content is lower in subjects with high salivary flow rates (Christensen, 1986). High concentrations of thiocyanate have been reported to lower the threshold for bitterness (Ehrenberg and Güttes, 1949, cited in Amerine et al., 1965). The role of the bitter electrolyte potassium chloride has not been explored. Caffeine has been detected in the saliva of coffee consumers, but was not correlated to bitterness sensitivity; isohumulones, the bitterants of hops, were not detected in the saliva of beer consumers (Tanimura, 1994). Interestingly, however, Tanimura and Mattes (1993) found an inverse correlation between caffeine consumption, isohumulone consumption and bitter sensitivity.

It would appear that bitter perception is little affected by salivary bitterants, in contrast to the significant effect of salivary sodium concentration on perception of saltiness (O'Mahony and Heintz, 1981). Psychophysically, Fischer (1990) found that bitterness perception was affected by salivary flow rate status. Unilateral parotid saliva was collected using a modified Carlson-Crittenden annular vacuum cup as described in Fischer et al. (1994). The temporal parameters of time to maximum bitterness intensity (T_{max}), I_{max}, T_{tot} and the area under the *TI* curve differed significantly among the low, medium and high salivary flow groups (Fisher et al., 1994). Averaged across all stimuli, low-flow subjects (ca. 500 mg saliva/20 min) perceived I_{max} later and more intensely than the high-flow subjects (ca. 1400 mg saliva/2 min), and T_{tot} was significantly greater for the low-flow subjects (Fischer et al., 1994). These data are consistent with a model of TRC cleansing of bitter stimuli by saliva. Guinard et al. (1994), however, found no significant correlations between salivary flow rate status and I_{max} or T_{tot} for isohumulone bitterness in beer. However, as beer is a potent sialagog (Guinard et al., 1994), it could well be that flow-rate status effects were masked, similar to the low pH effects Fischer et al. (1994) observed.

AGE

That the sensitivity to taste declines with age has been well established (Murphy, 1986); the research indicates that bitter taste sensitivity is especially prone to deterioration. This is true of bitter threshold sensitivity (Weiffenbach et al., 1982) as well as for suprathreshold sensitivity (Murphy and Gilmore, 1989). As the scaling procedure used in the latter study was memory-intensive, Gilmore and Murphy (1989) repeated their suprathreshold assessments using a just-noticeable-difference forced-choice paradigm. The Weber ratios thus derived were significantly higher for the geriatric population than the young subjects (Gilmore and Murphy, 1989). Schiffman (1993) has summarized detection thresholds of a number of bitter compounds in young and elderly subjects; on average, geriatric subjects had thresholds for bitterants seven times greater than those found for the young subjects. Cowart et al. (1994) have recently reported on compound-specific declines in bitter sensitivity; they observed that whereas quinine exhibited age-related differential sensitivity, urea did not. This is in agreement with Schiffman's (1993) finding, where urea had an elderly/young threshold ratio value of 1.12, indicating that loss of sensitivity to urea was minimal over the life span.

CONCLUSIONS

Reducing bitterness through improved understanding of perceptual mechanisms, rather than through hit-or-miss utilization of food additives, development of masking agents or modification of food systems, is an ambitious goal, that can only be achieved through a successful integration of interdisciplinary approaches. Progress in understanding the mechanism of bitterness perception depends on linking knowledge of factors affecting bitterness perception, such as those reviewed herein, with basic studies of taste reception and transduction, and the structure-activity relationships of bitter compounds and bitter inhibitors.

ACKNOWLEDGEMENTS

The author wishes to thank Dr. A. C. Noble for valuable editorial comments, and Ms. Susanne Keilhorn for library research assistance.

REFERENCES

Akabas, M. H., Dodd, J. and Al-Awqati, Q. 1988. A bitter substance induces a rise in intracellular calcium in a subpopulation of rat taste cells. *Science,* 242(4881): 1047–1050.

Amerine, M. A., Pangborn, R. M. and Roessler, E. B. 1965. *Principles of Sensory Evaluation of Food,* New York: Academic Press, Inc. pp. 602.

Anliker, J. A., Bartoshuk, L., Ferris, A. M. and Hooks, L. D. 1991. Children's food preferences and genetic sensitivity to the bitter taste of 6-*n*-propylthiouracil (PROP), *Am. J. Clin. Nutr.,* 54(2): 316–320.

Arnold, R. A. and Noble, A. C. 1978. Bitterness and astringency of grape seed phenolics in a model wine solution, *Am. J. Enol. Vitic.,* 29(3): 150–152.

Arnold, R. A., Noble, A. C and Singleton, V. L. 1980. Bitterness and astringency of phenolic fractions in wine, *J. Agric. Food Chem.,* 28(3): 675–678.

Azen, E. A., Lush, I. E. and Taylor, B. A. 1986. Close linkage of mouse genes for salivary proline-rich proteins (PRPs) and taste, *Trends Genet.,* 2: 199–200.

Bartoshuk, L. M., Dateo, G. P., Vandenbelt, D. J., Buttrick, R. L. and Long, L., Jr. 1969. Effects of *Gymnema sylvestre* and *Synsepalum dulcificum* on taste in man, In: *Olfaction and Taste III,* Pfaffman, C., ed., New York: Rockefeller University Press, pp. 436–444.

Bartoshuk, L. M. 1975. Taste mixtures: Is mixture suppression related to compression? *Physiol. Behav.,* 14(5): 643–649.

Bartoshuk, L. M. 1979. Bitter taste of saccharin related to the genetic ability to taste the bitter substance 6-*n*-propylthiouracil, *Science,* 205(4409): 934–935.

Bartoshuk, L. M., Rifkin, B., Marks, L. E. and Hooper, J. E. 1988. Bitterness of KCl and benzoate: Related to genetic status for sensitivity to PTC/PROP, *Chem. Senses,* 13(4): 517–528.

Bartoshuk, L. M. 1992. The biological basis of food perception and acceptance, Abstracts of the *Advances in Sensory Science: Rose Marie Pangborn Memorial Symposium,* August 2–6, Järvenpää, Finland, p. A11.

Bartoshuk, L. M. and Beauchamp, G. K. 1994. *Chemical Senses, Annu. Rev. Psychol.,* 45: 419–449.

Bartoshuk, L. M., Duffy, V. B. and Miller, I. J. 1994. PCT/PROP tasting: Anatomy, psychophysics, and sex effects, *Physiol. Behav.,* 56(6): 1165–1171.

Beets, M. G. J. 1978. *Structure-Activity Relationships in Human Chemoreception,* London: Applied Science Publishers, Ltd., pp. 408.

Beidler, L. M. 1961. Taste receptor simulation, *Prog. Biophys. Phys. Chem.,* 12: 107–151.

Belitz, H.-D. and Wieser, H. 1985. Bitter compounds: Occurrence and structure-activity relationships, *Food Rev. Intl.,* 1(2): 271–354.

Blakeslee, A. F. 1932. Genetics of sensory thresholds: Taste for phenyl thio carbamide, *Proc. Natl. Acad. Sci.,* 18(1): 120–130.

Burns, D. J. W. and Noble, A. C. 1985. Evaluation of the separate contributions of viscosity and sweetness of sucrose to perceived viscosity, sweetness and bitterness of vermouth, *J. Text. Studies,* 16: 365–381.

Calviño, A. M. 1984. Effects of concentration and temperature on gustatory persistence, *Percept. Motor Skills,* 58(2): 647–650.

Calviño, A. M., García-Medina, M. R. and Cometto-Muñiz, J. E. 1990. Interactions in caffeine-sucrose and coffee-sucrose mixtures: Evidence of taste and flavor suppression, *Chem. Senses,* 15(5): 505–519.

Calviño, A. M., García-Medina, M. R., Cometto-Muñiz, J. E. and Rodríguez, M. B. 1993. Perception of sweetness and bitterness in different vehicles, *Percept. Psychophys.,* 54(6): 751–758.

Christiansen, C. M. 1986. Importance of saliva in diet-taste relationships, In: *Interaction of the Chemical Senses with Nutrition,* Kare, M. R. and Brand, J. G., eds., Orlando, FL: Academic Press, Inc., pp. 3–24.

Cliff, M. and Heymann, H. 1993. Development and use of time-intensity methodology for sensory evaluation: A review, *Food Res. Internatl.,* 26(5): 375–385.

Collings, V. B. 1974. Human taste response as a function of locus of stimulation on the tongue and soft Palate, *Percept. Psychophys.,* 16(1): 169–174.

Cowart, B. J. 1987. Oral chemical irritation: Does it reduce perceived taste intensity? *Chem. Senses,* 12(3): 467–479.

Cowart, B. J., Yokomukai, Y. and Beauchamp, G. K. 1994. Bitter taste in aging: Compound-specific decline in sensitivity, *Physiol. Behav.,* 56(6): 1237–1241.

Cummings, T. A. and Kinnamon, S. C. 1992. Apical K^+ channels in necturus taste cells, *J. Gen. Physiol.,* 99: 591–613.

Dawes, C. 1984. Stimulus effects on protein and electrolyte concentrations in parotid saliva, *J. Physiol.,* 346: 579–588.

Delcour, J. A., Vandenberghe, M. M., Corten, P. F. and Dondeyne, P. 1984. Flavor thresholds of polyphenolics in water, *Am. J. Enol. Vitic.,* 35(3): 134–136.

DeSimone, J. A., Heck, G. L. and Bartoshuk, L. M. 1980. Surface active taste modifiers: A comparison of the physical and psychophysical properties of gymnemic acid and sodium lauryl sulfate, *Chem. Senses,* 5(4): 317–330.

Diamant, H., Oakley, B., Ström, L., Wells, C. and Zotterman, Y. 1965. A comparison of neural and psychophysical resposnes to taste stimuli in man, *Acta Physiol. Scand,* 64(1/2): 67–74.

Ehrenberg, R. and Güttes, H. J. 1949. Über die wirkung von rhodaniden und sulfaten auf die schwellenwertes des geschmacks, *Arch. Ges. Physiol.,* 251: 644–671.

Erickson, R. P. 1984. Definitions: A matter of taste. In: *Taste, Olfaction and the Central Nervous System: A Festschrift in Honor of Carl Pfaffman,* Pfaff, D. W., ed., New York: The Rockefeller University Press, pp. 129–150.

Faurion, A. 1987. Physiology of sweet taste. In: *Progress in Sensory Physiology,* Ottoson, D., ed., Berlin: Springer-Verlag, pp. 129–201.

Faurion, A. and Vayssettes-Courchay, C. 1990. Taste as a highly discriminative system: A hamster intrapapillar single unit study with 18 compounds, *Brain Res.*, 512(2): 317–332.

Fischer, R., Griffin, F., England, S. and Garn, S. M. 1961. Taste thresholds and food dislikes, *Nature* 191(4795): 1328.

Fischer, U. 1990. The influence of ethanol, pH, and phenolic composition on the temporal perception of bitterness and astringency, and parotid salivation, M.Sc. Thesis, University of California, Davis.

Fischer, U., Boulton, R. B. and Noble, A. C. 1994. Physiological factors contributing to the variability of sensory assessments: Relationship between salivary flow rate and temporal perception of gustatory stimuli, *Food Qual. Pref.*, 5: 55–64.

Fischer, U. and Noble, A. C. 1994. The effect of ethanol, catechin concentration, and pH on sourness and bitterness of wines, *Am. J. Enol. Vitic.*, 45(1): 6–10.

Fox, A. L. 1931. Six in ten "tasteblind" to bitter chemical, *Sci. News Lett.*, 19(253): 249.

Fox, A. L. 1932. The relationship between chemical constitution and taste, *Proc. Natl. Acad. Sci.*, 18(1): 115–120.

Frank, M. E. 1984. On the neural code for sweet and salty taste. In: *Taste, Olfaction and the Central Nervous System: A Festschrift in Honor of Carl Pfaffman,* Pfaff, D. W., ed., New York: The Rockefeller University Press, pp. 107–128.

Frijters, J. E. R. and Schifferstein, H. N. J. 1994. Perceptual interactions in mixtures containing bitter tasting substances, *Physiol. Behav.*, 56(6): 1243–1249.

Fuller, W. D. and Kurz, R. J. 1991. Ingestibles containing substantially tasteless sweetness inhibitors as bitter taste reducers or substantially tasteless bitter inhibitors as sweet taste reducers, World Patent Cooperation Treaty, WO 91/18523.

Funakoshi, M. and Kawamura, Y. 1967. Relations between taste qualities and parotid gland secretion rate. In: *Olfaction and Taste II,* Hayashi, T., ed., Oxford: Pergamon Press, Ltd., pp. 281–287.

Gent, J. F. and Bartoshuk, L. M. 1983. Sweetness of sucrose, neohesperidin dihydrochalcone, and saccharin is related to genetic ability to taste the bitter substance 6-*n*-propylthiouracil, *Chem. Senses,* 7: 265–272.

Gilmore, M. M. and Murphy, C. 1989. Aging is associated with increased Weber ratios for caffeine, but not for sucrose, *Percept. Psychophys.*, 46(6): 555–559.

Glanville, E. V. and Kaplan, A. R. 1965. Food preference and sensitivity of taste for bitter compounds, *Nature,* 205(4974): 851–853.

Glendinning, J. I. 1994. Is the bitter rejection response always adaptive? *Physiol. Behav.*, 56(6): 1217–1227.

Green, B. G. and Frankmann, S. P. 1987. The effect of cooling the tongue on the perceived intensity of taste, *Chem. Senses,* 12(4): 609–619.

Gaudagni, D. G., Maier, V. P. and Turnbaugh, J. H. 1974. Some factors affecting sensory thresholds and relative bitterness of limonin and naringin, *J. Sci. Food Agric.*, 25(1): 1199–1205.

Guinard, J.-X., Pangborn, R. M. and Lewis, M. J. 1986. Effect of repeated ingestion on temporal perception of bitterness in beer, *Am. Soc. Brew. Chem. J.*, 44(1): 28–32.

Guinard, J.-X., Hong, D. Y., Zoumas-Morse, C., Budwig, C. and Russell, G. F. 1994. Chemoreception and perception of bitterness of isohumulones, *Physiol. Behav.*, 56(6): 1257–1263.

Gurkan, S. and Bradley, R. M. 1988. Secretions of von Ebner's glands influence responses from taste buds in rat circumvallate papilla, *Chem. Senses,* 13(4): 655–661.

Hahn, H. and Günther, H. 1932. Über die reize und die reizbedingungen des geschmackssinnes, *Pflügers Arch. Ges. Physiol.*, 231: 48–67.

Hall, M. J., Bartoshuk, L. M., Cain, W. S. and Stevens, J. C. 1975. PTC taste blindness and the taste of caffeine, *Nature,* 253(5491): 442–443.

Hellekant, G. and Ninomiya, Y. 1994. Bitter taste in single chorda tympani taste fibers from chimpanzee, *Physiol. Behav.*, 56(6): 1185–1188.

Henkin, R. I. and Christiansen, R. L. 1967. Taste localization on the tongue, palate, and pharynx of normal man, *J. Appl. Physiol.*, 22(2): 316–320.

Jefferson, S. C. and Erdman, A. M. 1970. Taste sensitivity and food aversions of teenagers, *J. Home Econ.*, 62(8): 605–608.

Johnson, C., Birch, G. G. and MacDougall, D. B. 1994. The effect of the sweetness inhibitor 2(-4-methoxyphenoxy)propanoic acid (sodium salt) (Na-PMP) on the taste of bitter-sweet stimuli, *Chem. Senses,* 19(4): 349–358.

Kalmus, H. 1971. Genetics of taste. In: *Handbook of Sensory Physiology, Vol. IV, Chemical Senses Part 2, Taste,* Beidler, L. M., ed., New York: Springer-Verlag, pp. 165–179.

Kamen, J. M., Pilgrim, F. J., Gutman, N. J. and Kroll, B. J. 1961. Interactions of suprathreshold taste stimuli, *J. Exp. Psuychol.*, 62(4): 348–356.

Katsuragi, Y. and K. Kurihara. 1993. Specific inhibitor for bitter taste, *Nature,* 365(6443): 213–214.

Kennedy, L. M., Saul, L. R., Sefecka, R. and Stevens, D. A. 1988. Hodulcin: Selective sweetness-reducing principle for *Hovenia dulcis* leaves, *Chem. Senses,* 13(4): 529–543.

Kinnamon, S. C. 1988. Taste transduction: A diversity of mechanisms, *Trends Neurosci.*, 11(11): 491.

Kurihara, K., Kurihara, Y. and Beidler, L. M. 1969. Isolation and mechanism of taste modifiers; taste modifying protein and gymnemic acids. In: *Olfaction and Taste III,* Pfaffman, C., ed., New York: Rockefeller University Press, pp. 450–469.

Kurihara, K. 1971. Taste modifiers. In: *Handbook of Sensory Physiology, Vol. IV., Chemical Senses Part 2, Taste,* Beidler, L. M., ed., New York: Springer-Verlag, pp. 363–378.

Kurihara, K., Kamo, N. and Kobatake, Y. 1978. Transduction mechanism in chemoreception, *Adv. Biophys.*, 10: 27–95.

Kurihara, K., Kashiwayanagi, M., Yoshii, K. and Kurihara, Y. 1989. Molecular mechanisms of taste and odor reception, *Comm. Agric. Food Chem.*, 21(1): 1–50.

Kurihara, Y. 1992. Characteristics of antisweet substances, sweet proteins, and sweetness-inducing proteins, *Crit. Rev. Food Sci. Nutr.*, 32(3): 231–252.

Larson-Powers, N. and Pangborn, R. M. 1978. Paired comparison and time-intensity measurements of the sensory properties of beverages and gelatins containing sucrose or synthetic sweeteners, *J. Food Sci.*, 43: 41–46.

Lawless, H. T. 1979. Evidence for Neural Inhibition of bittersweet taste mixtures, *J. Comp. Physiol. Psychol.*, 93(3): 538–547.

Lawless, H. T. 1980. A comparison of different methods used to assess sensitivity to the taste of phenylthiocarbamide (PTC), *Chem. Senses,* 5(3): 247–256.

Lawless, H. T. and Stevens, D. A. 1984. Effects of oral chemical irritation on taste, *Physiol. Behav.*, 32: 995–998.

Lawless, H. T., Rozin, P. and Shenker, J. 1985. Effects of oral capsaicin on gustatory, olfactory and irritant sensations and flavor identification in humans who regularly or rarely consume chili pepper, *Chem. Senses,* 10(4): 579–589.

Lawless, H. T. 1986. Sensory interactions in mixtures, *J. Sens. Studies,* 1: 259–274.

Lawless, H. T. 1987. Gustatory psychophysics. In: *Neurobiology of Taste and Smell,* Finger, T. E. and Silver, W. L., eds., New York: John Wiley & Sons, pp. 401–420.

Lea, A. G. H. and Arnold, G. M. 1978. The phenolics of cider: Bitterness and astringency, *J. Sci. Food Agric.*, 29(5): 478–483.

Lea, A. G. H. and Arnold, G. M. 1983. Bitterness, astringency and the chemical composition of ciders. In: *Sensory Quality in Foods and Beverages: Definition, Measurement and Control,* Williams, A. A. and Atkins, R. K., eds., Chichester, England: Ellis Horwood, Ltd., pp. 203–211.

Leach, E. J. and Noble, A. C. 1986. Comparison of bitterness of caffeine and quinine by a time-intensity procedure, *Chem. Senses,* 11(3): 339–345.

Lee, W. E., III and Pangborn, R. M. 1986. Time-intensity: The temporal aspects of sensory perception, *Food Tech.*, 40(11): 71–78, 82.

Lewis, M. J., Pangborn, R. M. and Fuji-Yamashita, J. 1980. Bitterness of beer: A comparison of traditional scaling and time-intensity methods, *Inst. Brew., Proc. (Australia and New Zealand*), 17: 165–171.

Lindley, M. G. 1991. Phenoxyalkanoic acid sweetness inhibitors. In: *Sweeteners: Discovery, Molecular Design, and Chemoreception,* Walters, D. E., Orthoefer, F. T. and DuBois, G. E., eds., Washington, D.C.: American Chemical Society, pp. 251–260.

Mackey, A. O. 1958. Discernment of taste substances as affected by solvent medium, *Food Res.*, 23: 580–583.

Maes, F. W. 1984. A neural coding model for sensory intensity discrimination, to be applied to gustation, *J. Comp. Physiol. A,* 155(2): 263–270.

Martin, S. and Pangborn, R. M. 1970. Taste interaction of ethyl alcohol with sweet, salty, sour and bitter compounds, *J. Sci. Food Agric.*, 21(12): 653–655.

Mattes, R. and Labov, J. 1989. Bitter taste responses to phenylthiocarbamide are not related to dietary goitrogen intake in human beings, *J. Am. Diet. Assoc.*, 89(5): 692–694.

Mattes, R. 1994. Influence on acceptance of bitter foods and beverages, *Physiol. Behav.*, 56(6): 1229–1236.

Maurizi, M. and Cimino, A. 1961. L'influenza delle variazioni termiche sulla sensibilitá gustativa, *Boll. Mal. Orecchio, Gola, Naso,* 79: 626–634.

McBurney, D. H., Collings, V. B. and Glanz, L. H. 1973. Temperature dependence of human taste responses, *Physiol. Behav.*, 11: 89–94.

McBurney, D. H. 1974. Are there primary tastes for man? *Chem. Senses Flav.*, 1(1): 17–28.

Mela, D. J. 1989. Bitter taste intensity: The effect of tastant and thiourea taster status, *Chem. Senses,* 14(1): 131–135.

Miller, I. J., Jr. and Whitney, G. 1989. Sucrose octaacetate-taster mice have more vallate taste buds than non-tasters, *Neurosci. Lett.*, 100(1–3): 271–275.

Miller, I. J., Jr. and Reedy, F. E., Jr. 1990. Variations in human taste bud density and taste intensity perception, *Physiol. Behav.*, 47(6): 1213–1219.

Moskowitz, H. R. and Arabie, P. 1970. Taste intensity as a function of stimulus concentration and solvent viscosity, *J. Text. Studies,* 1(4): 502–510.

Murphy, C. 1986. Taste and smell in the elderly. In: *Clinical Measurement of Taste and Smell,* Meiselman, H. L. and Rivlin, R. S., eds., New York: Macmillan, pp. 343–371.

Murphy, C. and Gilmore, M. M. 1989. Quality-specific effects of aging on the human taste system, *Percept. Psychophys.*, 45(2): 121–128.

Naim, M., Seifert, R., Nünberg, B., Grünbaum, L. and Shultz, G. 1994. Some taste substances are direct activators of G-proteins, *Biochem. J.* 297: 451–454.

Ninomiya, Y. and Funakoshi, M. 1989. Peripheral neural basis for behavioral discrimination between glutamate and the four basic taste substances in mice, *Comp. Biochem. Physiol.*, 92A(3): 371–376.

Ninomiya, Y., Kajiura, H., Naito, Y. Mochizuki, K., Katsukawa, H. and Torii, K. 1994. Glossopharyngeal denervation alters responses to nutrients and toxic substances, *Physiol. Behav.*, 56(6): 1179–1184.

Noble, A. C. 1994. Bitterness in wine, *Physiol. Behav.*, 56(6): 1251–1255.

Noble, A. C. 1995. Application of time-intensity procedures for the evaluation of taste and mouthfeel, *Am. J. Enol. Vitic.*, 46(1): 128–133.

Norgren, R. 1984. Central neural mechanisms of taste. In: *Handbook of Physiology, Section 1: The Nervous System, Vol. III: Sensory Processes, Part 2,* Geiger, S. R., ed., Bethesda, MD: American Physiological Society, pp. 1087–1128.

O'Mahony, M. and Heintz, C. 1981. Direct magnitude estimation of salt taste intensity with continuous correction for salivary adaptation, *Chem. Senses,* 6(2): 101–112.

Pangborn, R. M. 1960. Taste interrelationships, *Food Res.*, 25(2): 245–256.

Pangborn, R. M., Trabue, I. M. and Szczesniak, A. S. 1973. Effect of hydrocolloids on oral viscosity and basic taste intensities, *J. Text. Studies,* 4(2): 221–241.

Pangborn, R. M., Gibbs, Z. M. and Tassan C. 1978. Effect of hydrocolloids on apparent viscosity and sensory properties of selected beverages, *J. Text. Studies,* 9(4): 415–436.

Pangborn, R. M., Lewis, M. J. and Yamashita, J. F. 1983. Comparison of time-intensity with category scaling of bitterness of iso-α-acids in model systems and in beer, *J. Inst. Brew.*, 89: 349–355.

Peleg, H. and Noble, A. C. 1995. Perceptual properties of benzoic acid derivatives, *Chem. Senses,* 20: 393–400.

Peryam, D. R. 1963. Variability in taste perception, *J. Food Sci.*, 28: 734–740.

Pfaffman, C., Bartoshuk, L. M. and McBurney, D. H. 1971. Taste psychophysics. In: *Handbood of Sensory Physiology, Vol. IV, Chemical Senses Part 2, Taste,* Beidler, L. M., ed., New York: Springer-Verlag, pp. 75–101.

Pfaffman, C., Frank, M. and Norgren, R. 1979. Neural mechanisms and behavioral aspects of taste, *Ann. Rev. Psychol.*, 30: 283–325.

Prescott, J., Allen, S. and Stephens, L. 1993. Interactions between oral chemical irritation, taste and temperature, *Chem. Senses,* 18(4): 389–404.

Pyle, H. 1986. Effect of ethanol, tartaric acid and sucrose on perceived intensity of caffeine in model solution, M.Sc. Thesis, University of California, Davis.

Rankin, K. M. and Marks, L. E. 1992. Effects of context on sweet and bitter tastes:

Unrelated to sensitivity to PROP (6-*n*-propylthiouracil), *Percept. Psychophys.*, 52(5): 479–486.

Robichaud, J. L. and Noble, A. C. 1990. Astringency and bitterness of selected phenolics in wine, *J. Sci. Food Agric.*, 53: 343–353.

Rossi, J. A., Jr. and Singleton, V. L. 1966. Flavor effects and adsorptive properties of purified fractions of grape-seed phenols, *Am. J. Enol. Vitic.*, 17: 240–246.

Roy, G. M. 1990. The applications and future implications of bitterness reduction and inhibition in food products, *Crit. Rev. Food Sci. Nutr.*, 29(2): 59–71.

Schifferstein, H. N. J. and Frijters, J. E. R. 1991. The perception of the taste of KCl, NaCl and quinine HCl is not related to PROP-sensitivity, *Chem. Senses,* 16(4): 303–317.

Schifferstein, H. N. J. and Frijters, J. E. R. 1992. Two-stimulus versus one-stimulus procedure in the framework of functional measurement: A comparative investigation using quinine HCl/NaCl mixtures, *Chem. Senses,* 17(2): 127–150.

Schifferstein, H. N. J. 1994. Contextual effects in the perception of quinine HCl/NaCl mixtures, *Chem. Senses,* 19(2): 113–123.

Schiffman, S. S. and Erickson, R. P. 1980. The issue of primary tastes versus a taste continuum, *Neurosci. Biobehav. Rev.*, 4(2): 109–117.

Schiffman, S. S., Cahn, H. and Lindley, M. G. 1981. Multiple receptor sites mediate sweetness: Evidence from cross adaptation, *Pharm. Biochem. Behav.*, 15(3): 377–388.

Schiffman, S. S. 1993. Perception of taste and smell in elderly persons, *Crit. Rev. Food Sci. Nutr.*, 33(1): 17–26.

Schmale, H., Holtgreve-Grez, H. and Christiansen, H. 1990. Possible role for salivary gland protein in taste reception indicated by homology to lipophilic-ligand carrier proteins, *Nature,* 343: 366–369.

Scott, T. R. and Giza, B. K. 1990. Coding channels in the taste system of the rat, *Science,* 249(4976): 1585–1587.

Shallenberger, R. S. and Acree, T. E. 1971. Chemical structure of compounds and their sweet and bitter taste. In: *Handbook of Sensory Physiology, Vol. IV, Chemical Senses Part 2, Taste,* Beidler, M., ed., New York: Springer-Verlag, pp. 221–277.

Shepherd, G. M. 1991. Sensory transduction: Entering the mainstream of membrane signaling, *Cell,* 67: 845–851.

Smith, A., June, H. and Noble, A. C. 1995. The effect of viscosity on temporal perception of bitterness and astringency. Chemical Senses Day XII, Dec. 2, 1994 (*Abstract J. Sens. Studies,* pp. 10, 116).

Snyder, L. H. 1931. Inherited taste deficiency, *Science,* 74(1910): 151–152.

Spielman, A. I. 1988. Purification and characterization of the proline-rich proteins from rabbit parotid saliva, Dissertation, University of Toronto, Toronto.

Spielman, A. I. 1990. Interaction of saliva and taste, *J. Dent. Res.*, 69(3): 838–843.

Tanimura, S. and Mattes, R. D. 1993. Relationships between bitter taste sensitivity and consumption of bitter substances, *J. Sens. Studies,* 8(1): 31–41.

Tanimura, S. 1994. "Determination of caffeine and isohumulones in saliva by HPLC analysis, *J. Agric. Food Chem.*, 42(8): 1629–1631.

Thorngate, J. H., III. 1992. Flavan-3-ols and their polymers in grapes and wines: Chemical and sensory properties, Dissertation, University of California, Davis.

Thorngate, J. H., III and Noble, A. C. 1995. Sensory evaluation of bitterness and astringency of 3R(−)-Epicatechin and 3S(+)-Catechin, *J. Sci. Food Agric.*, 67(4): 531–535.

Weiffenbach, J. M., Baum, B. J. and Burghauser, R. 1982. Taste thresholds: Quality specific variation with human aging, *J. Gerontol.*, 37(3): 372–377.

Weiffenbach, J. M., Fox, P. C. and Baum, B. J. 1986. Taste and salivary function, *Proc. Natl. Acad. Sci.*, 83(10): 6103–6106.

Yokomukai, Y., Cowart, B. J. and Beauchamp, G. K. 1993. Individual differences in sensitivity to bitter-tasting substances, *Chem. Senses*, 18(6): 669–681.

CHAPTER 8

Bitterness Perception across the Life Span

CLAIRE MURPHY[1]
JILL RAZANI[2]

ALTHOUGH there is a great deal of overlap between the senses of taste and smell in the perception of food flavor, each makes a unique contribution to such perceptions. The sense of taste alerts the brain to the presence of sweet, sour, bitter and salty substances present in the oral cavity. The sense of smell signals the presence of volatiles that produce the subtleties and complexities of food flavors. Odors reach the olfactory system through the nose or through the rear of the oral cavity. The olfactory contributions to the flavor of foods can be illustrated by blocking the nostrils before ingestion. Thus, in extreme cases in which the nasal airflow is blocked it may be impossible to identify the food or beverage without its olfactory component. The other chemosensory component that adds to our perception of food is the trigeminal. The trigeminal sense signals other sensations in the oral and nasal cavities: warmth, cold, irritation, pungency. Therefore, sensations derived from such foods as hot peppers or raw onions are carried by the trigeminal system.

Impairment of human chemosensory function as a result of aging has been reported at many levels. This chapter briefly reviews some of the literature that demonstrates these effects and highlights the results of studies with significant emphasis on bitter perception. A more complete understanding of age-associated changes in taste and olfaction may provide some insight into factors involved in dietary selection by the elderly.

[1]Department of Psychology, San Diego State University, 5500 Campanile Dr., San Diego, CA 92182-0551, U.S.A.
[2]SDSU/UCSD Joint Doctoral Program in Clinical Psychology, 6363 Alvarado Court, Suite 101, San Diego, CA 92120, U.S.A.

TASTE THRESHOLD

Previous research has shown a modest decline in taste sensitivity with aging. However, the question is, does aging affect all taste qualities equally, or are some affected more than others? Although the literature seems to show conflicting results regarding this question (see Murphy, 1986, for a detailed review), some conclusions may still be gleaned from previous research. However, to put these conflicting results in perspective we must first discuss plausible reasons for the apparent disagreement. First, a wide variety of experimental procedures were used in past studies. These methodological differences may account for some of the discrepancy. Second, the samples of subjects in these studies may have differed on a number of critical factors. The general health status of the subjects is one example of factors that need to be taken into account when comparing results across studies.

In an earlier study by Murphy (1979), detection thresholds for eight tastants (sucrose, NaCl, critic acid, tartaric acid, HCl, caffeine, quinine sulfate and magnesium) were measured in seven age groups with mean ages 6, 12, 19, 23, 31, 52 and 73 years. Three representatives of the sour quality and three representatives of the bitter quality were chosen in order to investigate the possibility that aging differentially affects perception of bitter substances that do not cross-adapt (Murphy, 1976). The results indicated that sensitivity varied as a function of age and tastant. Further, a significant age-by-tastant interaction was found. These data suggested that the magnitude of the change was on the order of one log step from adolescence to old age, and that perception of some substances was affected more than others.

Quality-specific age effects have also been demonstrated by other investigators. Weiffenbach, Baum and Burghauser (1982) investigated age-associated taste sensitivity for the "basic" qualities using the sucrose, NaCl, quinine sulfate and citric acid. Figure 1 illustrates their significant findings: decreased sensitivity for NaCl and quinine sulfate in the elderly. No age-associated differences for sucrose and citric acid were found (see Figure 2). Interestingly, they found a gender effect for citric acid, with males showing less sensitivity than females. In a more recent study, Weiffenbach, Cowart and Baum (1986) found similar results for sucrose and NaCl. That is, a scatter plot of the data revealed that NaCl thresholds increase as a function of age, while the slope for sucrose thresholds remains flat.

Cowart (1989) examined a total of 137 subjects, ranging in age from 19 to 87 years, on a battery of taste and smell tests, including detection threshold tests as well as intensity ratings for taste and smell stimuli. Anal-

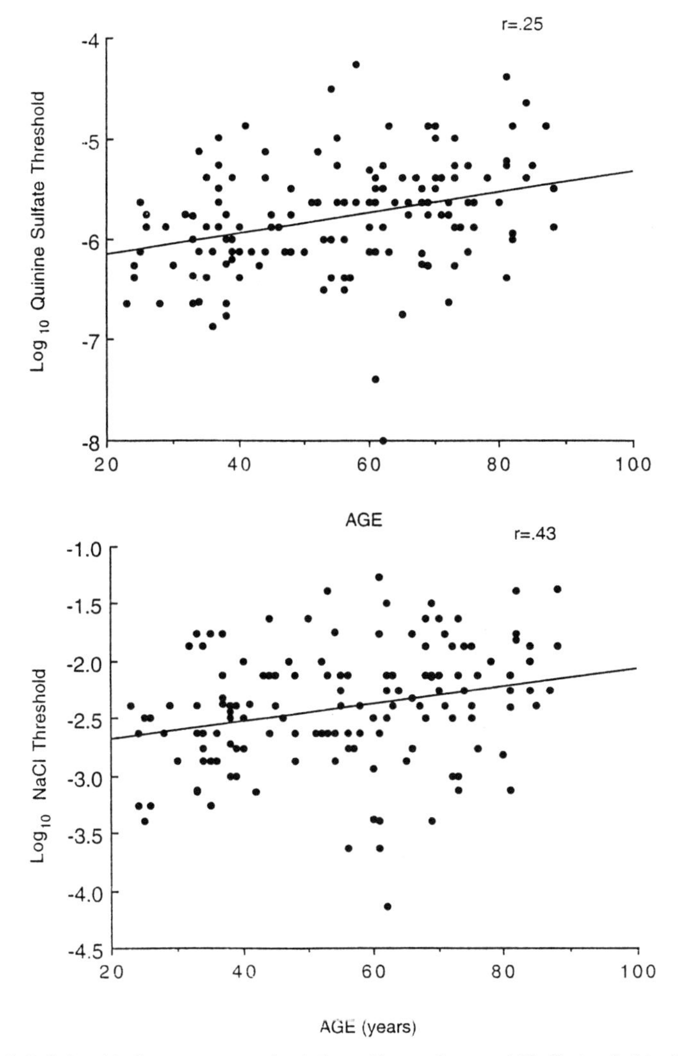

Figure 1 Relationship between age and quinine sulfate and age and NaCl thresholds. [Adapted from Weiffenbach, J. M., Baum, B. J. and Burghauser, R. 1982. *Journal of Gerontology, 37,* 372–377.]

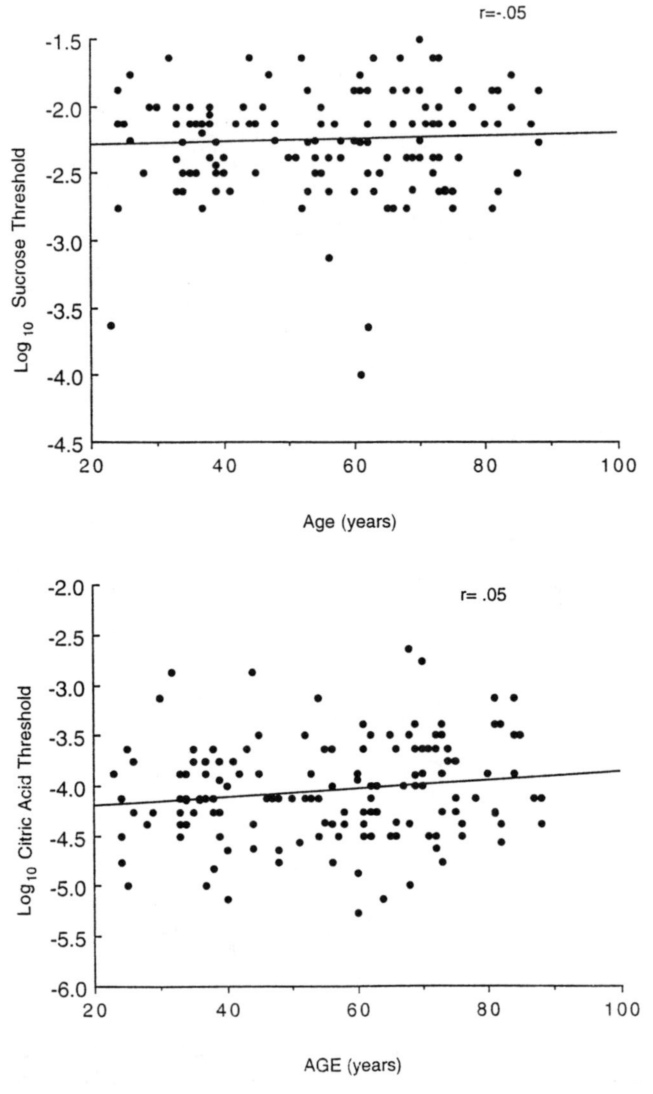

Figure 2 Relationship between age and citric acid and age and sucrose thresholds. [Adapted from Weiffenbach, J. M., Baum, B. J. and Burghauser, R. 1982. *Journal of Gerontology, 37,* 372–377.]

ysis of covariance revealed that the age covariate was not significant for sucrose, but approached significance for both NaCl and quinine sulfate.

In a more recent study, Schiffman et al. (1994) compared young and elderly subjects' taste thresholds for 12 different bitter compounds. Their results revealed higher thresholds for the elderly than for the young for six of the 12 compounds, with the difference in phenylthiocarbamide (PTC) thresholds approaching significance. More interestingly, the age-related threshold loss tended to be greater for the least lipophilic compounds. The fact that Schiffman et al. (1994) did not find age-associated changes in all 12 bitter compounds studied may suggest compound-specific decline in bitter sensitivity. This idea is supported by a recent study by Cowart, Yokomukai and Beauchamp (1994). These investigators examined threshold and suprathreshold sensitivity for two bitter compounds: urea and quinine sulfate. Their results were consistent with previous findings, indicating loss of sensitivity with age for quinine sulfate. However, no such differences were found for urea between the young and elderly. This study lends additional support for the existence of multiple bitter taste transduction mechanisms in humans and the idea that these mechanisms may be differentially affected by age.

These more recent studies suggest not only that thresholds for the four taste qualities are affected differentially by aging, but also that salty and bitter are affected more than sweet and sour qualities. This observation suggests the need for further study of the physiological mechanisms that underlie the psychophysical observation and points to the importance of early events in taste transduction and the role the aging process may play in their efficiency. In a more practical sense, differential effects of aging on taste modalities will contribute to altered taste perception in the elderly and may affect the acceptance or palatability of various foods.

SUPRATHRESHOLD INTENSITY

Age-related changes in suprathreshold taste intensity perception have also been demonstrated (see Murphy, 1986, for a more complete review). Psychophysical techniques such as magnitude estimation and magnitude matching have made it possible to investigate the effects of aging on the ability to distinguish and track different stimulus concentrations. With magnitude matching, subjects are required to estimate the magnitude of a series of stimuli from one modality as well as a series of stimuli on a continuum of another modality using the same scale (Marks and Stevens, 1980). This differs from magnitude estimation in which subjects are required to assign numbers that reflect stimulus intensity to a series of stimuli from a single modality that vary in concentration (Stevens, 1957).

The largest age-related losses in suprathreshold taste appear to be associated with the perception of bitterness and the smallest with the perception of sweetness (Cowart, 1989; Weiffenbach, Cowart and Baum, 1986; Murphy and Gilmore, 1989). Enns, Van Itallie and Grinker (1979) used magnitude estimation to examine perceived sweetness for sucrose in fifth-graders, college students and elderly individuals. They found steeper slopes for the children, but no differences in perceived sweetness from young adulthood to old age. Similarly, Hyde and Feller (1981) showed no differences in suprathreshold intensity ratings for sucrose and NaCl for elderly relative to young subjects, but did report lower ratings for caffeine and citric acid by the elderly. Cowart (1989) showed similar results in a life-span study of the four taste qualities. She found that psychophysical functions generated by the oldest group were always flatter than those generated by the other groups, with the flattening in function for quinine sulfate being the greatest across the age groups.

Weiffenbach, Cowart and Baum (1986) reported poorer performance by elderly subjects based on intraclass correlations. Intraclass correlation coefficients were calculated as the proportion of total variance of the judgments that is accounted for by variation in stimulus strength. This coefficient, thus, reflects the repeatability of the judgments made within a testing session. This measure showed a quality-specific difference between young and elderly, with significant decrements for bitter, salty, and bitter but not sweet.

Cowart (1989), using the method of magnitude estimation, examined intensity ratings for quinine sulfate, sodium chloride, sucrose and citric acid for 137 individuals divided into three age groups: 19–35 years, 45–59 years, 65–79 years. The results are presented in Figures 3 and 4. Figure 3 shows that the bitter compound produced the greatest difference between the age groups, with the greatest difference visible at the two highest concentrations found for the oldest group. Figure 4 presents the similar differences found when the oldest group was compared to the other two groups for both NaCl and citric acid. Once again, no such age differences were found for sucrose.

In another recent study, we used magnitude matching to assess this effect on the perceived intensity of the single tastants sucrose, caffeine, sodium chloride and citric acid. Perceived intensity for individual tastants in the following two-component mixtures was also assessed: sucrose/caffeine, sucrose/citric acid and sucrose/sodium chloride (Murphy and Gilmore, 1989). Stimuli were aqueous solutions of sucrose (0.0, 0.15, 0.30, 0.60 M), caffeine (0.0, 0.0025, 0.005, 0.01 M), citric acid (0.0, 0.0015, 0.003, 0.006 M) and NaCl (0.0, 0.10, 0.20, 0.40 M) and two-component mixtures of these stimuli. These stimulus ranges were of approximately equal

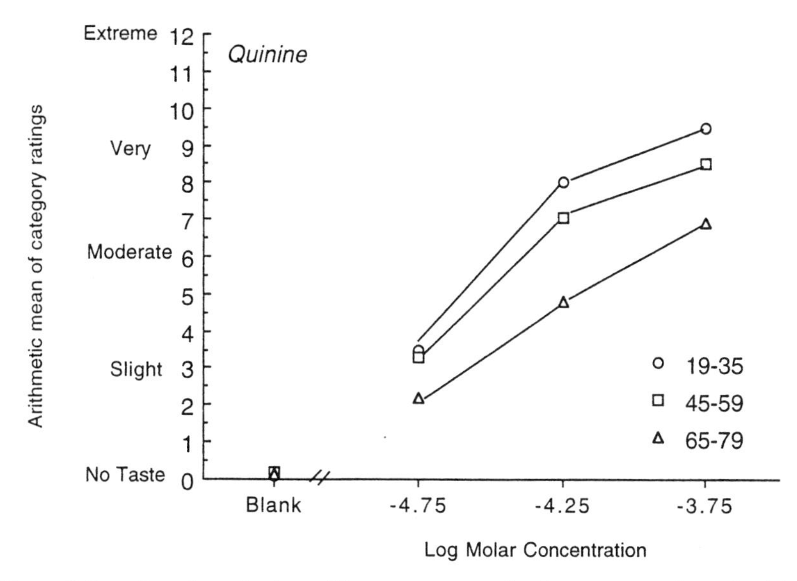

Figure 3 Intensity ratings for three different age groups for three concentrations of quinine sulfate. [Adapted from Cowart, B. J. 1989. Relationships between taste and smell across the adult life span. In Murphy, C., Cain, W. S. and Hegsted, D. M. (Eds.), *Nutrition and the chemical senses in aging* (pp. 39–55). New York: New York Academy of Sciences.]

subjective intensity for college students. Weights (20, 50, 100, 200, 400, 750 g) were used as the calibration continuum.

The results of this study showed no differences in intensity ratings for sweet and salty unmixed components, but revealed that the elderly subjects found bitter as less intense than the young (see Figure 5). In addition, age differences were found for the sour substance, but not for the sweet substance (Figure 6). Furthermore, the eta^2 values suggest the effects for bitter are greater (20% of the variance accounted for by age) than those for sour (10% of the variance accounted for by age). One of more interesting findings of this study is that quality-specific intensity scaling was also found in the binary mixture. Figure 7 shows that the elderly subjects judged bitter to be less intense in the sweet-bitter mixture than the young did, whereas saltiness of sweet-salty and sourness of sweet-sour mixtures remained relatively stable across the ages. Thus, both for unmixed and mixed components, the elderly subjects' intensity judgments for bitterness were significantly lower than those of the young subjects.

As evidenced from these studies and others, magnitude estimation and magnitude matching (Marks and Stevens, 1980) are currently the two methods most commonly employed for studying the differences in per-

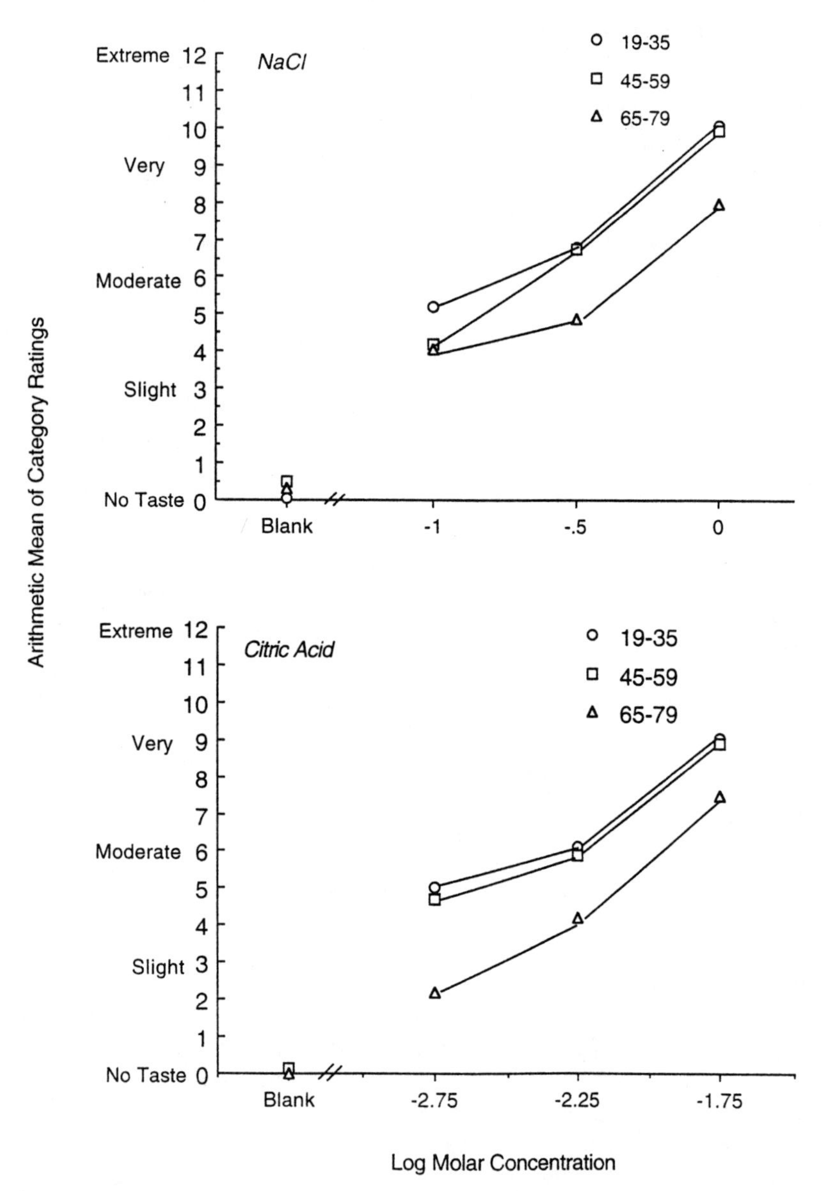

Figure 4 Intensity ratings for three different age groups for three concentrations of citric acid and NaCl. [Adapted from Cowart, B. J. 1989. Relationships between taste and smell across the adult life span. In Murphy, C., Cain, W. S. and Hegsted, D. M. (Eds.), *Nutrition and the chemical senses in aging* (pp. 39–55). New York: New York Academy of Sciences.]

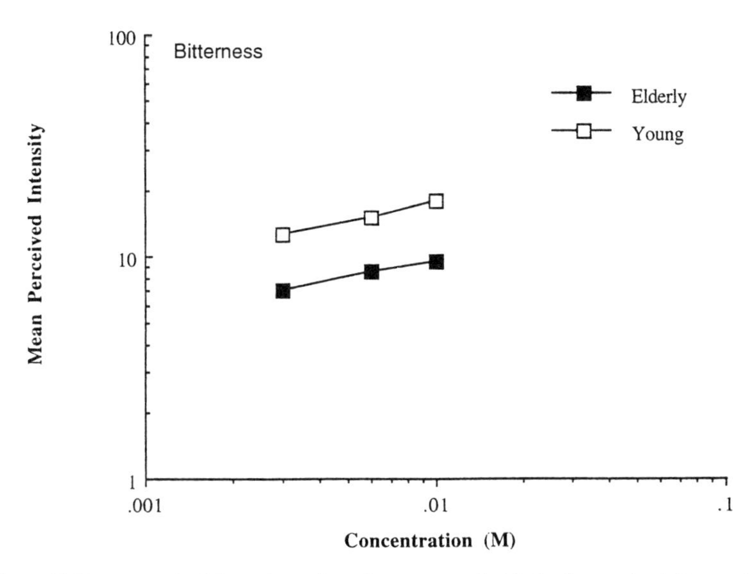

Figure 5 Mean perceived intensity ratings for young and elderly for unmixed bitter stimuli. [Adapted from Murphy, C. and Gilmore, M. M. 1989. Quality-specific effects of aging on the human taste system. *Perception and Psychophysics, 45,* 121–128.]

ceived intensity of suprathreshold stimulus concentrations. However, one must be especially aware of the contextual bias that would be expected to play a particularly large role in magnitude-matching experiments, since subjective magnitudes may vary with age depending on the sensory system being compared. That is, subjects' intensity judgments of stimuli from two modalities depended on the relative perceived intensity levels of the series of stimuli being compared; thus, when using scaling methods, care should be taken to select stimuli whose levels of subjective magnitude are similar.

These contextual effects on the sensory system may be even more difficult to control in scaling experiments where judgments of young and elderly are being compared, since aging may produce more variability in subjective magnitude ratings. Magnitude matching may, thus, underestimate the effects of age on suprathreshold intensity perception. Contextual bias may have operated when elderly and young persons judged intensity of bitter and sweet in the magnitude-matching experiment described above (Murphy and Gilmore, 1989). The differences between young and elderly subjects may have been underestimated since we know from the threshold data that their contextual worlds were probably different. Thus, while the conclusions that age-related impairment in bitter is relatively greater than that of sweet are undoubtedly true, this particular method may be underestimating the size of such differences.

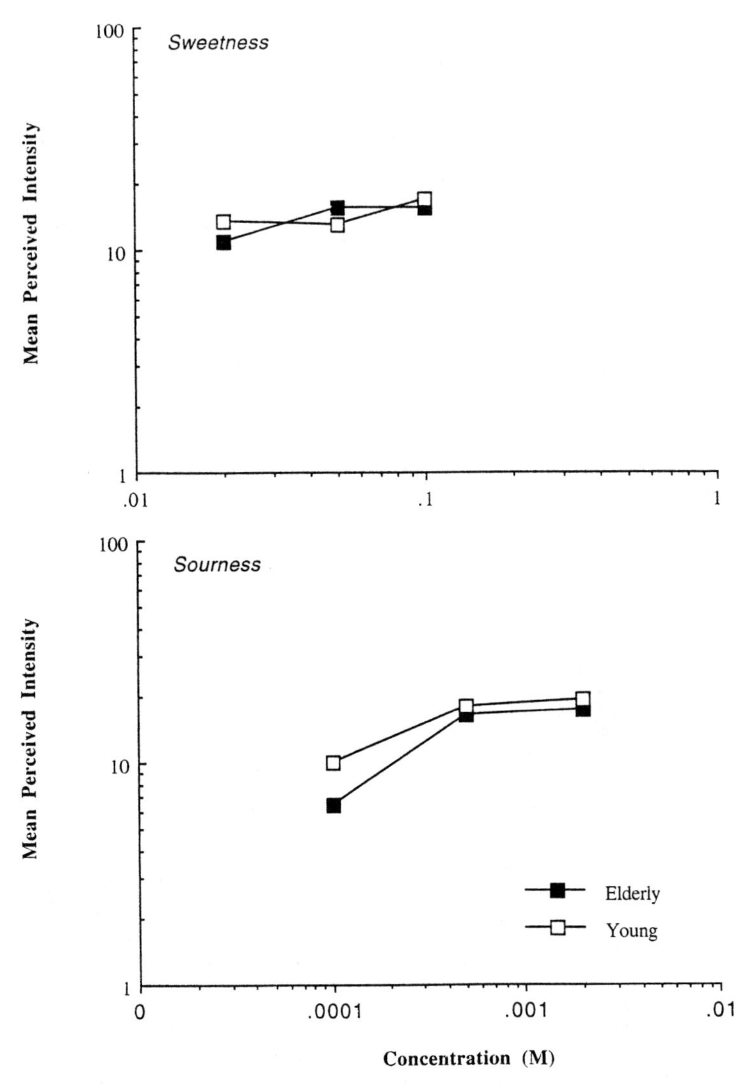

Figure 6 Mean perceived intensity ratings for young and elderly for unmixed sweet and sour stimuli. [Adapted from Murphy, C. and Gilmore, M. M. 1989. Quality-specific effects of aging on the human taste system. *Perception and Psychophysics, 45,* 121–128.]

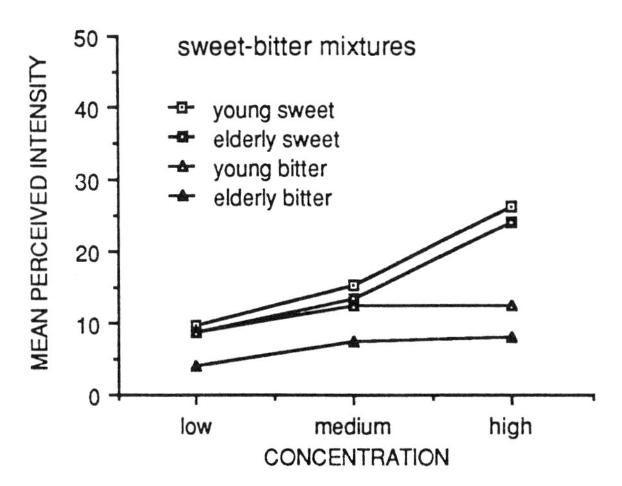

Figure 7 Mean perceived intensity ratings for young and elderly for bitter-sweet mixtures. [Adapted from Murphy, C. and Gilmore, M. M. 1989. Quality-specific effects of aging on the human taste system. *Perception and Psychophysics, 45,* 121–128.]

Magnitude matching with elderly subjects has other limitations, namely, the problem of calibration. Given that we do not yet completely understand the mechanisms underlying the effects of the aging process on sensory systems, excluding individuals who show impairment on the calibration continuum will exclude important subjects and may bias the results. For example, if hearing impairment is the calibration modality to which judgments of taste intensity are to be matched, then one will be required to exclude subjects whose auditory sensitivity does not meet a certain criterion. However, banning such subjects from participation will lead to an interesting subset of the elderly population being excluded.

Certain cognitive factors must also be taken into account when using magnitude-matching methods. To ensure the validity of the results obtained when using magnitude matching, one must be absolutely certain that subjects are using a unitary scale, since this is a cognitively demanding task and may be difficult for some elderly subjects.

An alternative approach is the use of Weber ratios to investigate questions of intensity discrimination between stimuli differing in concentration. This procedure, used to obtain just noticeable differences (JNDs), does not meet with the same contextual bias problems found with magnitude matching and it controls for criterion bias with a forced-choice task. The subjects' task is only to choose the strongest of two stimuli. This type of design is ideal for situations in which sensory or cognitive impairment in subjects of interest limits the use of magnitude matching.

WEBER RATIOS

Since scaling techniques place significant memory demands on elderly individuals, Gilmore and Murphy (1989) attempted to simplify the process of assessing suprathreshold taste function by determining an individual's just noticeable difference (JND): an indicator of how much of a physical stimulus is needed in order to obtain a perceptible difference in taste sensation. The subject's only requirement was to choose the strongest of two stimuli. By forcing a choice, the effects of criterion on responses were minimized.

The ratio of the JND to a particular standard concentration results in a Weber Ratio (WR). The average WR for taste is approximately 0.20, hence one must increase the stimulus 20% before a subject will perceive a noticeable difference in taste sensation. We were interested in whether elderly individuals required a greater increase in the physical stimulus in order to get a perceptible difference in sensation. We were also interested in whether or not age-associated effects would be quality-specific: whether the size of the WR would be dependent upon the stimulus tested. Since we had shown (Murphy and Gilmore, 1989) greater age-related differences in the perception of bitter stimuli than in the perception of sweet stimuli, these two tastants were examined in the first study.

In this experiment each subject tasted 30 pairs of stimuli associated with each of the eight standards: (1) 0.15 M sucrose, (2) 0.30 M sucrose, (3) 0.60 M sucrose, (4) 0.0025 M caffeine, (5) 0.005 M caffeine and (6) 0.01 M caffeine. Each standard was compared to six comparison stimuli.

The results of this study revealed no differences between the WRs of the young and elderly for sucrose at any of the standard concentration levels. This suggests that the same increase in stimulus concentration is needed for both groups in order to perceive a noticeable difference. However, differences were found between the WRs of the young and elderly for caffeine, particularly at the highest standard concentration. A mean WR of 2.5 was found for the elderly at the highest concentration level, suggesting that a 250° change in concentration level was required in order for the elderly to be able to perceive a noticeable difference. In contrast, a WR of 0.25 was observed for the young, suggesting that a 25% increase in concentration was necessary in order to perceive a noticeable difference. The findings from this study led us to investigate the same phenomenon, in terms of WR, for sour and salty taste qualities.

In the subsequent study (Nordin, et al., submitted), we again employed the method of constant stimuli for measuring the suprathreshold discrimination ability of young and elderly subjects for citric acid and NaCl. Each subject also performed a recognition threshold task in order to ensure that all subjects were able to taste the concentration range of the stimuli. Again, three standard concentrations were used for each taste quality and each standard was compared to six comparison stimuli. In the same forced-choice fashion, the subject's task was to identify the stimuli that had the stronger taste.

The results of this study once again supported the hypothesis that age-related changes in intensity perception are quality-specific. Results indicated differences in WRs between young and elderly women for citric acid but not for NaCl. In comparing eta^2 values for bitter and sour qualities (the two qualities that showed age effects), it was found that the effect size was larger for caffeine (0.17) than for citric acid (0.16).

WRs may help us understand why elderly subjects do not perceive taste stimuli in the same manner that young subjects do. WRs as an index of taste sensitivity may be particularly useful in assessment of individuals with impaired cognitive status, such as Alzheimer's disease patients, or with individuals who have impairments that would ordinarily prevent them from performing magnitude-matching tasks.

CHEMOSENSORY PREFERENCE

Is preference for chemosensory stimuli affected by age? Changes in flavor preference would be expected given the changes in smell and taste in the elderly. Although considerable proprietory information may exist, unfortunately, there is very little published literature to address this crucial question and guide our understanding of the relationships between chemosensory function, appetite and nutrition.

Very early reports suggested a greater preference for sour than for sweet taste in juice samples in older subjects (Laird and Breen, 1939). Preference for sweet and salty taste qualities have also been studied in children and adults. Desor, Green and Maller (1975) reported an increase in preference for sweet and saltiness for 9- and 15-year-olds than for adults. However, they did not find differences among adults up to 64 years old. Unfortunately, the lack of information about the age distribution of adult subjects in the study makes the results difficult to interpret.

Enns, Van Itallie and Grinker (1979) reported that female adults had higher preference for sucrose (in aqueous solution) than the elderly. Dye and Koziatek (1981) studied sucrose preference in young and elderly non-diabetic men. Their results suggested greater pleasantness ratings by the nondiabetic men as concentrations increased over the range of 0.125 M–1 M sucrose. The 0.25 M sucrose concentration was rated as the most pleasant stimulus and the concentrations stronger than 0.25 M were rated as less pleasant. Pangborn, Braddock and Stone (1983) found that older (36- to 66-year-old) subjects preferred greater amounts of salt added to chicken broth than younger (17- to 32-year-old) subjects.

Studies of taste preference across the adult life span have also been conducted in our laboratory (Murphy and Withee, 1986). In this study we investigated age-associated changes in pleasantness of various concentrations of tastants in aqueous solution and of the same tastants in complex mixtures. One hundred young, 100 middle-aged and 100 old adults scaled the pleasantness of sucrose, citric acid and NaCl, each presented in four concentrations in water and the same four concentrations in appropriate beverage bases.

Two measures of hedonics were used: (1) the peak preferred concentration (i.e., the single most preferred concentration in the series) and (2) the pleasantness judgment (i.e., an equivalent of magnitude estimate of the pleasantness or unpleasantness of a given stimulus).

The results showed that the samples were less pleasant overall in an aqueous base than in a beverage base and concentration significantly affected ratings. Both sucrose and NaCl were judged to be more pleasant in the beverage base, but the background had more effect on NaCl than on sucrose. Furthermore, salt was rated as more pleasant by the elderly than by the other age groups, regardless of the diluent. When presented in water, salt became more unpleasant as concentration increased. However, in beverage base, mid-range salt concentrations were preferred to the lowest by both middle-aged and elderly raters.

Sucrose pleasantness ratings increased with the first three concentrations, but decreased at the fourth, thus producing an invert U-shaped function. Conversely, pleasantness for both NaCl and citric acid decreased with increased concentration at every level. Also, sucrose and NaCl in high concentrations were rated as more pleasant by elderly than by younger participants.

The results of this study may have direct implications for the elderly's well-being, since preference for and dietary selection of salty and sugary foods may have negative health consequences. Older people have increased incidence of hypertension and diabetes, as well as increases in systolic blood pressure and blood sugar and many of these people need to avoid the highly palatable high salt and high sugar foods.

The olfactory system seems to be more affected by the aging process than the taste system (Stevens, Bartoshuk and Cain, 1984; Murphy, 1982a, 1985). Thus, age-associated changes in olfactory perception may be partially responsible for the differences in pleasantness judgments made by the young, middle-aged and elderly subjects in these experiments. Data from the age-related olfactory studies suggest that it may not be the high salt concentrations that make a beverage base more pleasant: it is equally possible that the higher salt concentration may be compensating for the reduction of olfactory stimulation in the overall flavor complex.

Literature on age-related changes in odor preference is also sparse; however, because of the great amount of overlap between the senses of taste and smell, some attention must be given to this topic. Moncrieff (1966) was one of the first investigators to report changes in odor preference with aging. Cain and Johnson (1978) have shown that simply being exposed to a particular odor can produce changes in pleasantness perception. Murphy (1983) examined young and elderly subjects' pleasantness ratings for a common odor, menthol. Subjects were repeatedly presented with the stimuli within the same session. The results indicated that the elderly did not alter their pleasantness ratings over presentation, whereas young subjects did. The elderly may respond in a similar manner in the context of daily life. That is, older individuals' responses to odors in the environment or in foods and beverages may be less influenced by repeated exposure.

CONCLUSION

The results of a majority of studies suggest age-related differences in threshold and suprathreshold sensitivity. However, these changes are not uniform for the four taste qualities. In fact, most studies indicate that age-related effects are larger for bitter compounds followed by salty and sour, with the sweet taste quality being affected the least by aging. In addition, even within a certain taste quality, there appear to be differential age-associated changes in sensitivity. This is particularly the case for bitter sensation.

There are reports of the loss of taste papillae, buds and cells in aged humans (Schiffman, 1986). The psychophysical results suggesting differential age effects on human taste qualities further suggest that membrane and receptor function may play an important role in taste dysfunction in the elderly. Modern neurobiological techniques may soon begin to shed light on this problem that has not yielded to more gross examination.

It is not surprising to find that eating enjoyment is affected by the quality-specific taste losses associated with aging. The majority of the studies reviewed agree that the ability to perceive bitterness diminishes with age, and this tastant serves as an underlying component in numerous foods and, in particular, many beverages. The use of herbs and spices that contain bitter flavors can be added as flavor enhancers. Selecting a diet that compensates for sensory losses may improve the palatability of foods for elderly people.

ACKNOWLEDGEMENT

Dr. Murphy's research and the preparation of this chapter were supported by NIH grants AG04085 and AG08203 from the National Institute on Aging.

REFERENCES

Cain, W. S. and Johnson, F. 1978. Lability of odor pleasantness: Influence of mere exposure. *Perception, 1,* 459–465.

Chalke, H. D. and Dewhurst, J. R. 1957. Accidental coal-gas poisoning. *British Medical Journal, 2,* 915–917.

Cowart, B. J. 1989. Relationships between taste and smell across the adult life span. In Murphy, C., Cain, W. S. and Hegsted, D. M. (Eds.), *Nutrition and the chemical senses in aging* (pp. 39–55). New York: New York Academy of Sciences.

Cowart, B. J., Yokomukai, Y. and Beauchamp, G. K. 1994. Bitter taste in aging: Compound-specific decline in sensitivity. *Physiology and Behavior, 56,* 1237–1241.

Desor, J. A., Green, L. S. and Maller, O. 1975. Preferences for sweet and salty tastes in 9- and 15-year-old and adult humans. *Science, 190,* 686–687.

Doty, R. L., Shaman, P., Applebaum, S. L., Giberson, R., Siksorski, L. and Rosenberg, L. 1984. Smell identification ability: Changes with age. *Science, 226,* 1441–1443.

Dye, C. J. and Koziatek, D. A. 1981. Age and diabetes effects on threshold and hedonic perception of sucrose solutions. *Journal of Gerontology, 36,* 310–315.

Enns, M. P., Van Itallie, T. B. and Grinker, J. A. 1979. Contributions of age, sex and degree of fatness on preferences and magnitude estimation for sucrose in humans. *Physiology and Behavior, 22,* 999–1003.

Fordyce, I. D. 1961. Olfaction tests. *British Journal of Industrial Medicine, 18,* 213–215.

Gilmore, M. M. and Murphy, C. 1989. Aging is associated with increased Weber ratios for caffeine, but not for sucrose. *Perception and Psychophysics, 46,* 555–559.

Hyde, R. J. and Feller, R. P. 1981. Age and sex effects on taste of sucrose, NaCl, citric acid and caffeine. *Neurobiology of Aging, 2,* 315–318.

Joyner, R. E. 1963. Olfactory acuity in an industrial population. *Journal of Occupational Medicine, 5,* 37–42.

Kimbrell, G. M. and Furchtgott, E. 1963. The effect of aging on olfactory threshold. *Journal of Gerontology, 18,* 364–365.

Laird, D. A. and Breen, W. J. 1939. Sex and age alterations in taste preferences. *Journal of the American Diet. Association, 15,* 549–550.

Marks, L. E. and Stevens, J. C. 1980. Measuring sensation in the aged. In Poon, L. W. (Ed.), *Aging in the 1980's: Psychological issues* (pp. 592–598). Washington, DC: American Psychological Association.

Minz, A. I. 1968. Condition of the nervous system in old men. *Zeitschrift fur Alternsforschung, 21*(3), 271–277.

Moncrieff, R. W. 1966. *Odour preferences.* Wiley: New York.

Murphy, C. 1976. Gustatory absolute thresholds as a function of age: An investigation into the mechanism for coding bitter. Doctoral Dissertation, University of Massachusetts, Amherst, MA.

Murphy, C. 1979. The effects of age on taste sensitivity. In Han, S. S. and Coons, D. H. (Eds.), *Special Senses in Aging* (pp. 21–33). Ann Arbor, MI: University of Michigan Institute of Gerontology.

Murphy, C. 1982a. Effects of aging on food perception. *Journal of American College of Nutrition, 1,* 128–129.

Murphy, C. 1982b. Effects of exposure and context on hedonics of olfactory-taste mixtures. In *Selected sensory methods: Problems and approaches to measuring hedonics,* Kuznicki, Johnson and Rutkiewic (Eds.), *ASTM STP 773,* Amer. Soc. Testing & Materials, Philadelphia, PA (pp. 60–70).

Murphy, C. 1983. Age-related effects on the threshold, psychophysical function, and pleasantness of menthol. *Journal of Gerontology, 38,* 217–222.

Murphy, C. 1985. Cognitive and chemosensory influences on age-related changes in the ability to identify blended foods. *Journal of Gerontology, 40,* 47–52.

Murphy, C. 1986. Taste and smell in the elderly. In Meiselman, H. L. and Rivlin, R. S. (Eds.), *Clinical measurement of taste and smell* (pp. 343–371). New York: Macmillan.

Murphy, C. and Cain, W. S. 1986. Odor identification: The blind are better. *Physiology and Behavior, 37,* 177–180.

Murphy, C. and Withee, J. 1986. Age-related differences in the pleasantness of chemosensory stimuli. *Psychology and Aging, 1,* 312–318.

Murphy, C. and Gilmore, M. M. 1989. Quality-specific effects of aging on the human taste system. *Perception and Psychophysics, 45,* 121–128.

Murphy, C., Cain, W. S. and Bartoshuk, L. M. 1977. Mutual action of taste and olfaction. *Sensory Processes, 1,* 204–211.

Murphy, C., Cain, W. S., Gilmore, M. M. and Skinner, R. B. 1991. Sensory and semantic factors in recognition memory for odors and graphic stimuli: Elderly versus young persons. *American Journal of Psychology, 104,* 161–192.

Murphy, C., Gilmore, M. M., Seery, C. S., Salmon, D. P. and Lasker, B. R. 1990. Olfactory thresholds are associated with degree of dementia in Alzheimer's disease. *Neurobiology of Aging, 11,* 465–469.

Murphy, C., Nunez, K., Withee, J. and Jalowayski, A. A. 1985. The effects of age, nasal airway resistance and nasal cytology on olfactory threshold for butanol. *Chemical Senses, 10,* 418.

Nordin, Razani, Markison and Murphy (submitted).

Pangborn, R. M., Braddock, K. S. and Stone, L. J. 1983. Ad lib mixing to preference vs. hedonic scaling: Salts in broths and sucrose in lemonade. *Association for Chemoreception Sciences Annual Meeting,* Sarasota, FL.

Schemper, T., Voss, S. and Cain, W. S. 1981. Odor identification in young and elderly persons: sensory and cognitive limitations. *Journal of Gerontology, 18,* 446–452.

Schiffman, S. S., Moss, J. and Erickson, R. P. 1976. Thresholds of food odors in the elderly. *Exper. Aging Res., 2,* 389–398.

Schiffman, S. S. 1979. Changes in taste and smell with age: Psychophysical aspects. In Ordy and Brizzee (Eds.), *Sensory systems and communication in the elderly.* Raven Press: New York.

Schiffman, S. S. 1986. Age-related changes in taste and smell and their possible causes. In Meiselman and Rivlin (Eds.), *Clinical measurement of taste and smell* (pp. 326–342). New York: Macmillan.

Schiffman, S. S., Gatlin, L. A., Frey, A. E., Heiman, S. A., Stagner, W. C. and Cooper, D. C. 1994. Taste perception of bitter compounds in young and elderly persons: Relation to lipophilicity of bitter compounds. *Neurobiology of Aging, 15,* 743–750.

Stevens, J. C. and Mark, L. E. 1980. Cross-modality matching functions generated by magnitude estimation. *Perception and Psychophysics, 27,* 379–389.

Stevens, J. C. and Cain, W. S. 1987. Old-age deficits in the sense of smell as gauged by thresholds, magnitude matching, and odor identification. *Psychology and Aging, 2,* 36–42.

Stevens, J. C., Plantinga, A. and Cain, W. S. 1982. Reduction of odor and nasal pungency associated with aging. *Neurobiology of Aging, 3,* 125–132.

Stevens, J. C., Bartoshuk, L. M. and Cain, W. S. 1984. Chemical senses and aging: Taste versus smell. *Chemical Senses, 9,* 167–179.

Stevens, S. S. 1957. On the psychophysical law. *Psychological Review, 64,* 153–181.

Strauss, E. L. 1970. A study on olfactory acuity. *Annals of Otolaryngology, Rhinology, and Laryngology, 79,* 95–104.

Venstrom, D. and Amoore, J. E. 1968. Olfactory threshold in relation to age, sex or smoking. *Journal of Food Science, 33,* 264–265.

Weiffenbach, J. M., Baum, B. J. and Burghauser, R. 1982. Taste thresholds: quality specific variation with human aging. *Journal of Gerontology, 37,* 700–706.

Weiffenbach, J. M., Cowart, B. J. and Baum, B. 1986. Taste intensity perception in aging. *Journal of Gerontology, 41,* 460–468.

CHAPTER 9

Suppression of Bitterness by Sodium: Implications for Flavor Enhancement

P. A. S. BRESLIN[1]
G. K. BEAUCHAMP[1]

INTRODUCTION

When two compounds that elicit different taste qualities are mixed in solution, the mixture will often yield a taste sensation that is less intense than the simple sum of the component tastes. In two-component mixtures, each taste quality is usually perceived as less intense than when it is tasted separately (Bartoshuk, 1975; Kamen et al., 1961; Kemp and Beauchamp, 1994; McBride, 1989). However, binary combinations of certain taste stimuli may result in asymmetrical changes. For example, Schifferstein and Frijters (1992a) recently reported that when quinine hydrochloride (QHCl), which usually elicits a bitter taste, is mixed in solution with sodium chloride (NaCl), which usually elicits a salty taste, the saltiness of the NaCl is relatively unaffected, while the bitterness of QHCl is suppressed from 50–70% (*see also* Frijters and Schifferstein, 1994), depending upon the concentrations involved (cf., Kroeze, 1982; Kemp and Beauchamp, 1994).

It is not known how NaCl decreases the bitterness elicited by QHCl although there is evidence that the decrease occurs peripherally (Bartoshuk, 1979, 1980; Bartoshuk and Seibyl, 1982; Kroeze and Bartoshuk, 1985). Also unknown is whether the taste profile of other bitter-salty mixtures would exhibit the same characteristics. There is growing evidence

This chapter is updated from "Suppression of Bitterness by Sodium," *Chemical Senses,* 20:609–623, 1995, by permission of Oxford University Press.

[1]Monell Chemical Senses Center, 3500 Market Street, Philadelphia, PA 19104-3308, U.S.A.

that multiple transduction pathways are associated with bitter taste and that a single compound, such as QHCl, does not stimulate all of them equally (see below for references). For example, a single transduction process is believed to account for the wide variation in sensitivity across individuals to the bitter compounds propylthiouracil (PROP) and phenylthiocarbamide (PTC), which varies independently of quinine's bitterness in the same subjects (Fischer and Griffin, 1963; Mela, 1989; Schifferstein and Frijters, 1991). McBurney et al. (1972) have also reported asymmetrical cross-adaptation among bitter compounds, a result consistent with multiple bitter transduction sequences. Yokomukai et al. (1993) and Cowart et al. (1994) have obtained data consistent with those of McBurney et al. (1972) (*see also* Lawless, 1979). Taken together, these studies suggest that there may be at least three classes of bitter transduction sequences in humans, one implicated in the transduction of PTC- and PROP-like compounds, one sensitive to quinine- and caffeine-like compounds, and one sensitive to urea- and magnesium sulphate ($MgSO_4$)- like compounds.

The goal of Experiment 1 of this chapter was to compare the interactions of NaCl with several bitter-eliciting compounds that may have different transduction sequences. In addition, more detailed studies were conducted with urea, a bitter compound that was effectively suppressed by NaCl. Specifically, Experiments 2 and 3 explored the effects of anion and cation substitution on bitter suppression, respectively.

GENERAL METHODS

SUBJECTS

Subjects between the ages of 21 and 30 were paid to participate in 12 studies after giving their informed consent. All were employees of the Monell Center. The number of subjects (12–27) in each study is given in Table 1; some subjects participated in more than one study. Each subject was coded with a random number.

STIMULI

The bitter agents and salts that were used are listed in Table 1. All solutions were prepared with deionized water. Solutions were kept at 5 °C in a dark cold-room and were replaced at least every two weeks. Prior to testing, the stimuli were brought to room temperature with the aid of a water bath.

TABLE 1. The Experimental Design of All 12 Studies in Experiments 1, 2 and 3.

Experiment	No. Subjects	Bitter Compound (concentrations)	Salt (concentrations)
1A	21	QHCl (0, 0.1, 1 mM) S*	NaCl (0, 0.1, 0.3, 0.5 M) F
1B	18	Caffeine (0, 1.25, 10 mM) S	NaCl (0, 0.1, 0.3, 0.5 M) F
1C	12	$MgSO_4$ (0, 0.3, 0.5 M) S	NaCl (0, 0.1, 0.3, 0.5 M) F
1D	15	Amiloride (0, 70, 100 μM) F	NaCl (0, 0.1, 0.3, 0.5 M) F
1E	27	KCl (0, 0.05, 0.10, 0.20 M) F	NaCl (0, 0.05, 0.10, 0.20 M) F
1F	20	Urea (0, 0.5, 1.0 M) F	NaCl (0, 0.1, 0.3, 0.5 M) F
2A	14	Urea (0, 0.5, 1.0 M) F	SA (0, 0.1, 0.3, 0.5 M) S
2B	15	Urea (0, 0.5, 1.0 M) F	SG (0, 0.1, 0.3, 0.5 M) A
2C	14	QHCl (0, 0.1, 1 mM) S	SA (0, 0.1, 0.3, 0.5 M) S
3A	14	Urea (0, 0.5, 1.0 M) F	KCl (0, 0.1, 0.3, 0.5 M) F
3B	12	Urea (0, 0.5, 1.0 M) F	LiCl (0, 0.1, 0.3, 0.5 M) F
3C	12	Urea (0, 0.5, 1.0 M) F	LALA (0, 0.1, 0.3, 0.5 M) S

*Abbreviations: QHCl = quinine hydrochloride; LALA = L-arginine-L-aspartate; SA = sodium acetate; SG = Sodium gluconate; F = Fisher Co.; S = Sigma Co.; A = Aldrich Co.

INTENSITY MATCHING

The range of concentrations for the bitter stimuli in each series was selected so that preceived bitter intensities were matched across series, as determined by pretesting, except for potassium chloride (KCl), which would be prohibitively salty when matched to quinine for bitterness.

The matching procedure was as follows. Two concentrations of quinine hydrochloride (QHCl), 0.1 and 1.0 mM, were selected as the medium and high levels of bitter sensation and all other bitter compound concentrations were selected to match these two in bitterness (except for KCl). Twenty individuals served as subjects and were run individually. To match the medium level of bitterness, each subject was presented with four pairs of solutions. One member of each pair was the medium level of QHCl and the other was a concentration of a different compound thought by the experimenter to appear close to the same level of bitterness. After sampling from both cups, a subject was asked to identify which one had the more bitter solution. This continued for all four pairs of cups. Subjects were instructed to rinse their mouth four times with room temperature deionized water before testing and twice between each sampling. If several (about five) subjects perceived either the QHCl or the other bitter compound as consistently more bitter on all four trials, then the concentration of the other bitter compound was adjusted appropriately and the testing restarted. When all five subjects did not select one of the compounds on four out of four trials, then all twenty subjects were tested with the four trial procedure. A tally was kept of how many times each compound had been selected as more bitter. If either the QHCl or the other bitter compound was selected, on average, as more bitter 55% of the time or more by the twenty subjects, then the entire procedure was repeated with a new test concentration.

The next test concentration was either half or double the former (as needed), and subsequent steps were halfway between those two steps, again moving up or down as needed. When neither compound was perceived as more bitter on more than 55% of the trials, the two compounds were considered to be matched on average for bitter intensity, since binomial variance dictates that a two-alternative procedure with 55% selection and 80 trials has a standard error of ±5.6 percentage points. The same procedure was used to match the higher QHCl; concentration, 1.0 mM.

These intensity matches were obtained from twenty subjects and were an average response across the whole sample population. Therefore, each individual subject may have found that any pair of solutions for the two compounds were *not* matched in intensity, when, *on the average,* all twenty subjects found the two solutions comparable in bitterness intensity (*see,* e.g., Cowart et al., 1994; Yokomukai et al., 1993). Matches were not calculated at the individual level as this would be prohibitively time consuming when trying to intensity match six different compounds for each subject.

PROCEDURE

Each study consisted of judgments of the bitterness and saltiness of all possible combinations of three or four concentrations of a bitter compound with four concentrations of a salt, which resulted in a $3 \times 4 = 12$ solution matrix for all but KCl-NaCl mixtures, which were tested in a $4 \times 4 = 16$ solution matrix. In every case, one concentration of the bitter agent and the salt was 0.0 M.

The method of magnitude estimation was used to obtain ratings of the perceived intensities of saltiness and bitterness from every solution sampled. All subjects were familiar with the method, and no modulus was given. Subjects were instructed to rate only the saltiness and the bitterness of each solution and to ignore any other qualities.

Within any bitter-salt mixture series (containing only one salt and one bitter agent), each solution ($n = 12$ or $n = 16$) was sampled twice. Subjects were instructed to rinse and expectorate with deionized water four times over a period of roughly 2 min prior to testing. The solutions were presented in random order, without replacement. The two salty ratings and the two bitter ratings for each solution were arithmetically averaged to yield single ratings of saltiness and bitterness. Subjects were required to rinse twice thoroughly with deionized water during the approximate 60-sec interstimulus interval. All samples were delivered in 10 ml volumes in polystyrene medicine cups.

STANDARDIZATION OF DATA

Experiment 1: To eliminate the variance produced by idiosyncratic number usage in the magnitude estimation task, the saltiness and bitterness ratings were standardized to the grand arithmetic mean of the saltiness ratings of NaCl in water for all subjects in all studies involving the use of NaCl. Each subject's individual mean saltiness rating was divided into the grand mean and the quotient was used as the multiplicative standardization factor for that individual's saltiness *and* bitterness ratings. This procedure equated mean ratings of salt in water across subjects while maintaining the individual relations between saltiness and bitterness for each subject.

Experiments 2 and 3: Because only one study in Experiments 2 and 3 included NaCl, whereas urea was used in all but one study (QHCl and NaAcetate, discussed below), the data from Experiments 2 and 3 were standardized to the urea data shown in Figure 3(a) (urea and NaCl). Each subject's individual mean urea bitterness rating from each study was divided into the grand arithmetic mean for the bitterness of urea in water without salt. The resulting multiplicative standardization factor was used for both the saltiness ratings *and* the bitterness ratings. In the study depicted in Figure 3(d) (QHCl and NaAcetate), neither NaCl nor urea was employed. Since QHCl was employed in both Figures 3(d) and 1(a) (QHCl and NaCl), the bitterness ratings of QHCl in water from data shown in Figure 1(a) were used to standardize the data from Figure 3(d) using a method parallel to that described above.

ANALYSIS

Since the distributions of standardized saltiness and bitterness ratings in each study were skewed in a manner that approximated a log-normal distribution, the data were transformed to logs before statistical analysis.

Because there were frequent reports of either zero saltiness or bitterness, 1.0 was added to all ratings prior to transformation. This addition of 1.0 to all ratings has a larger impact on smaller numbers in log coordinates, potentially resulting in less apparent suppression of bitterness at weaker bitter agent concentrations. However, in practice, this was not the case, as weak concentrations were always affected more than the stronger bitter concentrations (see Results and figures below).

Data from each study were analyzed separately using a 2-way within-subjects analysis of variance (ANOVA) [*Concentration* (3 or 4 steps) × *Added Compound* (3 or 4 levels)]. The two measurements of quality (saltiness and bitterness) were also analyzed separately. When interaction effects were obtained, one-way ANOVAs were performed on the different levels of the mixture for each concentration step. Because we viewed water as belonging to the concentration continuum, some interaction effects may be due to the subject's perception of bitterness of the salt in water or residual bitterness sensations reported when water was presented. All pairwise comparisons were performed with the Tukey HSD method. Different compounds were not directly statistically compared, since tests of each compound were conducted neither on exactly the same subjects nor on a completely different set of subjects. To provide some basis for comparisons among compounds, however, the arithmetic mean percent suppression of the two bitter concentrations is presented in Figures 2, 4, and 6 with standard errors. Percent suppression of bitterness was calculated by dividing the bitterness of the bitter-salt mixture by the bitterness of unmixed bitter compound concentration and then subtracting this value from 1: $=\{1 - ((\text{bitter} + \text{salt})/\text{unmixed bitter})\}$ for each subject at each concentration and then taking the arithmetic mean ± SEM (note that this calculation does not take into account baseline bitterness levels of water or unmixed salt).

There was never a significant suppression of saltiness for the highest concentration of NaCl. Therefore, percent suppressions were not calculated for the saltiness of NaCl.

RESULTS

All main effects of mixture and interaction effects for all experiments are presented in Table 2. Because in all cases saltiness and bitterness varied directly with concentration ($p < 0.1$), the main effect of concentration has been omitted from the table. Pairwise comparisons (Tukey HSD) with significant differences are indicated in the figures with asterisks (see Figures 1, 3, and 5).

TABLE 2. The Summary of Statistics for Main Effects and Interactions: Summary of Statistical Results.*

Experiment	Test Solutions	Bitterness Main Effect	Bitterness Interaction	Saltiness Main Effect	Saltiness Interaction
1A	NaCl and quinineHCl	$F(3,60) = 3.91$, $P < 0.05$	$F(6,120) = 9.47$, $P < 0.0001$	$F(2,40) = 5.22$, $P < 0.01$	$F(6,120) = 7.83$, $P < 0.0001$
1B	NaCl and caffeine	$F(3,42) = 3.79$, $P < 0.05$	$F(6,84) = 8.00$, $P < 0.0001$	$F(2,28) = 0.966$	$F(6,84) = 6.29$, $P < 0.0001$
1C	NaCl and $MgSO_4$	$F(3,33) = 2.49$	$F(6,66) = 4.09$, $P < 0.01$	$F(2,22) = 5.71$, $P < 0.01$	$F(6,66) = 14.22$, $P < 0.0001$
1D	NaCl and amiloride	$F(3,42) = 3.00$, $P < 0.05$	$F(6,84) = 7.90$, $P < 0.0001$	$F(2,28) = 9.80$, $P < 0.001$	$F(6,84) = 3.84$, $P < 0.01$
1E	NaCl and KCl	$F(3,78) = 51.24$, $P < 0.0001$	$F(9,234) = 11.39$, $P < 0.0001$	$F(2,28) = 9.80$, $P < 0.001$	$F(9,234) = 4.73$, $P < 0.0001$
1F	Urea and NaCl	$F(3,57) = 45.96$, $P < 0.0001$	$F(6,114) = 6.48$, $P < 0.0001$	$F(2,38) = 5.60$, $P < 0.01$	$F(6,114) = 4.95$, $P < 0.001$

(continued)

TABLE 2. (continued).

	Test Solutions	Bitterness Main Effect	Bitterness Interaction	Saltiness Main Effect	Saltiness Interaction
Experiment					
2A	Urea and NaAcetate	$F(3,39) = 17.16$, $P < 0.0001$	$F(6,78) = 3.79$, $P < 0.01$	$F(2,26) = 17.66$, $P < 0.0001$	$F(6,78) = 3.13$, $P < 0.01$
2B	Urea and NaGluconate	$F(3,42) = 22.09$, $P < 0.0001$	$F(6,84) = 4.73$, $P < 0.001$	$F(2,28) = 11.33$, $P < 0.001$	$F(6,84) = 5.68$, $P < 0.0001$
2C	QuinineHCl and NaAcetate	$F(3,39) = 0.60$	$F(6,78) = 4.30$, $P < 0.001$	$F(2,26) = 0.49$	$F(6,78) = 1.29$
Experiment					
3A	Urea and KCl	$F(3,39) = 18.38$, $P < 0.0001$	$F(6,78) = 10.85$, $P < 0.0001$	$F(2,26) = 1.02$	$F(6,78) = 2.13$
3B	Urea and LiCl	$F(3,33) = 11.58$, $P < 0.0001$	$F(6,66) = 10.03$, $P < 0.0001$	$F(2,22) = 1.15$,	$F(6,66) = 2.28$, $P < 0.05$
3C	Urea and L-arginine-L-aspartate	$F(3,33) = 5.13$, $P < 0.01$	$F(6,66) = 6.33$, $P < 0.0001$	$F(2,22) = 1.58$	$F(6,66) = 2.28$, $P < 0.05$

*Pairwise Tukey HSD statistics (see Method) with significant results are shown in figures.

EXPERIMENT 1

Mixture of NaCl and Various Bitter Compounds

Overview

Consistent with others (see Introduction), we found that most, but not all, of the bitter-tasting compounds were suppressed by NaCl. The extent of that suppression differed among the bitter coumpounds, even though they had been, on average, matched for intensity. For example, NaCl suppressed the bitterness of urea, 76 ± 6% (mean ± SE), but only suppressed the bitterness of $MgSO_4$, 4 ± 26%. These differences appear small in a logarithmic plot, so accompanying figures of the percent suppressions were included (Figures 2, 4 and 6). In most mixtures, saltiness was affected less than bitterness. The specific results for each bitter compound are presented below; statistical results are summarized in Table 2.

Quinine HCl [Figures 1(a), 2(a)]

BITTERNESS

NaCl significantly suppressed the bitterness of quinine HCl, suppressing 41 ± 11% of the maximum bitterness sensation. The bitterness of 10^{-4} M QHCl was suppressed by the addition of all concentrations of NaCl, whereas the bitterness of 10^{-3} M QHCl was suppressed only by 0.3 and 0.5 M NaCl.

SALTINESS

Only the addition of the highest QHCl concentration (10^{-3} M) suppressed the saltiness of the 0.1 M NaCl solution.

Caffeine [Figures 1(b), 2(b)]

BITTERNESS

The 0.3 and 0.5 M NaCl concentrations significantly suppressed the bitterness of caffeine. NaCl was able to suppress 55 ± 6% of the maximum bitterness sensation.

SALTINESS

Only the addition of the highest caffeine concentration (18 mM) suppressed the saltiness of the weakest NaCl concentration (0.1 M).

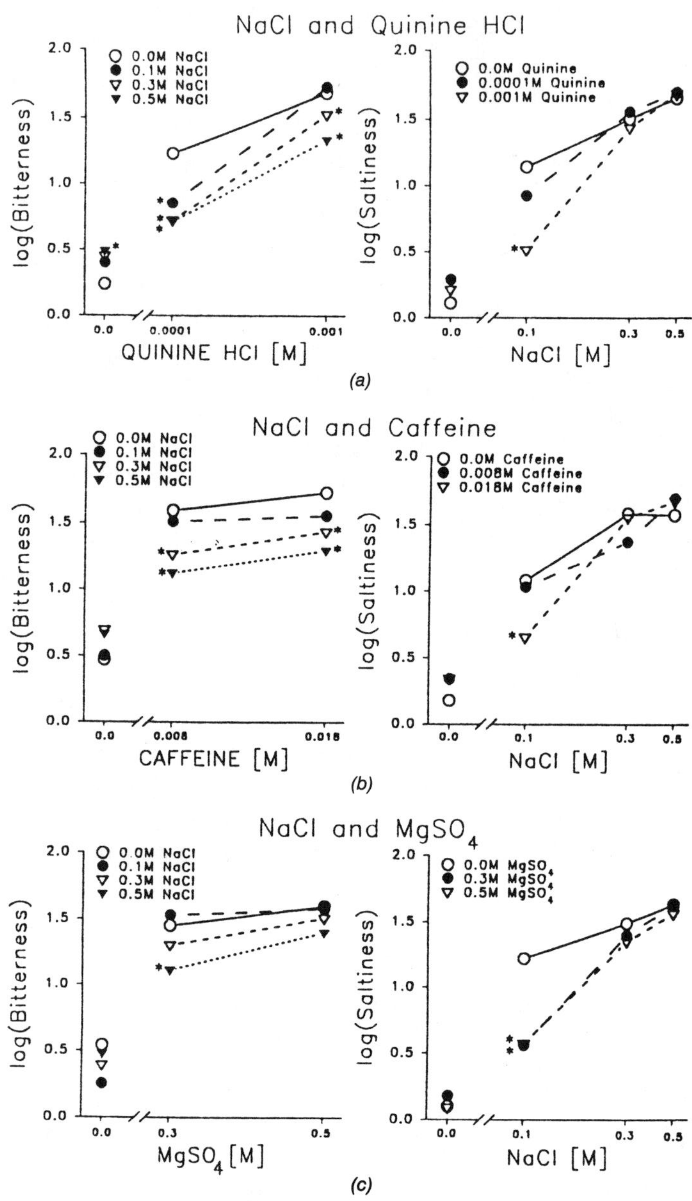

Figure 1 Graphs (a)–(e) depict the salt-bitter mixture interactions for NaCl and QHCl, NaCl and caffeine, NaCl and $MgSO_4$, NaCl and amiloride and NaCl and KCl, respectively. The left-hand column of panels shows the bitterness ratings for each study. The log-standardized bitterness ratings were plotted as a function of bitter compound concentration. The addition of varying amounts of NaCl to each level of the bitter compound is depicted by a separate curve for each sequential amount of NaCl that was added. The right-hand column of panels shows the saltiness ratings for each study. The log-standardized saltiness ratings were plotted as a function of NaCl concentration. The addition of varying amounts of bitter compound to each level of NaCl is depicted by a separate curve for each sequential amount of bitter compound that was added.

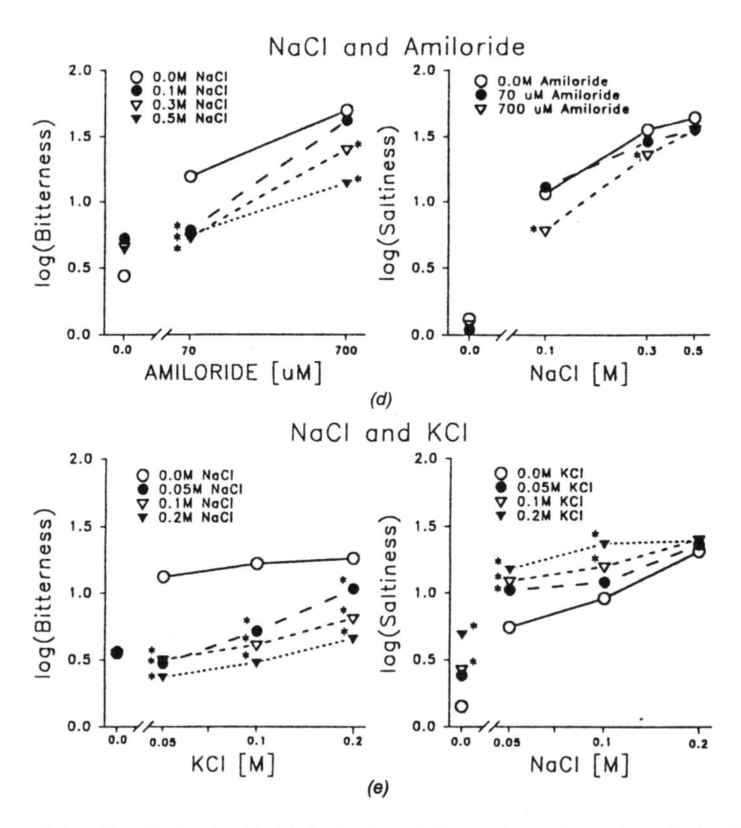

Figure 1 (continued) Graphs (a)–(e) depict the salt-bitter mixture interactions for NaCl and QHCl, NaCl and caffeine, NaCl and $MgSO_4$, NaCl and amiloride and NaCl and KCl, respectively. The left-hand column of panels shows the bitterness ratings for each study. The log-standardized bitterness ratings were plotted as a function of bitter compound concentration. The addition of varying amounts of NaCl to each level of the bitter compound is depicted by a separate curve for each sequential amount of NaCl that was added. The right-hand column of panels shows the saltiness ratings for each study. The log-standardized saltiness ratings were plotted as a function of NaCl concentration. The addition of varying amounts of bitter compound to each level of NaCl is depicted by a separate curve for each sequential amount of bitter compound that was added.

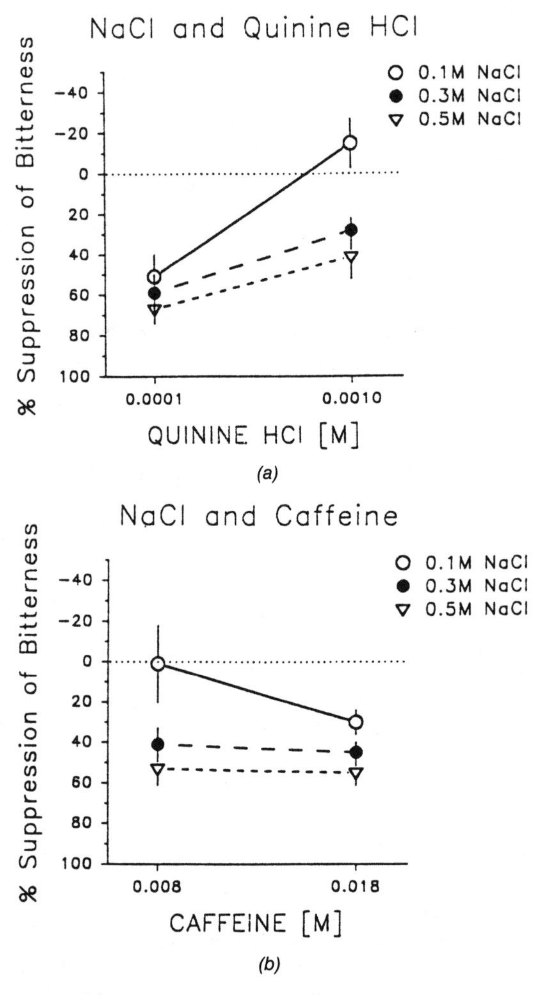

Figure 2 Graphs (a)–(e) are yoked to those of Figure 1 and follow the same layout, except only suppression of bitterness is shown. The percent suppression of bitterness is plotted as a function of the concentration of the bitter compound. Only three functions mixed with 0.1, 0.3, or 0.5 M added salt are shown, since the percent suppression is determined relative to no added salt.

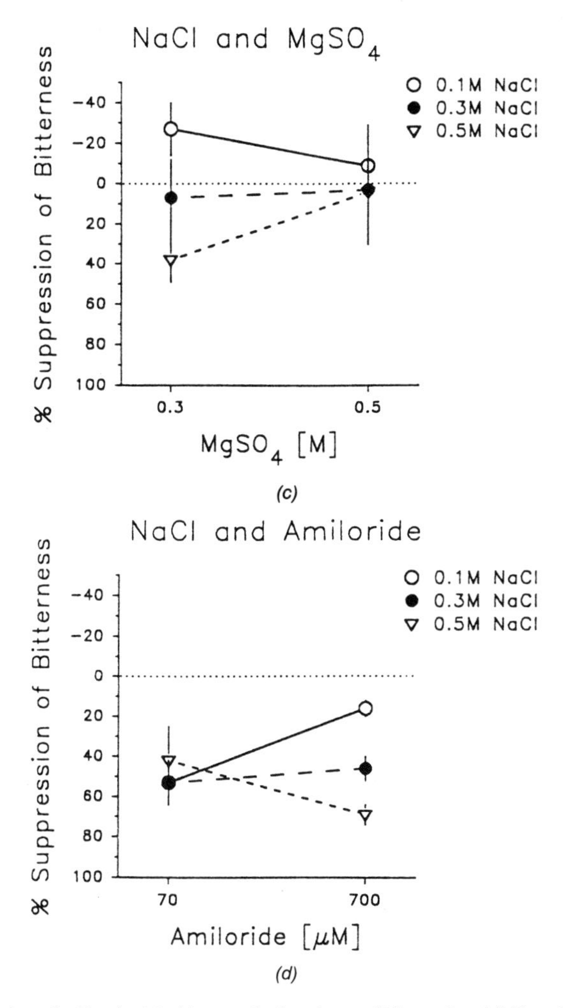

Figure 2 (continued) Graphs (a)–(e) are yoked to those of Figure 1 and follow the same layout, except only suppression of bitterness is shown. The percent suppression of bitterness is plotted as a function of the concentration of the bitter compound. Only three functions mixed with 0.1, 0.3, or 0.5 M added salt are shown, since the percent suppression is determined relative to no added salt.

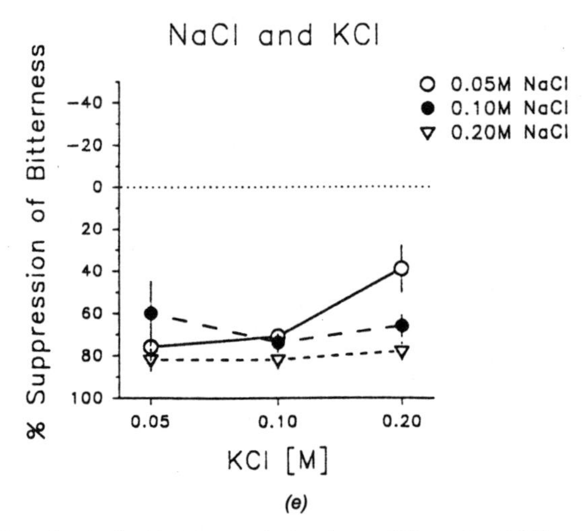

(e)

Figure 2 (continued) Graphs (a)–(e) are yoked to those of Figure 1 and follow the same layout, except only suppression of bitterness is shown. The percent suppression of bitterness is plotted as a function of the concentration of the bitter compound. Only three functions mixed with 0.1, 0.3, or 0.5 M added salt are shown, since the percent suppression is determined relative to no added salt.

$MgSO_4$ [Figures 1(c), 2(c)]

BITTERNESS

The addition of the strongest NaCl (0.5 M) decreased the bitterness of the lowest (0.3 M) concentration of $MgSO_4$. The NaCl was able to suppress only 4 ± 26% of the maximum bitterness sensation.

SALTINESS

The addition of both 0.3 and 0.5 M $MgSO_4$ suppressed the saltiness of 0.1 M NaCl, but not of 0.3 or 0.5 M NaCl. $MgSO_4$ was the only bitter agent tested that at an intermediate bitterness level (not the highest concentration) suppressed the saltiness of NaCl [cf. amiloride Figure 1(d)].

Amiloride [Figures 1(d), 2(d)]

BITTERNESS

The 0.3 and 0.5 M NaCl suppressed the bitterness of both amiloride concentrations, suppressing 69 ± 5% of the maximum bitterness sensation. All three levels of NaCl suppressed the bitterness of the lower amiloride (70 μM) concentration, whereas only the two high NaCl concentrations suppressed the bitterness of the 700 μM amiloride.

SALTINESS

The highest amiloride concentration (700 μM) suppressed the saltiness of 0.1 M NaCl and 0.3 M NaCl but not the saltiness of 0.5 M NaCl. Amiloride was the only bitter compound to suppress the saltiness of a higher NaCl concentration, 0.3 M NaCl.

KCl [Figures 1(e), 2(e)]

BITTERNESS

NaCl at all concentrations suppressed the bitterness of all concentrations of KCl. The maximum bitterness suppression achieved by NaCl on the highest KCl concentration eliminated 78 ± 4% of the bitterness. However, this maximum level of KCl bitterness was much lower than that obtained with the other bitter compounds employed, as noted previously.

SALTINESS

Saltiness ratings of KCl and NaCl mixture solutions were significantly greater than the saltiness ratings of the NaCl solutions alone. The 0.1 and 0.2 M KCl increased the saltiness of water, 0.05, and 0.1 M NaCl. Presumably because of its own inherent saltiness, KCl was the only bitter agent that when added to NaCl enhanced the saltiness of the NaCl mixture solution.

NaCl and Urea [Figures 3(a), 4(a)]

BITTERNESS

Both 0.3 and 0.5 M NaCl suppressed the bitterness of 0.5 and 1.0 M urea. The NaCl, when most effective, suppressed 76 ± 6% of the maximum bitterness sensation.

SALTINESS

The addition of the highest concentration of urea (1.0 M) suppressed the saltiness of 0.1 M NaCl, but no other concentrations of NaCl.

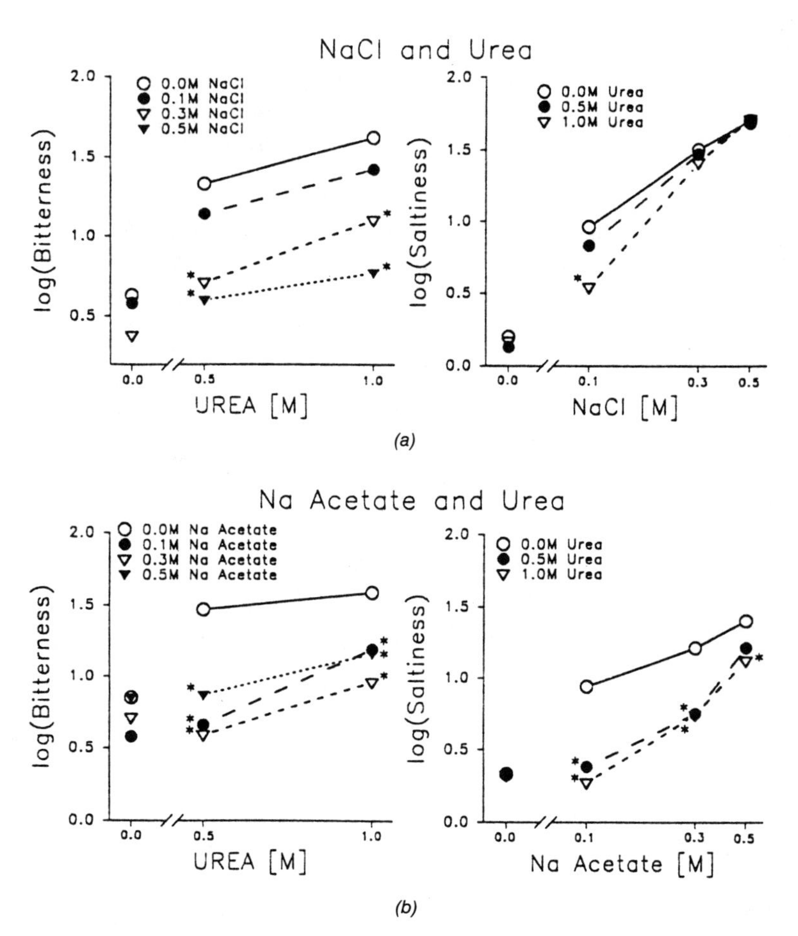

Figure 3 Graphs (a)–(d) depict the salt-bitter mixture interactions for NaCl and urea, NaAcetate and urea, NaGluconate and urea, and NaAcetate and QHCl, respectively. See the caption to Figure 1 for more details.

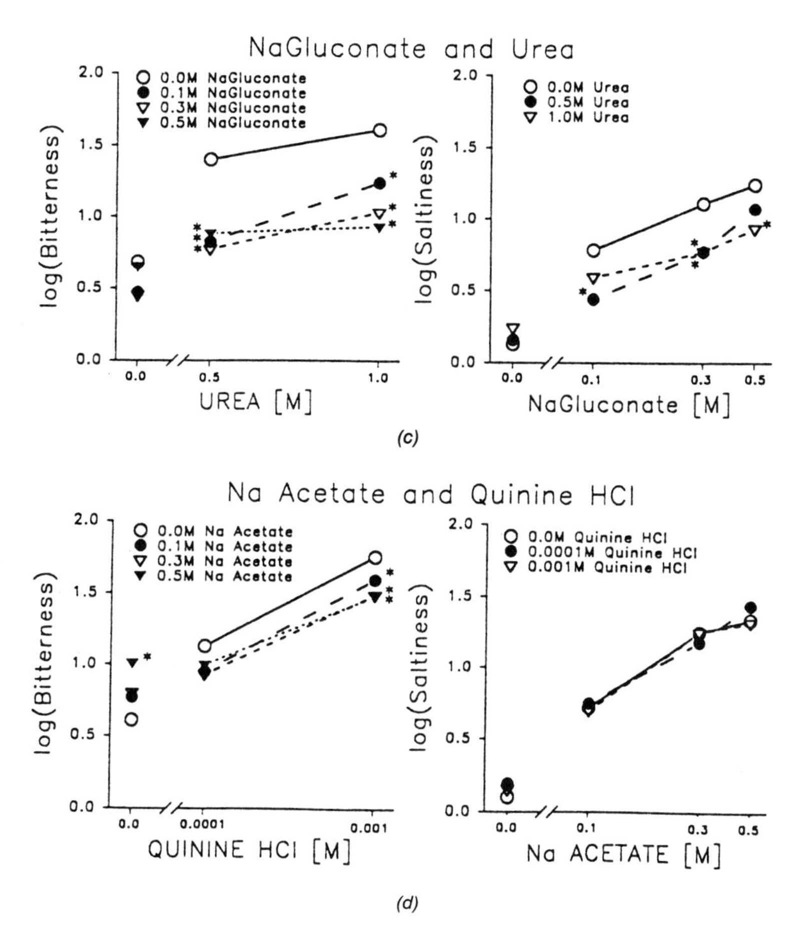

Figure 3 (continued) Graphs (a)–(d) depict the salt-bitter mixture interactions for NaCl and urea, NaAcetate and urea, NaGluconate and urea, and NaAcetate and QHCl, respectively. See the caption to Figure 1 for more details.

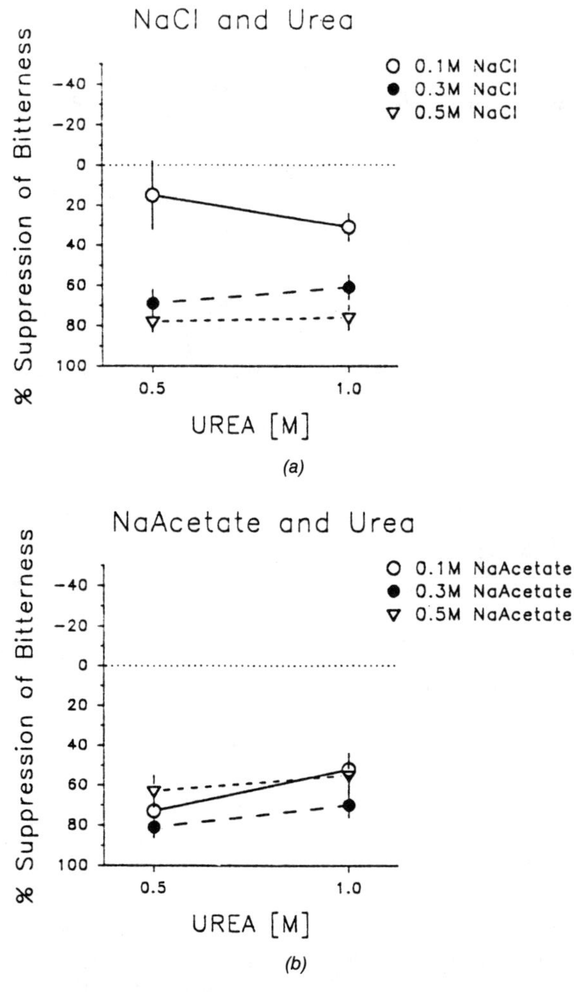

Figure 4 Graphs (a)–(d) are yoked to those of Figure 3 and follow the same layout, except only suppression of bitterness is shown. The panels are similar to those in Figure 2.

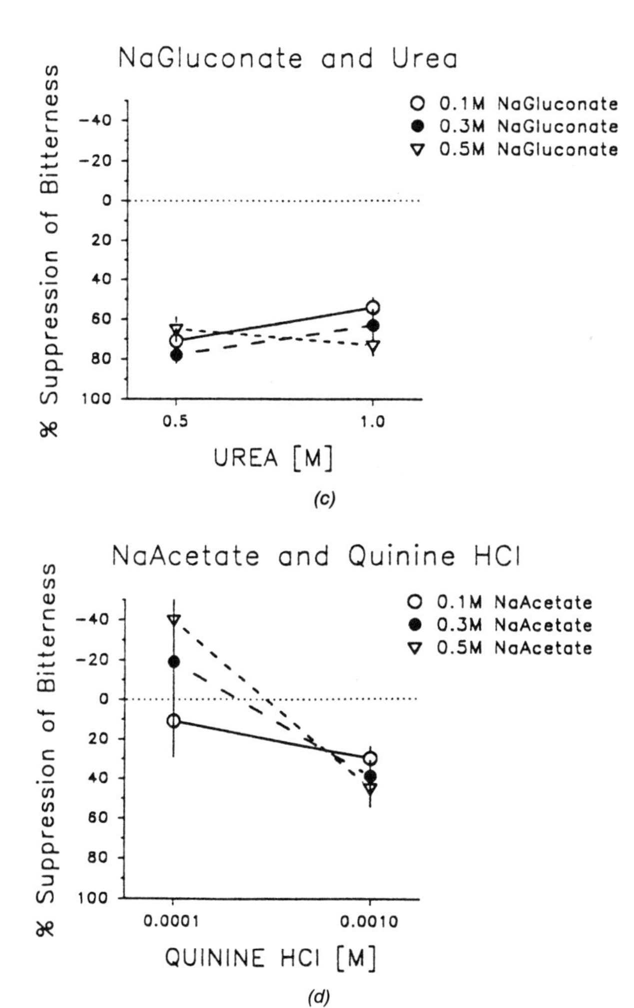

Figure 4 (continued) Graphs (a)–(d) are yoked to those of Figure 3 and follow the same layout, except only suppression of bitterness is shown. The panels are similar to those in Figure 2.

EXPERIMENT 2

Effects of Anions

Experiment 1 demonstrated an asymmetrical pattern of taste suppression for most bitter compounds: Bitterness was suppressed by NaCl, but there was less suppression of the saltiness by bitter compounds. We do not know why the suppression was asymmetrical, nor do we know why NaCl was effective as a bitter-suppressing agent. However, one strategy to elucidate the respective bitter-suppressing roles of the anion and the cation in salts is to hold one ion constant and vary the other. In Experiment 2 the anion was manipulated and in Experiment 3 the cation was manipulated. Because Experiment 1 showed that urea was a compound whose bitterness was most effectively suppressed by NaCl, it was selected as the main bitter stimulus in both Experiments 2 and 3. QHCl was also employed in one study of Experiment 2 in order to compare the effects of a nonchloride sodium salt on another bitter compound.

NaAcetate and Urea [Figures 3(b), 4(b)]

BITTERNESS

NaAcetate at all concentrations suppressed the bitterness of urea at all concentrations up to maximum suppression of 55 ± 9% of the highest bitterness sensation. Although the bitterness levels were relatively small, NaAcetate itself elicited the highest levels of bitterness of all the sodium salts (compare *y*-axis values for points over the 0 M urea in Figure 3).

SALTINESS

The 1.0 M urea suppressed the saltiness of NaAcetate at all concentrations. The addition of 0.5 M urea suppressed the saltiness of 0.1 M and 0.3 M NaAcetate, but not of 0.5 M NaAcetate.

NaGluconate and Urea [Figures 3(c), 4(c)]

BITTERNESS

NaGluconate at all concentrations suppressed the bitterness of all urea concentrations up to a maximum suppression of 73 ± 5% of the highest bitterness sensation.

SALTINESS

The addition of 0.5 M urea suppressed the saltiness of 0.1 M and 0.3 M NaGluconate but not of 0.5 M NaGluconate. In addition, saltiness supression was significant with 1.0 M urea only for the 0.3 and 0.5 M NaGluconate solutions.

NaAcetate and QuinineHCl [Figures 3(d), 4(d)]

BITTERNESS

The bitterness of 10^{-3} M QHCl was significantly suppressed by the addition of all concentrations of NaAcetate. Also, 0.5 M NaAcetate tasted bitter in water, again demonstrating that NaAcetate has some bitterness of its own. The NaAcetate, at its best, suppressed 41 ± 9% of the maximum bitterness sensation of QHCl.

SALTINESS

QHCl did not significantly affect the saltiness of NaAcetate.

EXPERIMENT 3

Effects of Cations

Experiment 2 revealed that NaAcetate suppressed the bitterness of both urea and QHCl when in mixture and that NaGluconate suppressed the bitterness of urea. The finding that the three different sodium salts (NaCl, NaAcetate and NaGluconate) were comparably effective at suppressing bitterness is consistent with the hypothesis that the anion was not as important as the sodium cation in the suppression of the bitterness of urea and QHCl by sodium salts. In Experiment 3 the cation and the anion were varied in L-ARG:L-ASP, which is a non-sodium/non-chloride salt that elicits a salty taste (Lee, 1992).

KCl and Urea [Figures 5(a), 6(a)]

BITTERNESS

KCl, when mixed with urea, significantly *increased* the bitterness over the levels from urea alone, showing 56 ± 18% enhancement of the maxi-

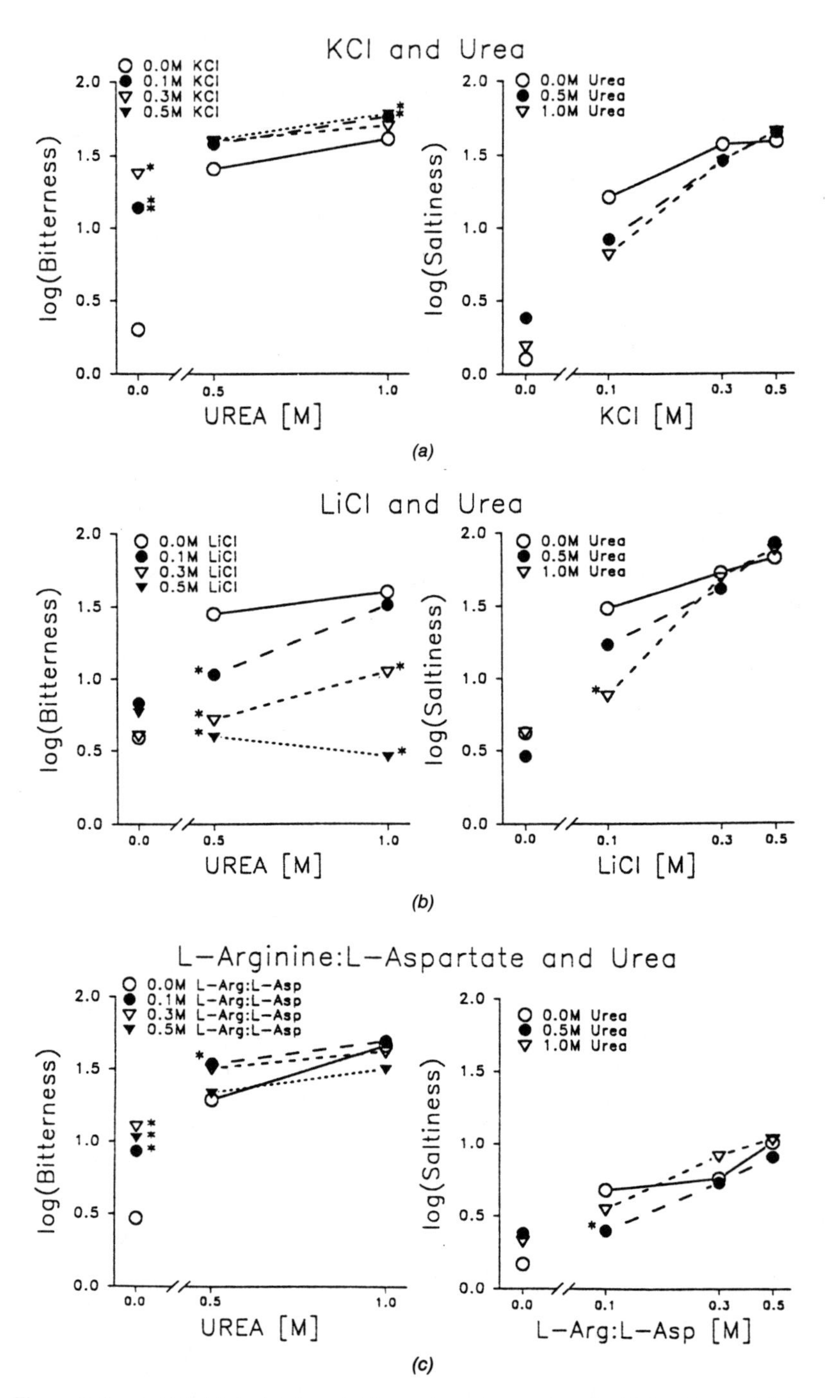

Figure 5 Graphs (a)–(c) depict the salt-bitter mixture interactions for KCl and urea, LiCl and urea, and L-arginine:L-aspartate acid and urea, respectively. See the caption to Figure 1 for more details.

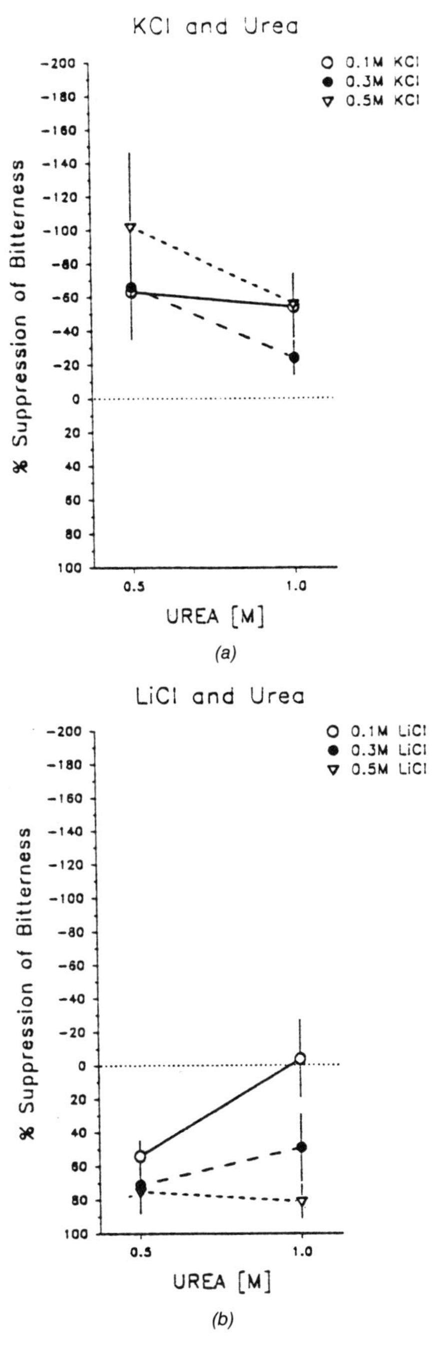

Figure 6 Graphs (a)–(c) are yoked to those of Figure 5 and follow the same layout, except only suppression of bitterness is shown. The large negative suppression values mean that the addition of the salt increased bitterness relative to the no-salt condition. The panels are similar to those in Figure 2.

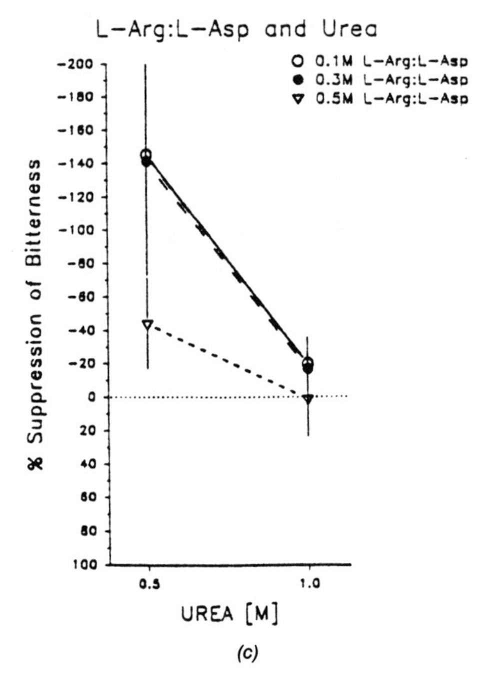

(c)

Figure 6 (continued) Graphs (a)–(c) are yoked to those of Figure 5 and follow the same layout, except only suppression of bitterness is shown. The large negative suppression values mean that the addition of the salt increased bitterness relative to the no-salt condition. The panels are similar to those in Figure 2.

mum bitterness sensation of urea alone. The KCl at all concentrations tasted bitter in water alone.

SALTINESS

Urea did not significantly affect the saltiness of KCl.

LiCl and Urea [Figures 5(b), 6(b)]

BITTERNESS

Both 0.3 and 0.5 M LiCl suppressed the bitter taste of urea at 0.5 and 1.0 M levels, showing 81 ± 10% suppression of the maximum bitterness sensation of urea alone. The 0.1 M LiCl significantly suppressed the bitterness 0.5 M urea but not the bitterness of 1.0 M urea.

SALTINESS

The addition of the highest urea concentration (1.0 M) suppressed the saltiness of 0.1 M LiCl but not that of other concentrations.

L-ARG:L-ASP and Urea [Figures 5(c), 6(c)]

BITTERNESS

The salt, L-arginine:L-aspartic acid (L-ARG:L-ASP), like KCl, also significantly *increased* the bitterness of urea. All concentrations of L-ARG:L-ASP tasted bitter, as well as salty, when alone in water. Also, the bitterness of the 0.5 M urea + 0.1 M L-ARG:L-ASP mixture was greater than that of the unmixed 0.5 M urea. The highest concentration of L-ARG:L-ASP barely altered the bitterness of 1.0 M urea by $1 \pm 22\%$.

SALTINESS

The addition of 0.5 M urea suppressed the saltiness of 0.1 M L-ARG:L-ASP but not that of any other concentration.

DISCUSSION

BITTER SUPPRESSION

Sodium chloride suppressed the bitter sensation elicited by several compounds when mixed in solution with them. The amount of bitterness suppression tended to vary directly with the concentration of NaCl, and inversely with the concentration of the bitter agent. These general observations are consistent with the findings of others (Bartoshuk and Gent, 1985; Bartoshuk and Seibyl, 1982; Bartoshuk, 1980, 1979, 1977, 1975; Kroeze and Bartoshuk, 1985; Kroeze, 1980; Schifferstein and Frijters, 1992a).

The degree of average bitterness suppression varied widely across bitter substances. For example, the bitterness of KCl, urea, and amiloride was suppressed up to about 78% by high concentrations of NaCl. In contrast, the suppression of the bitterness of quinineHCl and caffeine appeared to be weaker, roughly 48% suppression [compare Figures 2(a) and 2(b) with others], and previous reports had indicated that NaCl had no suppressing effect on the bitterness of caffeine (Kamen et al., 1961). NaCl, on average, had no significant effect on the bitterness of the high concentration of

$MgSO_4$ and although the bitterness of urea and QHCl used in these studies was judged approximately the same as that of $MgSO_4$, NaCl was more effective at suppressing the bitterness of the high concentration of urea ($\sim 76 \pm 6\%$ suppression) than of QHCl ($\sim 41 \pm 11\%$ suppression), as can be seen from a comparison of Figures 2(a) and 4(a). The baseline bitterness for water and/or the salt solutions when the bitter-tasting compound was not present was usually above zero, either due to lingering bitterness from previous trials or due to bitterness elicited by the salt, particularly for certain bitter-tasting salts such as for KCl, NaAcetate and L-ARG:L-ASP.

McBurney et al. (1972) have suggested, based on the results of cross-adaptation studies, that QHCl and urea may elicit bitter sensations through different transduction sequences, a conclusion consistent with findings reported by Yokomukai et al. (1993) and Cowart et al. (1994). The differential suppression of bitter we have observed would seem to provide further support for this hypothesis, although unlike McBurney et al. (1972) and Yokomukai et al. (1993) we did not observe similar patterns of responses to urea and $MgSO_4$.

To investigate whether suppression of the bitter taste of urea was due to the Na^+ ion, the Cl^- ion, or both, and whether the perceived saltiness of the compound was associated with its ability to suppress bitterness, the anion and/or cation of the salt stimulus was varied. If the presence of the chloride ion or the perceived saltiness were responsible for suppression of bitterness, then KCl should suppress the bitterness urea, since it has a strong salty taste as well as a weak bitter taste (Murphy et al., 1981). However, the KCl-urea mixtures were more bitter than was urea alone. To determine if there was a noticeable suppression of bitterness, despite the overall increase in bitterness from the added bitterness of KCl, we calculated the ratio of the actual bitterness rating to the urea + KCl mixture divided by the bitterness rating of the urea alone plus the bitterness rating of KCl alone $[(urea_{alone} + KCl_{alone})/(urea + KCl)_{mixed}]$. The total bitterness of any particular mixture was $92 \pm 6\%$ of the sum of the average bitterness of KCl alone plus the average bitterness of urea for the 1.0 M urea, and $85 \pm 17\%$ of their sum for the 0.5 M urea. Thus, relative to NaCl, KCl had a much weaker bitter-suppressing effect.

L-ARG:L-ASP is an ionic salt of the base, L-arginine and the acid, L-aspartic acid. It, too, was chosen because it elicits a mild salty taste (as well as a slight bitter taste) and has been implicated in the salt taste system as a compound that enhances the saltiness of NaCl, while containing neither sodium nor chloride (Lee, 1992). However, its effects on bitterness resembled those of KCl; it increased the bitterness of the solution when mixed with urea. Most likely the enhanced bitterness comes from the base, L-arginine, which is reported to have a slight bitter taste (Stone, 1967; Schiffman et al., 1975).

Thus, the two salty-tasting compounds (KCl and L-ARG:L-ASP) that contained neither sodium nor lithium had little efficacy suppressing the bitterness of urea. This suggests that the suppressing effects of NaCl (and LiCl) are due to their chemical properties acting in the periphery, rather than to their taste properties acting centrally. This conclusion is strongly supported by the results of Experiment 2, which indicated that the active component in the bitterness suppression of urea is the Na^+ ion, independent of the anion and the perceived saltiness of the salt. Specifically, NaCl, NaAcetate and NaGluconate were comparably effective in suppressing urea bitterness (Figures 3 and 4), even though these sodium salts were substantially different in saltiness. For example, 0.5 M NaCl was roughly 2 times as salty as 0.5 M NaAcetate, and 3-4 times as salty as 0.5 M NaGluconate (Figure 3), as has previously been reported (Bartoshuk, 1980; DeSimone, 1976; Kahlenberg, 1901; Weiffenbach and Ryba, 1993).

Figure 7 makes this bitter-suppression/saltiness comparison graphically. It depicts the mean saltiness ratings for all three concentrations of NaCl, NaAcetate, and NaGluconate in the top panel and their effectiveness as bitterness suppressors when mixed with 1.0 M urea in the bottom panel. At a given concentration, saltiness was independent of bitter-suppressing efficacy. A very similar distinction may be made between the bitter-suppressing efficacy of NaCl and NaAcetate on quinineHCl.

The results of several previous studies with QHCl also support the hypothesis that the suppression of bitterness by NaCl has a peripheral component (Bartoshuk, 1979, 1980; Bartoshuk and Seibyl, 1982; Kroeze and Bartoshuk, 1985). In the first of these studies, examining mixture suppression, Bartoshuk (1979) employed the strategy of adapting the subjects to one of the components of the mixture and then having the subjects taste the mixture. If the suppression is a central phenomenon, then adapting to one of the mixture components will prevent the adapted stimulus from eliciting a perceived taste and hence from suppressing. In general, peripheral adaptation to one component (e.g., HCl) tended to release the suppression of the other component (e.g., sucrose); however, there was one notable exception. Adaptation to NaCl did not release the suppression of QHCl in the NaCl-QHCl mixture, suggesting a peripheral locus of interaction.

Second, adaptation to a NaCl-QHCl mixture had no effect upon the bitterness of QHCl administered alone (Bartoshuk and Seibyl, 1982). This would most likely occur if the QHCl was not able to interact normally with transduction elements while in mixture with NaCl. Third, Kroeze and

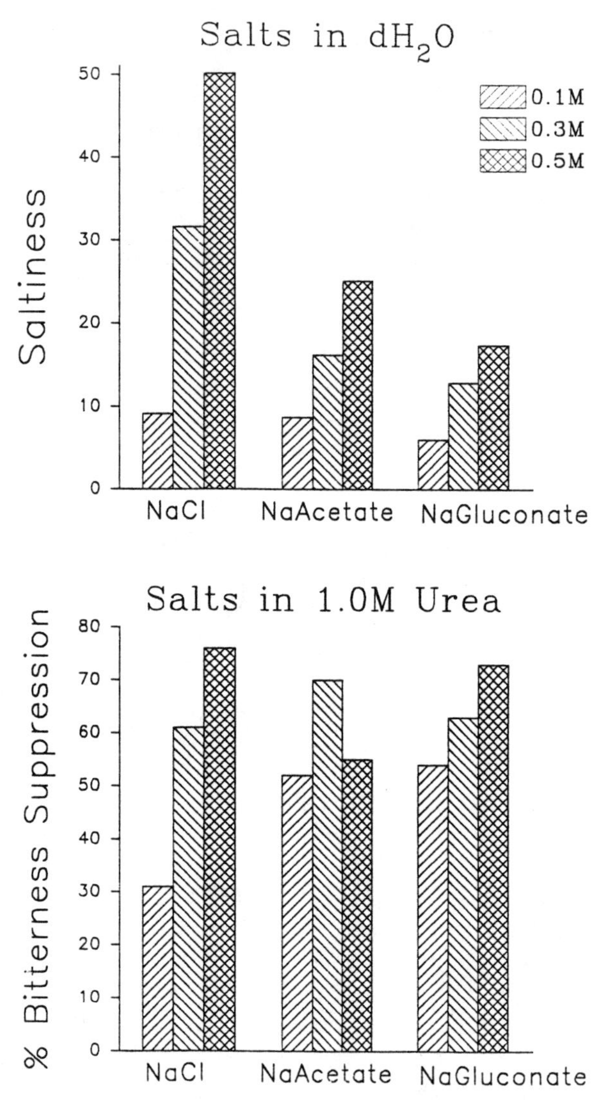

Figure 7 The top panel shows the mean standardized saltiness ratings of 0.1, 0.3, and 0.5 M NaCl, NaAcetate, and NaGluconate. The bottom panel depicts the percent suppression of the bitterness of 1.0 M urea by these three salts each at three concentrations.

Bartoshuk (1985) compared the mixture suppression of NaCl and QHCl when they were placed together on the same side of the tongue, and when they were placed simultaneously on opposite sides of the divided tongue. If the suppression were central, then equal suppression should occur in either case, but if the suppression were peripheral, then it should predominantly occur when the two were placed on the same side of the tongue. They found that much greater suppression occurred when NaCl and QHCl were placed together on the same side. In contrast, suppression in NaCl-sucrose occurred equally for the two situations.

If the suppression of several bitter compounds by salts is a peripheral phenomenon, how do sodium and lithium interact with the bitter transduction mechanism(s) to block bitter perception? An answer to this question will come when the bitter transduction mechanism(s) are better understood (*see* Brand and Shah, 1992; Kumazawa et al., 1986, 1988; Spielman et al., 1992). Because sodium's bitter-suppressing effects on these compounds were so varied, it is difficult to speculate which properties of the bitter compounds (polarity, charge, lipophilicity, etc.) or which aspect of the transduction sequence (ion channels, ion pumps, receptors, G proteins, enzymes, etc.) are involved in the differential suppression.

SALTINESS SUPPRESSION

As mentioned above, the mixture suppression was asymmetrical in that saltiness was suppressed little, relative to bitterness. On the average, $MgSO_4$ had a slight but significant tendency to decrease the saltiness of NaCl, especially at the 0.1 M NaCl level. The suppression of saltiness was less with other bitter stimuli (with the exception of amiloride) although several bitter compounds, including QHCl, caffeine, $MgSO_4$ and urea, partially suppressed the saltiness of 0.1 M NaCl, the lowest concentration tested. This asymmetry is best seen with the urea and NaCl mixtures [Figures 1(a), 2(a)]. NaCl reduced the bitterness of urea 76%, while urea had little affect on saltiness, suppression being significant only for the 0.1 M concentration. KCl was unique among the stimuli tested in that it enhanced the perceived saltiness of the solution when mixed with NaCl. That is, in addition to NaCl suppressing the bitterness of KCl, there was also a large increase in the saltiness of NaCl solution when mixed with KCl. This effect is most likely a simple summation of the independent salty tastes of NaCl and KCl.

The main differences between the effects of urea on the saltiness of NaCl, NaAcetate and NaGluconate were that urea had a stronger effect upon the latter two compounds. This may be due to the fact that NaAcetate and NaGluconate are perceived as less salty than NaCl. It appears that weak salty sensations (as from NaGluconate) may be suppressed by certain bitter compounds, but stong salt sensations (from NaCl at the same concentration as NaGluconate) are more difficult to suppress (Ossebaard and Smith, 1995). Thus, the suppression of saltiness in not dependent on low concentrations of sodium ions; rather, it is dependent on low intensities of perceived saltiness independent of number of sodium ions, suggesting a central locus for saltiness suppression. The suppression of strong sensations of saltiness was rarely seen. Indeed, *only* 700 nM amiloride was able to suppress the saltiness of the 0.3 M NaCl significantly, but not the saltiness of 0.5 M NaCl.

Since amiloride has received considerable attention as a salt taste suppressor (Schiffman et al., 1983; Desor and Finn, 1989; McCutcheon, 1992; Tennissen, 1992; Ossbaard and Smith, 1995), it warrants further comment. Although amiloride suppressed higher NaCl concentrations (0.3 M) than any other bitter compound employed, it did not completely eliminate saltiness even at the lowest concentration of NaCl. In view of the significant suppression shown by other bitter compounds here (e.g., $MgSO_4$), amiloride appeared to differ quantitatively rather than qualitatively from other bitter compounds in its salt-suppressing capacity. Surprisingly, the saltiness-suppressing capacity of amiloride was of much smaller magnitude than the ability of NaCl to suppress the bitterness of amiloride.

METHODOLOGICAL ISSUES

Throughout this study, mixture effects were examined by asking subjects to rate saltiness and bitterness simultaneously. The manner in which gustatory ratings are obtained can impact the responses given by subjects (Frank et al., 1993; Schifferstein, 1994a, 1994b; Schifferstein and Frijters, 1992b; Stillman, 1993). For example, rather than asking subjects to rate saltiness and bitterness, as was done in the present study, they could have been asked simply to rate bitterness alone and then separately to rate saltiness on another day, or subjects could have been asked to rate saltiness and bitterness in addition to simultaneously rating sweetness and sourness and/or other sensations. The method by which the rating is obtained can result in certain response biases that can affect measurements of sensation intensity (Frank et al., 1993). However, the point of the present discussion is to make *relative* comparisons of bitterness suppression by various salts among compounds matched for bitterness intensity, rather than absolute statements of magnitude of effect. Since all compounds were evaluated using the same methodology, any major differences among compounds should not be attributed to response biases as a result of the method. However, there is a chance that side-tastes may affect bitter intensity responses differentially as a function of how many qualities were rated. We have examined this idea and preliminary data (unpublished) suggest that three different instructions for rating sensations [e.g., rating one quality/session (saltiness or bitterness); rating two qualities/session (saltiness and bitterness); rating five qualities/session (saltiness, bitterness, sweetness, sourness, otherness)] all reveal the same general suppression effects of NaCl, NaAcetate, and NaGluconate on the bitterness of urea and quinineHCl, as those presented here.

SUMMARY

Overall, the bitterness of a large array of compounds was suppressed by the addition of sodium salts. In contrast, the saltiness of the different salts was suppressed little by the addition of bitter compounds. Saltiness was suppressed only when the levels of perceived saltiness were low. The suppression of bitterness appeared to be dependent upon the presence of the sodium ion. Neither the anion of the salt nor the level of perceived saltiness was a determinant of bitter suppression. The efficacy of the suppression of bitterness depended upon which compounds were eliciting the bitterness. The variability in the suppression of bitterness across bitter compounds provides additional evidence for multiple bitter transduction sequences.

We next wanted to determine how bitterness suppression by sodium salts would affect the taste of more complex stimulus mixtures. When in a complex mixture, salts can function in at least two conceptually distinct ways (Lynch, 1987). First, they impart a pleasant salty taste, particularly NaCl. Second, salts seem to modify the flavor of the food in a positive manner (e.g., soups, desserts, vegetables, breads). There is widespread belief in the food industry that salts act as flavor potentiators (i.e., to increase the intensity of other desirable flavors; Gillette, 1985). However, almost all published experimental studies indicate that NaCl either suppresses or has no effect on other flavors (Helleman, 1992; Kamen et al., 1961; Kemp and Beauchamp, 1994; Kroeze, 1977; Moskowitz, 1972; Pangborn and Chrisp, 1964; Pangborn and Trabue, 1964; Pangborn, 1962).

How can these contradictory observations be reconciled? We have postulated (Kemp and Beauchamp, 1994) that NaCl acts to *differentially* suppress flavors, such that off-tastes are more suppressed than palatable ones. Selective suppression could result in an enhancement of the positive flavors either *relative* to the negative ones or in an *absolute* sense. For example, in a mixture that is both bitter and sweet, bitterness and sweetness mutually suppress each other (Bartoshuk, 1975; Calvino et al., 1990; Kamen et al., 1961; Kroeze and Bartoshuk, 1985; Lawless, 1982). If a sodium-containing compound is added, it may suppress the bitterness much more than the sweetness, thereby releasing the sweetness from suppression by bitterness. The resultant mixture would taste sweeter than it would without the salt.

Using a model aqueous system, the following experiment was designed to test this selective suppression hypothesis. Urea was tested in mixtures with sucrose and NaAcetate. NaAcetate was used specifically because it does not have as strong a salty taste as does NaCl, thereby permitting a test of the flavor-modifying effects of sodium in the absence of a strong perceived saltiness. We hypothesized that in a three-component mixture containing urea, sucrose, and NaAcetate, the bitterness of urea would be suppressed more than the sweetness, and the mixture would, as a consequence, appear to have a relatively, and perhaps absolutely, heightened sweetness. Methods were similar to those described above. Results were consistent with this prediction lending support to the argument that salt sodium-containing compounds, including NaCl, may enhance food flavors by differentially suppressing other flavor components (Breslin and Beauchamp, 1995).

Although our simple three-component aqueous model system does not fully mimic the complex food systems in which salts purportedly act as flavor potentiators, it did provide sufficient scope to uncover at least one mechanism by which a salt may increase both the relative and the absolute intensity of palatable components of foods. This mechanism has not been previously described in mixture studies in large part because experimental work has tended to focus either on two-component mixtures in which phenomena such as differential suppression and release from suppression cannot be observed (e.g., Calvino et al., 1990; Kamen et al., 1961; Schifferstein and Frijters, 1992) or on complex foods when interpretations are difficult (e.g., Calvino et al., 1990; Hellemann, 1992; Pangborn and Chrisp, 1964; Pangborn and Trabue, 1964).

These data support the hypothesis that a key role of salts in foods – in addition to adding desired saltiness – is to potentiate flavors through differential suppression and release from suppression. People's desire for NaCl and other salts in foods as diverse as (often bitter) vegetables, oily foods and meats may be due in part to their ability to suppress off-flavors. If this is so, this would help explain one of the major reasons why it is so difficult to make low-sodium foods acceptable; not only do they not have a desirable salty taste, but off-flavors are more prominent than if sodium were present.

On a practical level, biophysical evidence (Brand and Shah, 1992; Gilbertson, 1993) suggests that it may be extremely difficult or impossible to develop a salty-tasting substitute for salt that contains no sodium. However, perhaps the differential flavor-suppressing effect of Na salts, if they do not occur by the same mechanism that sodium stimulates saltiness, could be duplicated by non-Na^+ substances. In the search for salt substitutes it would be wise to take into account the multiple sensory functions of salts in foods.

ACKNOWLEDGEMENTS

This research was supported by NIH Grants DC 00882 and DC 00100. Many thanks are due to Roberta George, Kenneth Gerald, Jeffrey Leen, Laila Burgos, and Karen Opdyke for their excellent technical assistance. We are grateful to Barry Green, Beverly Cowart, and Joseph Brand for their careful reading of this manuscript.

REFERENCES

Bartoshuk, L. M. 1975. Taste mixtures: Is mixture suppression related to compression? *Physiol. Behav., 14,* 643–649.

Bartoshuk, L. M. 1977. Psychophysical studies of taste mixtures. In: *Olfaction and Taste VI.* LeMagnen, J. and MacLeod, P. (Eds.). Information Retrieval Ltd., Press, Washington, D.C., 377–384.

Bartoshuk, L. M. 1979. Taste interactions in mixtures of NaCl with QHCl and sucrose with QHCl. *Society for Neuroscience Abstract, 9th Annual Meeting.*

Barotshuk, L. M. 1980. Sensory analysis of the taste of NaCl. In: *Biological and Behavioral Aspects of Salt Intake.* Kare, M. R., Fregly, M. J. and Bernard, R. A. (Eds.). Academic Press, New York, pp. 83–98.

Bartoshuk, L. M. and Gent, J. F. 1985. Taste mixtures: An analysis of synthesis. In: *Taste, Olfaction, and the Central Nervous System.* Pfaff, D. W. (Ed.). Rockefeller Univer. Press, New York, pp. 210–232.

Bartoshuk, L. M. and Seibyl, J. P. 1982. Suppression of QHCl in mixtures: Possible mechanisms. *AChemS Abstract, 4th Annual Meeting.*

Brand, J. G. and Shah, P. S. 1992. The transduction of taste and olfactory stimuli. In: *Physical Chemistry of Foods.* Schwartzberg, H. and Hartel, R. (Eds.). Marcel Dekker, New York, 517–540.

Breslin, P. A. S. and Beauchamp, G. K. 1995. Suppression of bitterness by sodium: Variation among bitter taste stimuli. *Chem. Sens., 20,* 609–623.

Breslin, P. A. S. and Beauchamp, G. K. 1995. Sodium as a flavor potentiator: Selective suppression of tastes. *Chemical Senses, 20,* 609–623.

Calvino, A. M., Garcia-Medina, M. R. and Cometto-Muniz, J. E. 1990. Interactions in caffeine-sucrose and coffee-sucrose mixtures: Evidence of taste and flavor suppression. *Chem. Sens., 15,* 505–519.

Cowart, B. J., Yokomukai, Y. and Beauchamp, G. K. 1994. Bitter taste in aging: Compound-specific decline in sensitivity. *Physiol. Behav., 56,* 1237–1242.

DeSimone, J. A. and Price, S. 1976. A model for the stimulation of taste receptor cells by salt. *Biophys. J., 16,* 869–880.

Desor, J. A. and Finn, J. 1989. Effects of amiloride on salt taste in humans. *Chem. Sens.*, *14*, 793–803.

Fischer, R. and Griffin, F. 1963. Quinine dimorphism: A cardinal determinant of taste sensitivity. *Nature*, *200*, 343–347.

Frank, R. A., van der Klaauw, N.J. and Schifferstein, H. N. J. 1993. Both perceptual and conceptual factors influence taste-odor and taste-taste interactions. *Percept. and Psychophys.* *54*, 343–354.

Frijters, J. E. R. and Schifferstein, H. N. J. 1994. Perceptual interactions in mixtures containing bitter tasting substances. *Physiol. Behav.*, *56*, 1243–1249.

Gilbertson, T. A. 1993. The physiology of vertebrate taste reception. *Curr. Opin. Neurobiol.*, *3*, 532–539.

Gillette, M. 1985. Flavor effects of sodium chloride. *Food Technol.*, *39*, 47–56.

Hellemann, U. 1992. Perceived taste of sodium chloride and acid mixtures in water and bread. *J. Food. Sci. Tech.*, *27*, 201–211.

Kamen, J. M., Pilgrim, F. J., Gutman, N. J. and Kroll, B. J. 1961. Interactions of suprathreshold taste stimuli. *J. Exp. Psychol.*, *62*, 348–356.

Kahlenberg, L. 1901. The action of solutions on the sense of taste. *Bull. Univ. Wisc.*, *2*, 3–31.

Kemp, S. E. and Beauchamp, G. K. 1994. Flavor modification by sodium chloride and monosodium glutamate. *J. Food Sci.*, *59*, 682–686.

Kroeze, J. H. A. 1977. Taste thresholds for bilaterally and unilaterally presented mixtures of sugar and salt. In: *Olfaction and Taste VI.* Le Magnen, J. and MacLeod, P. (Eds.). IRL Press, London, pp. 486.

Kroeze, J. H. A. 1980. Masking in two- and three-component taste mixtures. In: *Olfaction and Taste VII.* van der Starre, H. (Ed.). Information Retrieval Ltd., Press, Washington, D.C., 435.

Kroeze, J. H. A. 1982. The relationship between the side tastes of masking stimuli and masking binary mixtures. *Chem. Senses*, *7*, 23–27.

Kroeze, J. H. A. and Bartoshuk, L. M. 1985. Bitterness suppression as revealed by split-tongue taste stimulation in humans. *Physiol. and Behav.*, *35*, 779–783.

Kumazawa, T., Kashiwayanagi, M. and Kurihara, K. 1986. Contribution of electrostatic and hydrophobic interactions of bitter substances with taste receptor membrane to generation of receptor potentials. *Biochim. Biophys. Acta.*, *888*, 62–69.

Kumazawa, T., Nomura, T. and Kurihara, K. 1988. Liposomes as a model for taste cells: Receptor cites for bitter substances including N–C=S substances and mechanism of membrane potential change. *Biochemistry*, *27*, 1239–1244.

Lawless, H. 1979. The taste of creatine and creatinine. *Chem. Sens. & Flav.*, *4*, 249–258.

Lawless, H. 1982. Paradoxical adaptation to taste mixtures. *Physiol. Behav.*, *25*, 149–152.

Lee, T. 1992. US Patent #5145707, "Salt Enhancer."

Lynch, N. M. 1987. In search of a salty taste. *Food Technol.*, *41*, 82–86.

McBride, R. L. 1989. Three models of taste mixtures. In: *Perception of Complex Smells and Tastes*, Laing, D. G., Cain. W. S., McBride, R. L. and Ache, B. W. (Eds.). Academic Press, New York, pp. 265–282.

McBurney, D. H. 1969. Effects of adaptation on human taste function. In: *Olfaction and Taste III.* Pfaffmann, C. (Ed.). Rockefeller University Press, New York, 405–419.

McBurney, D. H., Smith, D. V. and Shick, T. R. 1972. Gustatory cross-adaptation: Sourness and bitterness. *Percept. and Psychophys., 11,* 228-232.

McCutcheon, N. B. 1992. Human psychophysical studies of saltiness suppression by amiloride. *Physiol. and Behav., 51,* 1069-1074.

Mela, D. J. 1989. Bitter taste intensity: The effect of tastant and thiourea taster status. *Chem. Sens., 14,* 131-135.

Moskowitz, H. R. 1972. Perceptual changes in taste mixtures. *Percept. Psychophys., 11,* 257-262.

Murphy, C., Cardello, A. V. and Brand, J. G. 1981. Tastes of fifteen halide salts following water and NaCl: Anion and cation effects. *Physiol. and Behav., 26,* 1083-1095.

Ossebaard, C. A. and Smith D. V. 1995. Effect of amiloride on the taste of NaCl, Na-gluconate and KCl in humans: Implications for Na^+ receptor mechanisms. *Chem. Sens., 20,* 37-46.

Pangborn, R. M. 1962. Taste interrelationships. III. Suprathreshold solutions of sucrose and sodium chloride. *J. Food Sci., 27,* 495-500.

Pangborn, R. M. and Chrisp, R. B. 1964. Taste interrelationships. VI. Sucrose, sodium chloride and citric acid in canned tomato juice. *J. Food Sci., 29,* 490-498.

Pangborn, R. M. and Trabue, I. M. 1964. Taste interrelationships. V. Sucrose, sodium chloride and citric acid in lima bean puree. *J. Food Sci., 29,* 233-240.

Schifferstein, H. N. J. and Frijters, J. E. R. 1991. The perception of the taste of KCl, NaCl and quinine HCl is not related to PROP-sensitivity. *Chem. Senses, 16,* 303-317.

Schifferstein, H. N. J. and Frijters, J. E. R. 1992a. Two-stimulus versus one-stimulus procedure in the framework of functional measurement: A comparative investigation using quinineHCl/NaCl mixtures. *Chem. Sens., 17,* 127-150.

Schifferstein, H. N. J. and Frijters, J. E. R. 1992b. Contextual and sequential effects on judgements of sweetness intensity. *Percept. Psychophys., 56,* 227-237.

Schifferstein, H. N. J. 1994a. Contextual effects in the perception of quinineHCl/NaCl mixtures. *Chem. Senses, 19,* 113-123.

Schifferstein, H. N. J. 1994b. Sweetness suppression in fructose/citric acid mixtures: A study of contextual effects. *Percept. Psychophys., 56,* 227-237.

Schiffman, S. S., Moroch, K. and Dunbar, J. 1975. Taste of acetylated amino acids. *Chemical Senses and Flavor, 1,* 387-401.

Schiffman, S. S., Lockhead, E. and Maes, F. W. 1983. Amiloride reduces the taste intensity of Na^+ aand Li^+ salts and sweeteners. *Proc. Nat. Acad. Sci., 80,* 6136-6140.

Spielman, A. I., Huque, T., Whitney, G. and Brand, J. G. 1992. The diversity of bitter taste signal transduction mechanisms. In: *sensory Transduction,* Corey, D. P. and Roper, S. D. (Eds.) Rockefeller University Press, 307-324.

Stillman, J. A. 1993. Context effects in judging taste intensity: A comparison of variable line and category rating methods. *Percept. and Psychophysic., 54,* 447-484.

Stone, H. 1967. Gustatory responses to L-amino acids in man (sic). In: *Olfaction and Taste II,* Hyashi, T. (Ed.). Pergamon Press, New York, 289-306.

Tennissen, A. M. 1992. Amiloride reduces intensity responses of human fungiform papillae. *Physiol. and Behav., 51,* 1061-1068.

Weiffenbach, J. and Ryba, N. 1993. Anions determine the taste intensity and perceived saltiness of three sodium salts. *Chem. Sens., 18,* 647-648.

Yokomukai, Y., Cowart, B. J. and Beauchamp, G. K. 1993. Individual differences in sensitivity to bitter-tasting substances. *Chem. Sens., 18,* 669-681.

CHAPTER 10

Development of a Low-Sodium Salt: A Model for Bitterness Inhibition

ROBERT J. KURTZ[1]
WILLIAM D. FULLER[2]

SALT (sodium chloride) has played a central role in the diet, economy and religion of man since ancient times. It was such an important commodity that the Romans built the Via Salaria from Ostia to Rome and Caesar's soldiers were paid part of their salary in this product. The Bible mentions salt more than 30 times and the Book of Job asks, "Can that which is tasteless be eaten without salt?"

Despite the attraction of man for salt, modern medical evidence suggests that high levels of sodium ion in the diet may be detrimental. Currently, the accepted value for the average daily intake of sodium by Americans is 4000 to 6000 mg, while the FDA recommended average daily intake is 2400 mg. The NLEA (Nutrition Labeling and Education Act of 1990) nutrition facts label on a 12 oz. single serving can of a popular beverage describes this product as having 980 mg of sodium, which is 41% of the recommended average daily intake. Although this value is high, it is not atypical of other single-serving foods such as soups. Chronic high sodium intake has been associated with hypertension and associated diseases, including heart and kidney dysfunction. The well-established consumption of high levels of sodium coupled with the current NLEA labeling requirements have led many companies to try to develop a good-tasting salt substitute for use by the food industry.

Of the three common sodium chloride (NaCl) analogues that are salty, lithium chloride (LiCl), potassium chloride (KCl), and ammonium chloride (NH_4Cl), lithium chloride is highly toxic and ammonium

[1]Bioresearch Inc., 727 Twenty-Third Street South, Suite 300, Arlington, VA 22202, U.S.A.
[2]Bioresearch Inc., 11189 Sorrento Valley Road #4, San Diego, CA 92121, U.S.A.

chloride has an unpleasant smell and taste and is potentially unstable. Therefore, only KCl has any feasibility as a NaCl substitute. Its acceptability, however, has been severely limited due to its pronounced bitter/metallic aftertaste. At present, sodium reduction is obtained by simple abstinence, by the substitution of potassium chloride for sodium chloride, by the use of "fillers" with sodium chloride to bulk up the sodium chloride, by the use of sodium chloride with a high surface area such as flake salts and by the use of mixtures of sodium chloride and potassium chloride. None of these methods has been widely accepted by the American public, and one review article on salt substitutes concluded that there appears to be no truly viable approach to a replacement for NaCl other than the masking of the bitter/metallic aftertaste of KCl (Erickson, 1990).

Sodium chloride is not the only source of sodium in the diet. Dietary sodium also comes from leavening agents (sodium bicarbonate and sodium carbonate), flavor enhancers (monosodium glutamate, sodium guanylate, sodium inosinate), preservatives (sodium benzoate, sodium nitrate, sodium sulfite) and other sources. Obviously, if these sodium sources could be replaced by their potassium analogues, a further reduction of dietary sodium could be realized. Based on the above, it became clear to us that one possible approach to allow a major reduction in the ingestion of sodium ions would involve the removal of the bitter/metallic taste from potassium salts without affecting their flavor or salty taste.

In the early 1980s, while collaborating with Professor Murray Goodman on the development of high-potency sweeteners and his model for sweet taste perception, we conducted an analysis of taste patterns in structurally similar molecules. This evaluation led us to conclude that there was probably a close relationship between bitter and sweet taste. Thus, for example, the literature indicated that whereas L-aspartyl-L-phenylalanine methyl ester (aspartame) is sweet, the corresponding L-aspartyl-D-phenylalanine methyl ester is bitter (Figure 1). Moreover, the closely related compound L-aspartyl-L-phenylalanine (AP) is tasteless. We reasoned that it was likely that these three similar molecules bind to the same or closely related receptor sites. However, aspartame triggers a sweet response whereas the LD stereoisomer induces a bitter sensation. Most importantly, it was reasonable to us that AP also interacted with the taste receptor but failed to induce any taste response.

Since the discovery of aspartame in the late 1960s, a wealth of literature has appeared describing a large number of new high-intensity sweeteners. Based on the numerous structure-taste relationships that were forthcoming, many models for the sweet taste receptor have been developed. The molecular requirements for sweet taste have been described and refined by Schallenberger and Acree (1967, 1969), Kier (1972; Holtje and Kier, 1974), Tinti and Nofre (1990), Tinti et al. (1980, 1981), Goodman (Douglas and

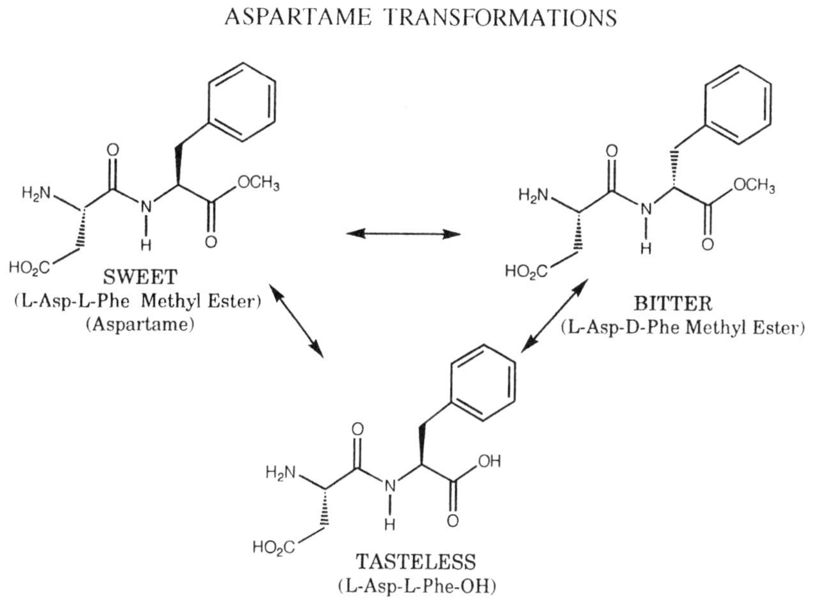

ASPARTAME TRANSFORMATIONS

SWEET
(L-Asp-L-Phe Methyl Ester)
(Aspartame)

BITTER
(L-Asp-D-Phe Methyl Ester)

TASTELESS
(L-Asp-L-Phe-OH)

Figure 1 Small changes in chirality or substitution can change a sweet compound (L-aspartyl-L-phenylalanine methyl ester) into a bitter compound (L-aspartyl-D-phenylalanine methyl ester) or a tasteless compound (L-aspartyl-L-phenylalanine). All compounds are tasted as salts; zwitterions for the methyl esters, zwitterion, monosodium salt for Asp-Phe. We refer to such changes in structure with an accompanying change in taste as a "transformation."

Goodman, 1990; Yamazaki et al., 1994; Benedetti et al., 1990; Feinstein et al., 1991; Fuller et al., 1985) and others (Van der Heijden et al., 1979, 1985; Mazur et al., 1969; Lelj et al., 1976; Temussi et al., 1978, 1984). In addition, Belitz et al. (1979) have described a model for bitter taste reception. These researchers have clearly shown that there is a close relationship between sweet and bitter taste, and that both taste sensations are probably elicited by the same or similar receptors.

We have used elements of these models, particularly those developed by Goodman, and Tinti and Nofre, to predict changes in the molecular topology, stereochemistry and substituents of known sweet or bitter compounds that would result in tasteless compounds. Thus, as shown from the literature, changing functional groups on an aromatic ring can lead to isomers that are sweet, bitter or tasteless (Figures 2 and 3). We refer to such molecular reorientation, or to the addition or removal of substituents from a parent molecule that results in a change of taste, as a transformation (Figures 1–3) (Kurtz and Fuller, 1993a, b). In this nomenclature, a compound that is substantially tasteless and blocks either the bitter or sweet taste of another substance is termed a tastand. In our model, tasteless mol-

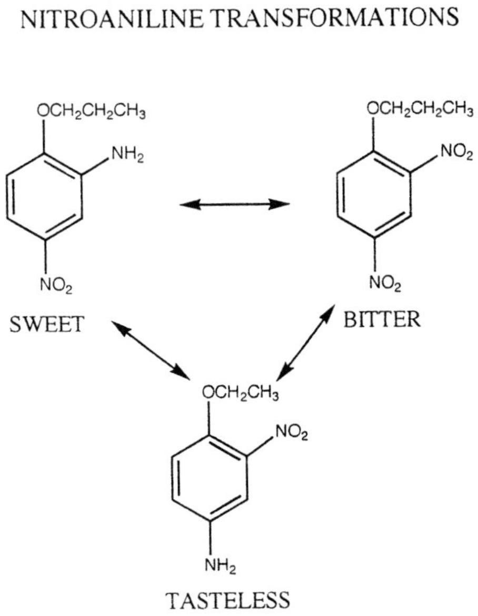

NITROANILINE TRANSFORMATIONS

$OCH_2CH_2CH_3$, NH_2, NO_2 — SWEET

$OCH_2CH_2CH_3$, NO_2, NO_2 — BITTER

OCH_2CH_3, NO_2, NH_2 — TASTELESS

Figure 2 Nitroaniline transformations.

SACCHARIN TRANSFORMATIONS

SWEET

O_2N — BITTER

$N—CH_3$ — TASTELESS

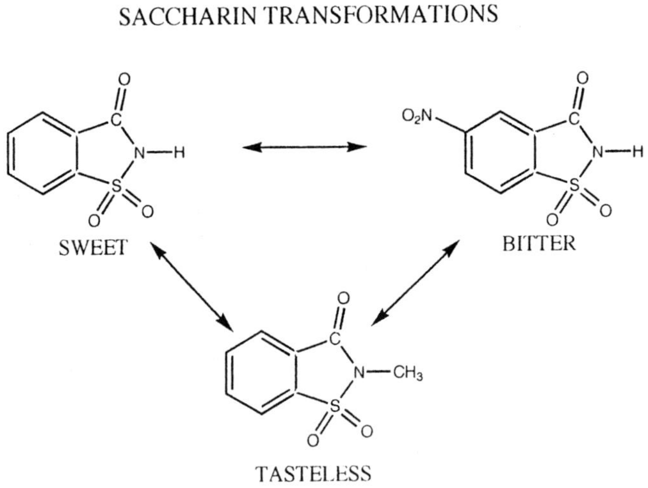

Figure 3 Saccharin transformations.

ecules that possess the molecular characteristics required for binding to a taste receptor but do not possess the "trigger" site to induce a taste response (Figure 4) could eliminate the taste of sweet and/or bitter compounds. We, therefore, began to search for compounds that could dissociate the bitter sensation associated with KCl from its saltiness.

Since our model assumed a close similarity of bitter and sweet receptors, we hypothesized that an essentially tasteless compound that could inhibit bitter taste might also inhibit sweet taste, and conversely, that an essentially tasteless compound that inhibited bitter taste might also inhibit sweet taste. Consequently, our first attempt to remove the bitterness from KCl involved the use of lactisole or 2-(4-methoxyphenoxy)propionic acid sodium salt (2,4-MPP, Figure 5), a known potent sweetness inhibitor (Lindley, 1990). When this compound is added to a 2.25% solution of KCl (w/v) in water, the bitterness of the KCl is markedly reduced. At a ratio of 2,4-MPP to KCl of 1/400 (0.25% relative to the KCl), the bitterness of the above solution is totally eliminated as judged by a random sample of taste volunteers using a sip-and-spit methodology.

The results of this first study were very encouraging. We had discovered a tasteless blocker for the bitter taste of KCl that was effective at 0.25% relative to KCl and that already had some FEMA approval. However, during the course of normal application, salt is added to various foods to achieve a desirable final taste. We, therefore, examined the effect of adding 2,4-MPP to KCl/NaCl mixtures and of applying the final concoction to popcorn or cooked potatoes. A mixture of KCl/NaCl/2,4-MPP (80:20:0.25 w/w/w) when tasted by itself had a salty taste approximating

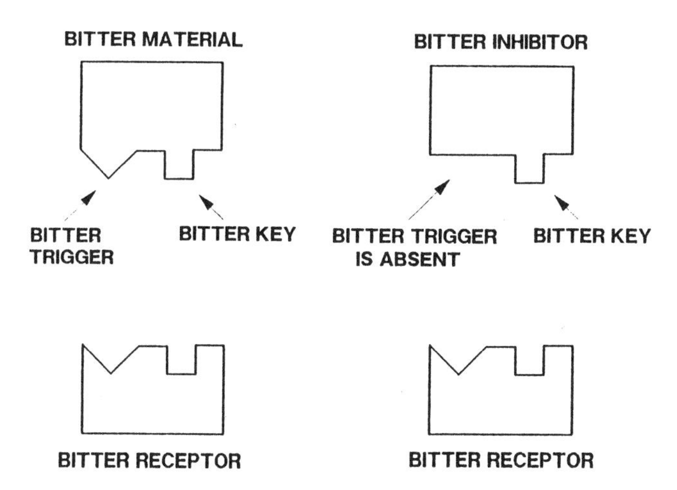

Figure 4 Schematic representation of the bitter taste receptor and interactions with a bitter compound and with a bitter inhibitor.

LACTISOLE

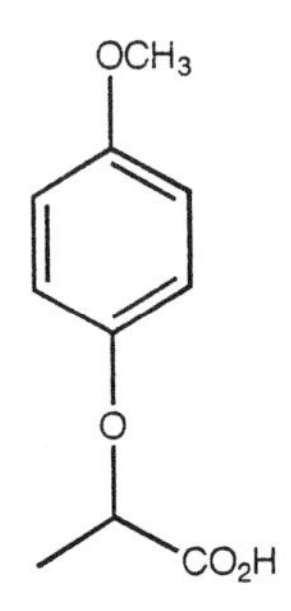

Figure 5 2-(4-Methoxyphenoxy)-propionic acid (lactisole).

that of NaCl. Interestingly, this desired taste was maintained when the mixture was used to "salt" popcorn. However, when the same mixture was used to salt cooked potatoes, the bitterness of KCl was manifest. Careful analysis showed that whereas popcorn displays virtually no sweetness, there is a significant sweet component associated with the taste of cooked potatoes. This difference is likely due to the sugar content of these goods. Popcorn has virtually no sugars whereas potatoes have a significant saccharide content. Thus, lactisole is too potent a sweetness inhibitor to be useful in foods that contain sugars or other sweet components. In addition, although lactisole blocked the bitterness of KCl, it was not very effective at blocking the bitter taste associated with quinine or caffeine. Figure 6 shows the relative inhibitory effect of lactisole on several sweet and bitter compounds.

Based on the above observation, it appears logical that the ideal molecule to eliminate the bitterness of KCl would interact strongly with the bitter site on the taste receptor(s) but would not be a strong sweetness inhibitor. Thus, we began the search for compounds that were essentially tasteless, that would inhibit bitter tastes with higher potency than sweet tastes and whose structure would seem to allow FEMA GRAS approval. Our search quickly led us to the di- and trihydroxybenzoic acids, several of which possess inhibitory properties for bitter taste (Figure 7). The salts of 2,4-dihdroxybenzoic acid (2,4-DHB) were quickly identified as good candidates since they did not inhibit sweetness at all and, in fact, are actually sweet at higher concentrations. In contrast, when present at about 0.5% relative to the amount (w/w) of KCl 2,4-DHB eliminates the bitterness of this salt. When a baked potato was salted with a composition composed of KCl/NaCl/sodium 2,4-DHB (80/20/0.5), the taste of the potato was very close to the taste of a potato salted with sodium chloride.

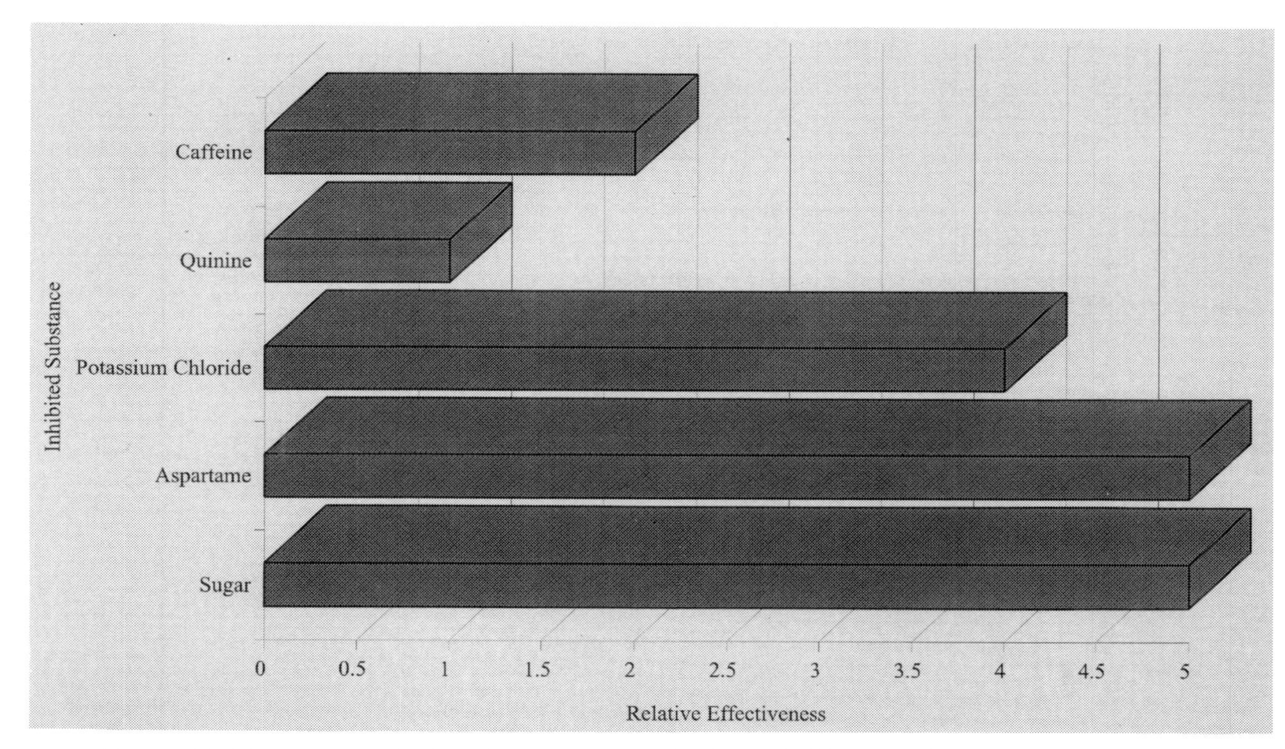

Figure 6 Inhibition properties of lactisole. The relative ability of lactisole to inhibit various bitter compounds is represented.

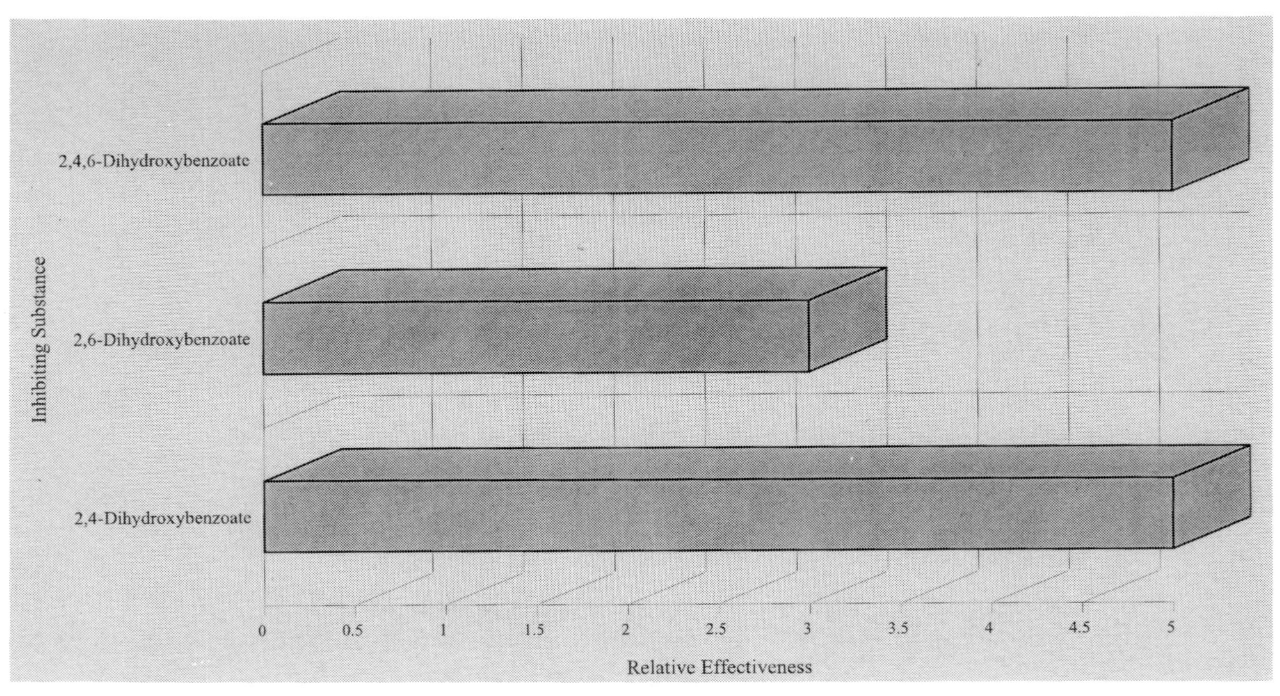

Figure 7 Inhibition of the bitter taste of KCl by various hydroxybenzoic acids.

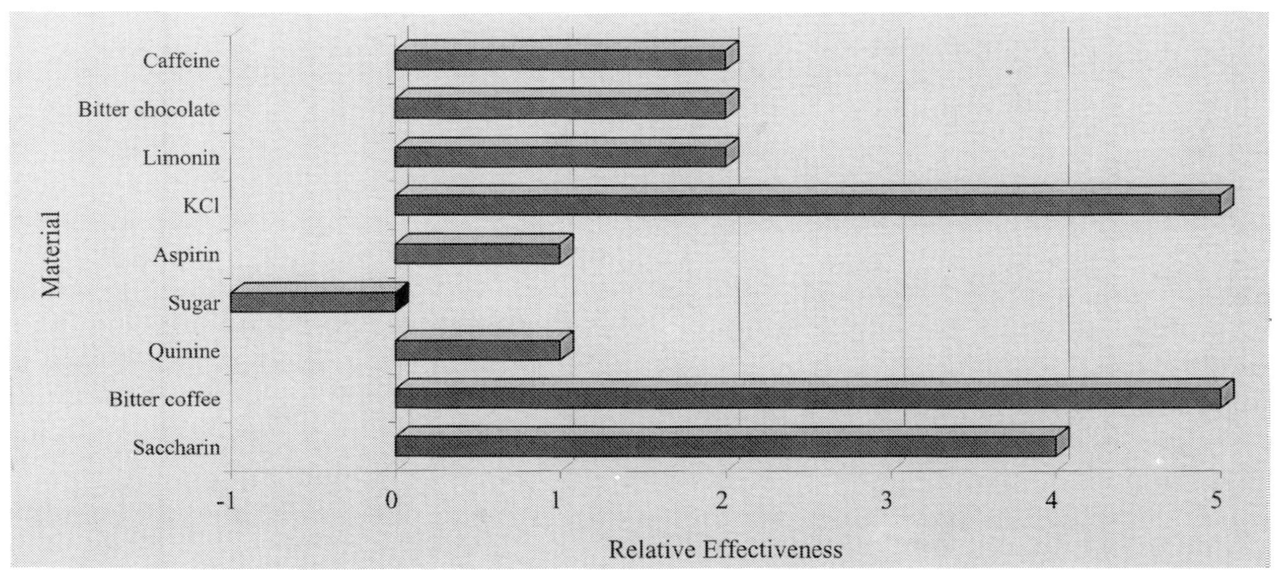

Figure 8 Inhibition of bitter compounds by 2,4-dihydroxybenzoate.

The bitter inhibition properties of sodium 2,4-dihydroxybenzoate are summarized in Figure 8. It should be noted that sugar is shown below the origin because at higher concentrations 2,4-DHB is actually sweet. 2,4-DHB is also very effective at eliminating the undesirable metallic, bitter aftertaste of saccharin. If about 25 mg of 2,4-DHB is added to a 12 oz. can of cola sweetened with saccharin, the taste is more sugar-like with virtually no aftertaste. Most significantly, at the concentrations tested, 2,4-DHB does not inhibit or enhance the sweetness of sugar. It, therefore, can be used to block the bitterness of KCl in food compositions and we have successfully applied it to potatoes, popcorn and other edibles.

Although in a specific food application 2,4-DHB is a very good tastand, there are several potential problems with it that must be kept in mind.

(1) It is prone to decarboxylation at low pH and, therefore, its usefulness is pH dependent.
(2) It is not heat stable; it decarboxylates at elevated temperatures losing its blocking capabilities.

We, therefore, decided to continue our search for a better compound for general use to block the undesirable taste of potassium chloride. At least 60 compounds have been identified that can inhibit bitter taste. Among these, a compound from BioResearch BR-613 is stable to high temperature, a wide pH range, and can be used to eliminate the bitterness associated with KCl, high-intensity sweeteners, and chocolate, as well as many other foods.

The findings reported herein support the suggestion that the mechanism of salt perception is likely different from the mechanism of sweet perception and bitter perception. As a consequence, we conclude that a low-sodium salt substitute is readily achieved using the tastand concept described in this chapter. Tastands can be used to debitter a variety of ingestibles and therefore will, we believe, find wide-ranging application in the food, beverage and pharmaceutical industries. The products of interest are covered in a recent patent. Other examples of tastands include taurine, β-alanine and β-amino ethyl phosphonic acid. Surfactants may offer a means to enhance the effectiveness of certain tastands.

REFERENCES

Ariyoshi, Y. 1979. Taste-giving peptides (molecular theories of sweet taste), in *Dev. Food Sci.*, 378, *Proceedings of the fifth Kyoto International Congress of Food Science and Technology,* New York, 1978, Elsevier North Holland.

Belitz, H.-D., Chen, W., Jungel, H., Treleano, R., Wieser, H., Gasteiger, J. and Marsili, M. 1979. Sweet and bitter compounds: structure and taste relationship, *Food Taste Chemistry, American Chemical Society,* 93–139.

Benedetti, E., Blasio, B. D., Pavone, V., Pedone, C., Fuller, W. D. and Mierke, D. F. 1990. Crystal structure of two retro-inverso sweeteners, *J. Am. Chem. Soc.*, *112*, 8909.

Douglas, A. J. and Goodman, M. 1990. Molecular basis of taste: a stereoisomeric approach, *A.C.S. Symposium Series, 450*, 128.

Erickson, D. 1990. Trick of the tongue, *Scientific American,* May 1990.

Feinstein, R. D., Polinski, A., Douglas, A. J., Beijer, M. G. F., Chadha, R. K., Benedetti, E. and Goodman, M. 1991. Conformational analysis of the dipeptide sweetener alitame and two stereoisomers by proton NMR, computer simulations, and X-ray crystallography, *J. Am. Chem. Soc.*, *113*(9), 3467–73.

Fuller, W. D., Goodman, M. and Verlander, M. S. 1985. A new class of amino acid based sweeteners, *J. Am. Chem. Soc.*, *107*, 5821.

Fuller, W. D. and Kurtz, R. J. 1992. World Patent Cooperation Treaty PCT Int. Appl. WO 92 10179 to Bioresearch Inc.

Goodman, M., Coddington, J., Mierke, D. F. and Fuller, W. D. 1987. A model for the sweet taste of stereoisomeric retro-inverso and dipeptide amides, *J. Am. Chem. Soc.*, *109*, 4712.

Goodman, M., Yamaza, T., Zhu, Y.-F., Benedetti, E. and Chadha, R. K. 1993. Structures of sweet and bitter peptide diastereomers by NMR, computer simulations, and X-ray crystallography, *J. Am. Chem. Soc.*, *115*, 428.

Holtje, H. D. and Kier, L. B. 1974. Sweet taste receptor studies using model interaction energy calculations, *J. Pharm. Sci.*, *63,* 1722.

Kier, L. B. 1972. A molecular theory of sweet taste, *J. Pharm. Sci.*, *61*, 1394.

Kurtz, R. J. and Fuller, W. D. 1993a. Specific eatable taste modifiers, Eur. Patent Appl. EP 0900657, assigned to Bioresearch Inc., Arlington, Va.

Kurtz, R. J. and Fuller, W. D. 1993b. Ingestibles containing substantially tasteless sweetness inhibitors as bitter taste reducers or substantially tasteless bitter inhibitors as sweet taste reducers, US Patent 5,232,735, assigned to Bioresearch Inc., Farmingdale N.Y.

Kurtz, R. J. and Fuller, W. D. 1991. Ingestibles containing substantially tasteless sweetness inhibitors as bitter taste reducers or substantially tasteless bitter inhibitors as sweet taste reducers, Eur. Patent Appl. EP 0911565, assigned to Bioresearch Inc., Arlington, Va.

Lelj, F., Tancredi, T., Temussi, P. A. and Toniolo, C. 1976. Interaction of α-L-aspartyl-L-phenylalanine methyl ester with the receptor site of the sweet taste bud, *J. Am. Chem. Soc.*, *98,* 6669.

Lindley, M. G. 1990. Phenylalkanoic acid sweetness inhibitors, *ACS Symposium Series, 450,* 251.

Mazur, R. H., Schlatter, J. M. and Goldkamp, A. H. 1969. Structure-taste relationships of some dipeptides, *J. Am. Chem. Soc.*, *91*, 2684.

Nofre, C. and Tinti, J. M. 1983. Sweetening agents, Eur. Patent Appl. 0,107,597, assigned to Universite Claude Bernard, Lyon, France.

Nofre, C., Tinti, J. M. and Ouar Chatzopoulos, F. 1986. Sweeteners derived from glycine and β-alanine, Eur. Patent Appl. 0,195,730, assigned to Universite Claude Bernard, Lyon, France.

Nofre, C., Tinti, J. M. and Ouar Chatzopoulos, F. 1987. Preparation of (phenylguanidino)- and ([1-(phenylamino)ethyl]amino) acetic acids as sweeteners, Eur. Patent Appl. 0,241,395, assigned to Universite Claude Bernard, Lyon, France.

Nofre, C. and Tinti, J. M. 1988. Preparation of *N*(heterocyclylcarbamoyl)-dipeptides as sweeteners, Eur. Patent Appl. 0,321,368, assigned to Universite Claude Bernard, Lyon, France.

Shallenberger, R. S. and Acree, T. E. 1967. Molecular theory of sweet taste, *Nature, 216,* 480.

Shallenberger, R. S. and Acree, T. E. 1969. Molecular structure of sweet taste, *J. Agric. Food Chem., 17,* 701.

Temussi, P. A., Lelj, F. and Tancredi, T. 1978. Three dimensional mapping of the sweet taste receptor site, *J. Med. Chem., 21,* 1154.

Temussi, P. A., Lelj, F., Tancredi, T., Castiglione-Merelli, M. A. and Pastore, A. 1984. Soft agonist receptor interactions: theoretical and experimental simulations of the active site of the receptor of sweet molecules, *Int. J. Quantum Chem., 26,* 889.

Tinti, J. M., Durozard, D. and Nofre, C. 1980. Sweet taste receptor, evidence of separate specific sites for carboxyl ion and nitrite/cyanide groups in sweeteners, *Naturwissenshaften, 67,* 193.

Tinti, J. M., Durozard, D. and Nofre, C. 1981. Studies on sweeteners requiring the simultaneous presence of both NO_2/CN and COO− groups, *Naturwissenshaften, 68,* 143.

Tinti, J. M. and Nofre, C. 1990. Why does a sweetener taste sweet? A new model, *ACS Symposium Series, 450*, 206.

Van der Heijden, A., Brussel, L. B. D. and Peer, H. G. 1979. Quantitative structure-activity relationship in sweet aspartyl dipeptide methyl esters, *Chem. Sense Flav., 4,* 141.

Van der Heijden, A., Van der Wel, H. and Peer, H. G. 1985. Structure-activity relationships in sweeteners. 1. Nitroanilines, sulfamates, oximes, isocoumarins and dipeptides, *Chem. Senses, 10,* 57.

Yamazaki, T., Benedetti, E., Kent, D. and Goodman, M. 1994. Conformational requirements for sweet-tasting peptides and peptidomimetics, *Angew. Chemie Int. Ed. Eng., 33*, 1437.

CHAPTER 11

The Use of Exopeptidases in Bitter Taste Modification

GRAHAM BRUCE[1]
DENISE PAWLETT[1]

INTRODUCTION

THE industrial use of enzymes for the modification of food proteins has very distant origins, although the historical use usually involved crude preparations where the active components, the enzymes, were not recognized as such. Nevertheless, the desirable effects were exploited and the use of the natural materials containing the enzymes was passed down to future generations. Many of these food conversions involved the use of proteolytic enzymes, and it is important to remember that even today the most typical application of proteolytic enzymes is *within* a food fermentation process rather than as selected and isolated enzymes. The deliberate use of selected and isolated industrial enzymes has developed largely over the last 50 years. It is convenient to think of two categories of enzyme process:

(1) Traditional processes using enzymes within live cultures alone or in combination with crude animal or plant extracts containing enzymes. Key enzymes and their role are not well understood.
(2) Modern industrial processes where extracted (but not pure) enzymes are used to achieve specific ends. Here, key enzymes are better understood, but owing to the complexity of the substrate understanding remains far from complete.

Both of these approaches rely on limited hydrolysis and a careful balance of endoproteases and exopeptidases to control flavor and produce a

[1]Imperial Biotechnology Ltd., Southbank Technopark, 90 London Road, London SE1 6LN, U.K.

desirable product. The difficulties of this control are evident in the frequency with which bitter flavors can occur, both in natural fermented processes and in modern industrial processes where specific enzymes are used.

Bitterness produced via enzymatic hydrolysis was recognized as far back as 1950 by Cuthbertson who noticed the objectionable taste in protein hydrolyzates destined for clinical feeding. Bitterness as a defect in cheese and other fermented processes is significantly less common but goes back as far as the origin of the processes themselves. Bitterness remains a major drawback, often restricting the commercial viability of products, and has resulted in an industry in itself to discover methods to actively debitter or mask the bitter flavors.

Examples of traditional processes dependent on proteolytic enzymes include the use of fungal cultures in oriental foods such as soy sauce and tempeh, and in the West the production of cheese that depends on enzymes from two separately added sources. The protease chymosin (originally minced calf fourth stomach) serves to coagulate the milk and begin the breakdown of casein, and the proteolytic enzymes (largely exopeptidases) from lactic acid bacteria are key to further flavor development. Another traditional process that also involves proteolytic enzymes from lactic acid bacteria is that of fermented sausage production. Until relatively recently the role of cultures in cheese and fermented sausage was thought to be mainly one of lactic acid production, essentially a tool to preserve nutritious foods. Over the last 10 years the importance of endo- and exopeptidases in flavor production and bitterness prevention has been recognized. A final example of the indirect use of enzymes is in the production of yeast autolyzates, which involves self-digestion of the yeast cell contents, but it sometimes augmented by the addition of protease. In all of these processes, the enzymes involved remain incompletely defined and the balance of the component enzymes is "accidental," relying on following a formula that has traditionally worked. Deviation from this formula often leads to undesirable results such as bitter off-flavors, and there is a need for a detailed knowledge of the enzyme processes in order to return to a quality product.

Commercial uses of proteolytic enzymes that have developed more recently fall into three broad categories. Proteins are hydrolyzed to improve: (1) functional properties such as gelling, whipping, foaming, solubility and stability; (2) flavor enhancement, particularly with meat, fish, dairy proteins and vegetable proteins; (3) increased digestibility for medical, dietary, infant and animal foods. Production of flavors from vegetable proteins is a growing area due to the concern over potentially carcinogenic chloro-compounds that can be produced by conventional acid hydrolysis. Bitterness is commonly encountered in all of these processes.

The following sections will deal with the chemical nature of the bitter flavor; the mechanisms that lead to these bitter species being produced; the action of exopeptidases; and some detailed examples that look at bitter taste removal and prevention. Examples are taken from protein hydrolysis, one where a bland product is desired and one to be used as a flavoring agent, and a fermented product, the production of cheese.

THE NATURE OF BITTERNESS

An up-to-date review of the chemical structure of bitterness is presented in Section I of this volume. Here we focus on bitter peptides and those characteristics that are important for understanding the action of exopeptidases in debittering.

After the enzymatic hydrolysis of a native protein, the resultant hydrolyzate displays a wide molecular weight distribution profile that ranges from unhydrolyzed protein to small peptides and free amino acids. The exact distribution will depend on the reaction conditions (time, temperature, pH), enzyme specificity and the nature and extent of denaturation of the substrate. Early work on enzymatic hydrolyzates showed that the bitter taste flavor was due to peptides rather than free amino acids (Murray and Baker, 1952). Specifically, bitterness is associated with peptides containing hydrophobic amino acids; leucine, isoleucine; proline, phenylalanine, tyrosine and tryptophan.

PREDICTION OF BITTERNESS AND THE "Q"-RULE

Working with single peptides, described as being bitter or nonbitter, Ney proposed a rule for predicting bitterness. Ney calculated the calorific values of individual amino acids, these are used to calculate the average hydrophobicity or Q-value of a peptide (Ney, 1971). If the Q-value exceeded a certain value [originally +1400 cal/mol but later revised to +1350 cal/mole (Ney and Retzlaff, 1986)], then that particular peptide would be bitter, and if the value was less then no bitterness should be present. This empirical correlation between the presence or absence of bitterness and the average hydrophobicity was termed the Q-rule. The relative hydrophobicity and calorific values of the amino acids are listed in Table 1 (values from Tanford, 1962).

The Q-rule was further substantiated by observations on the taste of a peptide of seven amino acids which was synthesized in a stepwise manner. The addition of isoleucine, a strongly hydrophobic amino acid to a nonbitter tetrapeptide resulted in a bitter taste. The bitterness disappeared when two further hydrophilic amino acids were added, which resulted in an average hydrophobicity below the bitterness threshold (Ney, 1979). Follow-

TABLE 1. Relative Hydrophobicity of Amino Acids.

Amino Acid	Hydrophobicity	Q-Value (cal/mol)
Alanine	0	730
Arginine	0	730
Asparagine	0	−10
Aspartic	0	540
Cysteine	+	–
Glutamic	0	550
Glutamine	0	−100
Glycine	0	0
Histidine	0	550
Isoleucine	+++	2970
Leucine	+++	2420
Lysine	+	1500
Methionine	+	1300
Phenylalanine	++	2650
Proline	++	2620
Serine	0	40
Threonine	0	440
Tryptophane	+++	3000
Tyrosine	++	2870
Valine	++	1690

ing a compilation of over 200 peptides which had a bitter taste, it was found that practically all the bitter-tasting peptides complied with the Q-rule (Guiqoz and Solms, 1976). The chain length of these bitter peptides varied, with the majority containing two to 12 amino acid residues, but with three peptides containing greater than 20 amino acid residues. The size limit for bitter peptides can be as high as 6000 daltons.

The link between amino acids with hydrophobic side-chains and the bitter taste sensation was independently proposed by Matoba and Hata (1972). This work concluded that hydrophobic amino acids exert the strongest bitterness when both their ends are blocked, e.g., by forming peptide linkages. The bitterness perceived was comparatively weaker when the amino acid was in a terminal position and weakest when it was free. Tri- and tetrapeptides tend to be more bitter than dipeptides, for example, Phe-Gly-Phe-Gly is at least 10-fold more bitter than the single dipeptide Phe-Gly (Shiraishi, 1973).

PEPTIDE, SEQUENCE, STRUCTURE AND BITTERNESS

Importantly, the intensity of bitterness is also dependent on the sequence of amino acids in the peptide. Shiraishi (1973) showed that the peptide Phe-Pro is more intensely bitter than Pro-Phe and Gly-Phe-Pro being more bitter than Phe-Pro-Gly. The presence of a basic amino acid at the N-terminal is an important contributor to bitterness (Shinoda 1986) and when leucine is located at the C-terminal of a peptide then that peptide will invariably be bitter. The conformation of the C-terminal of the peptide is particularly critical for bitterness intensity (Ishibashi, 1987, 1988a).

Larger peptides can retain some three-dimensional structure, which will also impact on the associated levels of bitterness. Upon formation of peptide bonds, the rotation about the C-N linkage is limited and confines the bond to a planar configuration. The most common structure formed is an alpha-helix; however, depending on pH and the nature of the side-chains present, a random coil configuration may also be formed. Helix formation can be inhibited by side-groups that have the same charge (hydrophobic) or are large and spatially inhibitory such as the proline ring structure (Shallenberger, 1993). When proline is present near the center of a peptide, then the bitterness of that peptide is increased (Ishibashi, 1988a). From studies on model derivatives, Ishibashi (1988b) concluded that bitterness required two separate reactive sites, a "binding site" and a "stimulating unit," to interact with the taste receptor. Consequently, modification of the steric properties of a peptide can alter the extent of the bitter response elicited by the peptide.

ENZYME HYDROLYSIS AND GENERATION OF BITTERNESS

The appearance of bitterness as a protein is hydrolyzed may be summarized biochemically as follows. Intact food proteins do not contribute significantly to flavor, their molecular size alone suggests that interaction with taste receptors is unlikely. Even for very hydrophobic proteins, such as casein, in the native, globular form the majority of the hydrophobic side-chains are concealed within the interior and are therefore not free to interact with taste receptors. The resulting flavor is therefore bland and not bitter.

As the intact protein is hydrolyzed by endoproteases, peptides of varying sizes are formed. The largest peptides will retain to some extent a degree of three-dimensional conformation and hydrophobic side-chains can still be hidden within either a U-shaped pocket or within aggregates of peptides (Adler-Nissen, 1986). This breakdown pattern explains why there have been so few large bitter peptides reported. Upon further hydrolysis, the number of small peptides increases and more hydrophobic side-chains will become exposed and thus available to interact with taste receptors.

PREDICTION OF BITTERNESS IN PROTEIN HYDROLYZATES

As an approach to predict the level of bitterness resulting from hydrolysis, Ney extrapolated his Q-rule (Ney, 1979). Ney proposed that the likelihood of bitterness as a result of protein hydrolysis could be estimated by calculating the average hydrophobicity of the resultant hydrolyzate. This extrapolation does not hold true and was first questioned by Adler-Nissen (1986). Although the more hydrophobic the protein, the higher the statistical probability of forming very hydrophobic, and hence bitter, peptides, it is *not* the average hydrophobicity of the whole mixture that is the cause of bitterness but the presence and concentration of specific hydrophobic peptides. There is now common consensus that the Q-rule holds for isolated peptides but that this rule cannot be extrapolated to assess the bitterness of a hydrolyzate.

The composition of a hydrolyzate ranges from free amino acids through peptides of varying chain length to unhydrolyzed native protein. This composition will be influenced by the enzyme (or enzymes) used and the extent of hydrolysis. To date, the average hydrophobicity of food proteins (Figure 1) remains a good predictor of the *likelihood* of encountering bitterness on hydrolysis. All food proteins can produce bitter peptides under certain hydrolysis conditions, but for those proteins with lower hydrophobicity (e.g., gelatin, fish), bitterness is easier to control or avoid.

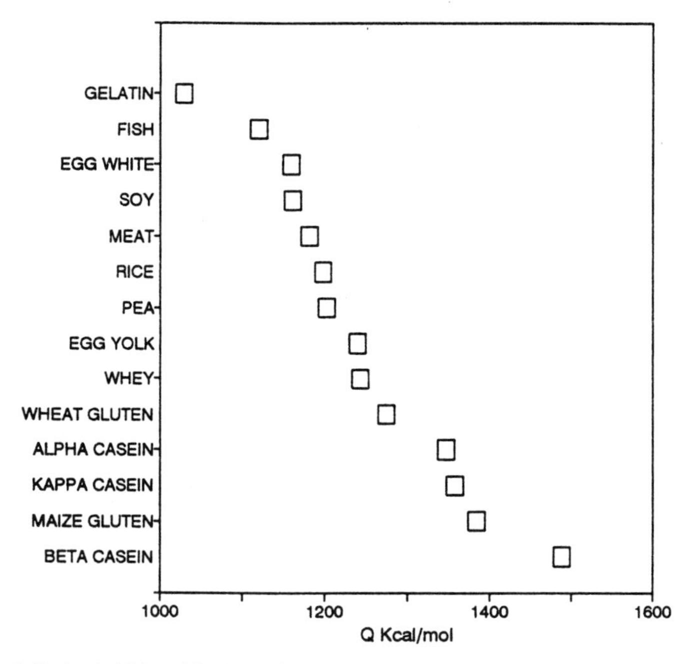

Figure 1 Hydrophobicity of food proteins. The hydrophobicity (Q-value kcal/mol) is a good indicator of the probability of producing bitterness on hydrolysis. Proteins with high Q-values will give the most bitter hydrolyzates. [Reproduced from *Industrial Enzymology* (1996), 2nd ed., ed. T. Godfrey and S. West, with the kind permission of Macmillan Press Ltd.]

Two remaining important points need to be made. Firstly, despite their hydrophobicity, bitter peptides must be slightly soluble in water to interact with the taste buds and elicit the bitter response. Secondly, the concentration of bitter peptides must exceed a certain threshold level before bitterness is detected by the human palate and impairs the overall flavor profile.

ENZYME SELECTION

The intensity of bitterness of protein hydrolyzates is clearly complex and difficult to predict. The final level of bitterness is a result of a particular combination of enzyme, substrate and reaction conditions. It is possible to limit bitterness by careful selection of the endoprotease and reaction conditions and remove any residual bitterness by simultaneous or subsequent hydrolysis with an exopeptidase.

The influence of the endoprotease can be predicted to some extent by its specificity and reference to the amino acid sequence of the substrate. For example, proteases such as pepsin, *Bacillus* neutral and *Bacillus* alkaline have some preference for hydrophobic amino acids at the carbonyl side and will tend to produce peptides with N-terminal hydrophobic amino acids. Proteases such as trypsin, papain and bromelain have a preference for lysine and arginine or aspartate and glutamate at the carbonyl side. These will, therefore, produce a very different spectrum of peptides and the presence of a hydrophilic amino acid at the N-terminal of the peptide will significantly affect the solubility and bitterness of the resultant hydrolyzate. Table 2 lists the substrate specificity of the commonly used industrial proteases.

This insight has, so far, had limited practical use for a number of reasons. Many of the industrial proteases have such broad specificity that accurate prediction of the hydrolysis sites is difficult and several have proteolytic side activities with different specificities. Secondly, the three-dimensional structure of the native protein will initially control the first bonds to be cleaved, and subsequent preferred cleavage sites can be made unavailable by the formation of aggregates of peptides through hydrogen bonding and hydrophobic interaction. Thirdly, the hydrolysis reaction rarely goes to completion.

Figure 2 illustrates the hydrolysis of casein with three different proteases. In general, bitterness is related to the degree of hydrolysis and tends to increase with increasing hydrolysis and eventually reaches a plateau. The method of limiting protein hydrolysis to low degrees of hydrolysis has been advocated as a method for controlling bitterness (Adler-Nissen, 1984), and can be acceptable particularly for proteins with lower

TABLE 2. Industrial Food Enzymes and Their Specificity.

Enzyme	General Specificity	Side Activities
Animal		
Chymosin	Glu-Ala Leu-Val	Pepsin
Pancreatin	Broad	Mixed, aspartic, serine, metallo, exopeptidase
Pepsin	Leu-Val Phe-Tyr	
Trypsin	Lys-Ala Arg-Gly	Chymotrypsin
Bacterial		
Bacillus subtilis	Broad His-Leu, Ala-Leu, Gly-Phe	
Bacillus cheniformis	Broad Gln-His, Leu-Tyr, X-Tyr, X-Trp	
Fungal		
Aspergillus oryzae	Very broad	Mixed, metallo, carboxy serine and exopeptidase
Aspergillus niger	Broad Leu-Val, Phe-Tyr	Mixed metallo, carboxy peptidase
Plant		
Papain	Broad Asn-Gln, Glu-Ala, Leu-Val, Phe-Tyr	Mixed
Bromelain	Lys-Arg, Tyr-Phe	

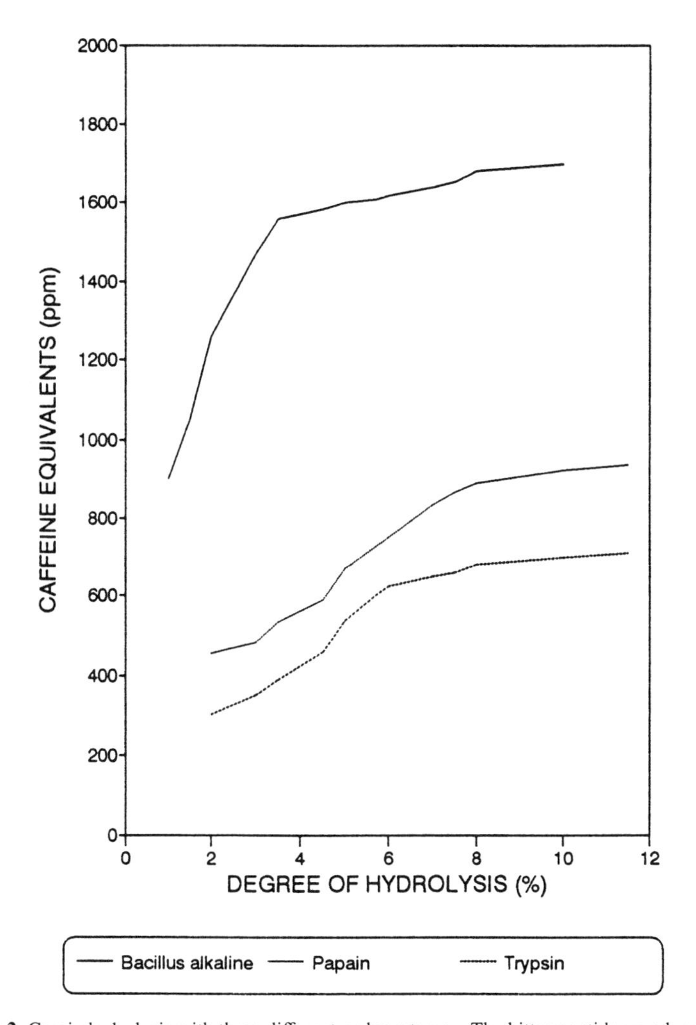

Figure 2 Casein hydrolysis with three different endoproteases. The bitter peptides produced during hydrolysis will depend on the specificity (cleavage site) of the endoprotease used. Hence, to some extent the level of bitterness can be reduced by protease selection. [Reproduced from *Industrial Enzymology* (1996), 2nd ed., ed. T. Godfrey and S. West, with the kind permission of Macmillan Press Ltd.]

Q-values. The disadvantages are restrictions in the final protein functionality, yield constraints and the need for good control measures to ensure that a product with acceptable bitterness is produced every time. For proteins with high Q-values such as casein, even very slight hydrolysis gives rise to unacceptable levels of bitterness. Casein hydrolyzates produced with the plant sulphydryl proteases tend to be less bitter than those produced with trypsin or *Bacillus* protease.

Very extensive protein hydrolysis (DH $> 30\%$) has been associated with a reduction in the level of bitterness and usually the generation of savory flavors. This is an indication of the action of exopeptidases which are often present as side activities in protease preparations, with endoprotease activity only bitterness does not reduce.

PEPTIDASES AND DEBITTERING

The use of peptidase enzymes to reduce or eliminate bitterness has increased significantly in recent years. Peptidases remove single or pairs of amino acids from the terminal of a peptide chain – carboxypeptidases acting from the C-terminal and aminopeptidases from the N-terminal. They are largely ineffective on intact proteins and are therefore always used after or in conjunction with an endoprotease.

The effectiveness of peptidases in debittering will largely depend on their specificity. However, because of the complex nature of the interaction of bitter peptides with taste receptors, removing only a few key residues from the terminal of a bitter peptide can produce a large reduction in bitterness. Peptidases able to cleave hydrophobic amino acid residues and proline are most valuable, but, as can be inferred from the discussion of the structure of bitter peptides, removal of a basic amino acid from the N-terminal will have a major impact on bitterness intensity.

Removal even of hydrophilic amino acids and proline from a peptide can alter its solubility or steric properties and hence affect bitterness. For example, BPl-a, a bitter peptide isolated from cheese.

H – ARG – GLY – PRO – PRO – PHE – ILE – VAL – OH

The three N-terminal amino acids are easily cleaved by a combination of leucine aminopeptidase and prolyl dipeptidyl aminopeptidase (Arg- + GlyPro). The resulting peptide (Pro-Phe-Ile-Val) is both less soluble and less bitter. It is not necessary to reduce the peptide to free amino acids for complete debittering although the action of peptidases will always result in an increase in the free amino acid content (Pawlett and Fullbrook, 1988).

High levels of free amino acids will impart savory or brothy flavors which, in applications such as dietary foods or dairy ingredients, are as undesirable as bitterness.

AMINOPEPTIDASES

Aminopeptidases are produced commercially from two sources; lactic acid bacteria and *Aspergillus oryzae.* Peptidases from *A. oryzae* may contain low levels of endoprotease as a side activity whereas in peptidase preparations from lactic acid bacteria, endoprotease activity is nondetectable. Leucine aminopeptidase is widely present as a side activity, often at significant levels, in fungal enzyme preparations, e.g., *Aspergillus oryzae* proteases.

Aminopeptidases cleave single or pairs of amino acids from the amino terminal of peptide chains (Figure 3). One of the most common aminopeptidases is aminopeptidase N, previously referred to as leucine aminopeptidase, although it has its greatest reaction rate against lysine. This enzyme has been widely isolated from lactic acid bacteria but variations in terms of specificity have been reported (Tan, et al. 1993; Pritchard and Coolbear, 1993). This enzyme is quite different in terms of specificity, pH and temperature optima from a leucine aminopeptidase from *Aspergillus oryzae* (Nakadai, et al., 1973). The aminopeptidases currently available are listed in Table 3 together with their source and characteristics.

The main commercial sources of peptidases are: (1) mixtures of aminopeptidase from lactic acid bacteria produced by Imperial Biotechnology Ltd. under the trade names Accelase®, Savorase® and Debitrase®; (2) *Aspergillus oryzae* aminopeptidase, which is produced relatively endoprotease free by Imperial Biotechnology Ltd. (Debitrase® DBS50) and Rohm Gmbh (Corolase™ 7093); (3) preparations of *Aspergillus oryzae* protease, which are widely commercially available and contain varying levels of aminopeptidase as a side activity. Carboxypeptidases are not commercially available but are present as side activities in pancreatin, some fungal proteases, yeast and *Penicillium* spp.

Unlike the majority of bulk commercial enzymes *L. lactis* aminopeptidases are produced intracellularly and have to be extracted from the cell. These preparations contain mixtures of aminopeptidase. Conventional mutation selection techniques have allowed the production of different strains which overproduce either general aminopeptidase or prolyl dipeptidyl aminopeptidase, other aminopeptidases (aminopeptidase C, dipeptidase and tripeptidase) are present as side activities. A disadvantage of the enzymes from lactic acid bacteria is their thermolability with a maximum use temperature of 40 °C. In contrast, *A. oryzae* leucine aminopeptidase is

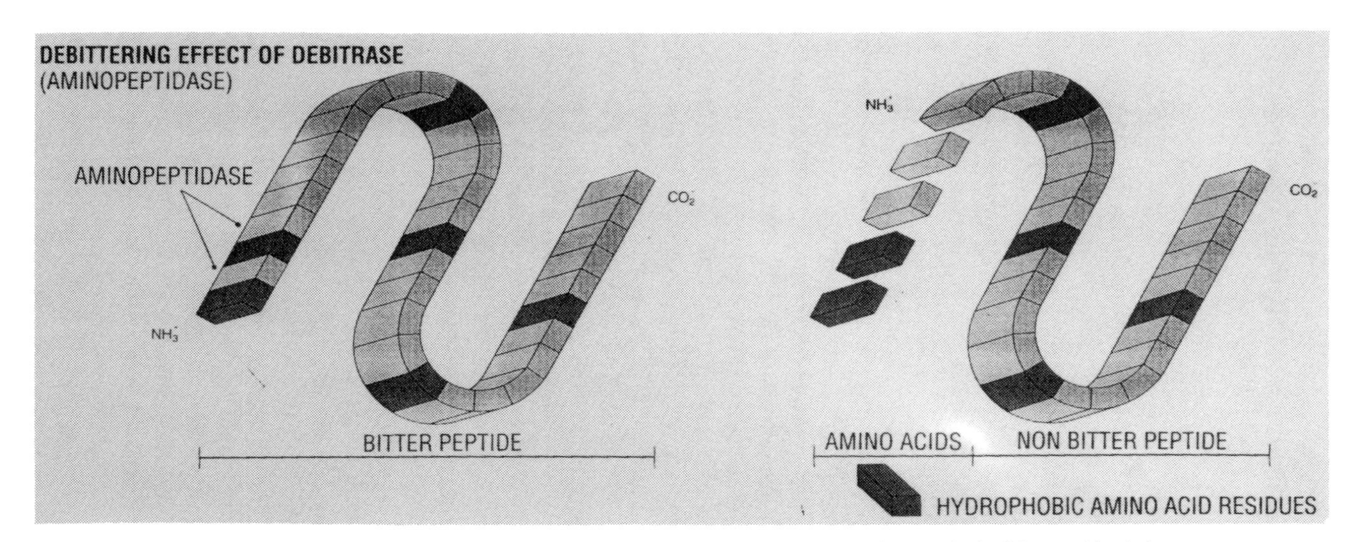

Figure 3 Aminopeptidases remove single or pairs of amino acids from the amino terminal of the peptide chain.

TABLE 3. Aminopeptidases.

Enzyme	Type	pH Range	Maximum Temperature	Primary Specificity	Source	Commercial Availability
Aminopeptidase N (general aminopeptidase)	Metallo	5–8.5	40°C	Lys, Arg, Leu, Met	*Lactococcus* *Pediococcus* *Lactobacillus* *Micrococcus*	Accelase Savorase Debitrase
Aminopeptidase M	Metallo	6.5–8.5	50°C	Leu	Pig kidney (microsimal)	No
Leucine aminopeptidase	Metallo	6–8.0	60°C	Leu	*Aspergillus*	Debitrase Corolase 7093 Savorase and as side activity in protease preparations
Aminopeptidase C	Thicol	6–8.0	40°C	Lys, Arg, Leu, His, Phe	*Lactococcus* *Lactobacillus*	Accelase (side activity)
Aminopeptidase A	Metallo	6.5–8.5	55°C	Asp, Glu	*Lactococcus*	No
X-prolyl dipeptidyl Peptidase (Pep X)	Serine	5.0–9.0	45°C	Argpro-, alapro-, glypro-	*Lactococcus* *Lactobacillus*	Debitrase
Prolidase	Metallo	6.5–8.5	40°C	Dipeptides X-pro	*Lactococcus* *Lactobacillus* *Penicillium* Pig kidney	As side activity only
Tripeptidase	Metallo	6.5–8.5	50°C	Tripeptides	*Lactococcus* *Lactobacillus*	As side activity only
Dipeptidase	Metallo	6.5–8.5	50°C	Dipeptides	*Lactococcus* *Lactobacillus*	As side activity only

very thermostable and can be used at 60°C. However, aminopeptidases from *L. lactis* have a broader pH optimum than the leucine aminopeptidase from *A. oryzae* (Figure 4).

Neither the general aminopeptidase or leucine aminopeptidase will cleave proline residues. The presence of a prolyl dipeptidyl aminopeptidase in preparations of *L. lactis* aminopeptidase provides a particularly powerful debittering effect. This enzyme hydrolyzes dipeptides with the sequence X-Pro from the amino terminal of a polypeptide (Zevaco, et al., 1990). To date no activity has been identified which will release proline from the N-terminal of a polypeptide.

USE OF PEPTIDASES TO DEBITTER PROTEIN HYDROLYZATES

Frequently, the aim when preparing protein hydrolyzates for nutritional purposes is to arrive at a bland final product. However, the generation of large amounts of free amino acids leads to meaty and brothy flavors which are undesirable. By making intelligent choices of the proteases and peptidases used, the level of these off-flavors can be minimized. The other extreme of protein hydrolysis is when meat or vegetable proteins are hydrolyzed to produce flavor components. Here the aim is to produce the maximum amount of free amino acids and it is necessary to select endoproteases and exopeptidases with very broad specificity (Taylor et al., 1991).

To compare the debittering effectiveness of several aminopeptidases, a standard, bitter protein hydrolyzate was used (Pawlett and Bruce, 1996). Casein was the protein of choice, as this is the most hydrophobic food protein and notoriously the most difficult to debitter. A hydrolyzate was prepared using *Bacillus subtilis* neutral protease, which has no exopeptidase side activity. The hydrolysis was stopped after a relatively low degree of hydrolysis, 5% (measured using the TNBS method for free amino nitrogen) had been obtained; this means that large bitter peptides are still present. This substrate therefore represents the most difficult debittering task a food chemist is likely to face and is a useful model for evaluating the debittering efficiency of various peptidases. A trained sensory panel assessed the hydrolyzate relative to caffeine standards. A 1% solution was equivalent in bitterness to a 1700 ppm caffeine solution, the threshold of bitterness detection was 200 ppm. The mixture of peptidases from *L. lactis* was able to effectively debitter the hydrolyzate with an increase in the degree of hydrolysis of around 8%; this is a reflection of the high specificity for hydrophobic amino acids and proline. The *A. oryzae* leucine aminopeptidase required a significantly higher dose rate to produce a similar reduction in bitterness. The increased dose rate gives a greater increase

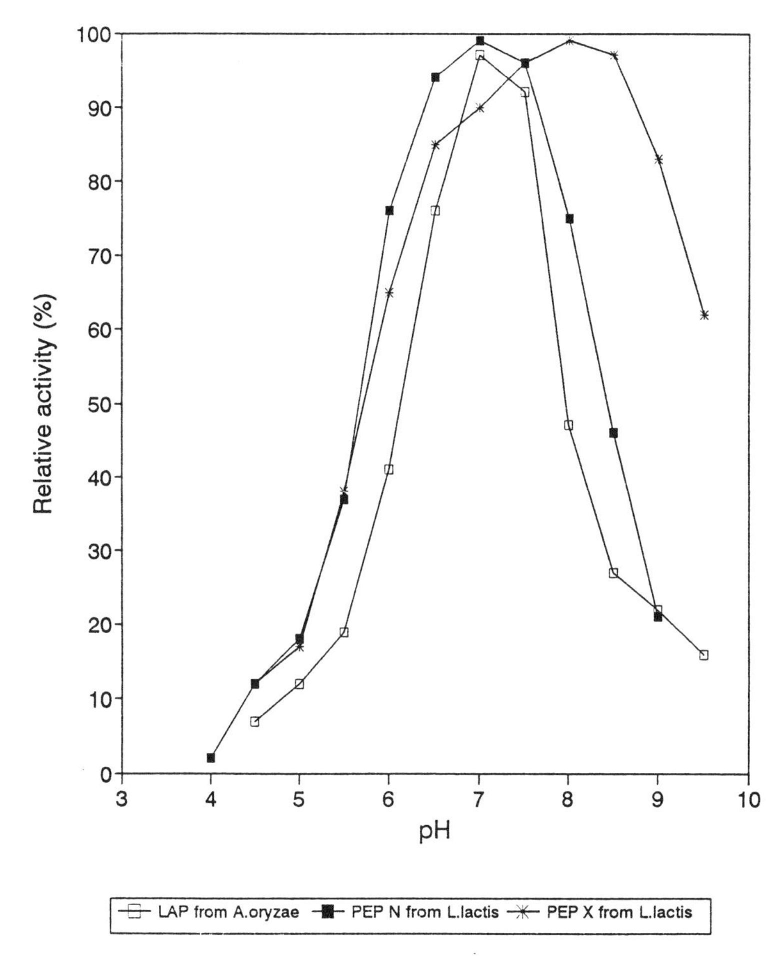

Figure 4 pH and activity of leucine aminopeptidase (LAP) from *Aspergillus oryzae,* general aminopeptidase (Pep N), and prolyl dipeptidyl aminopeptidase (Pep X) from *Lactococcus lactis.*

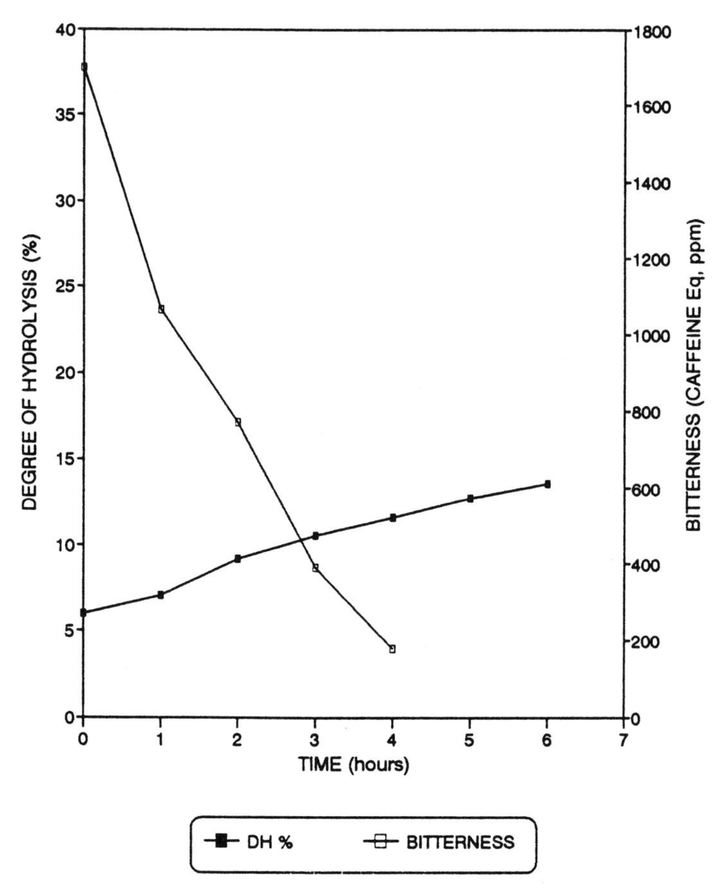

Figure 5 Secondary hydrolysis of a standard bitter casein hydrolyzate with aminopeptidases from *Lactococcus lactis* (dose rate, 2 LAP units/g protein, activity measured against leucine paranitroanilide), showing effective debittering for a relatively small increase in degree of hydrolysis. Bitterness threshold is 200 ppm caffeine equivalents. [Reproduced from *Industrial Enzymology* (1996), 2nd ed., ed. T. Godfrey and S. West, with the kind permission of Macmillan Press Ltd.]

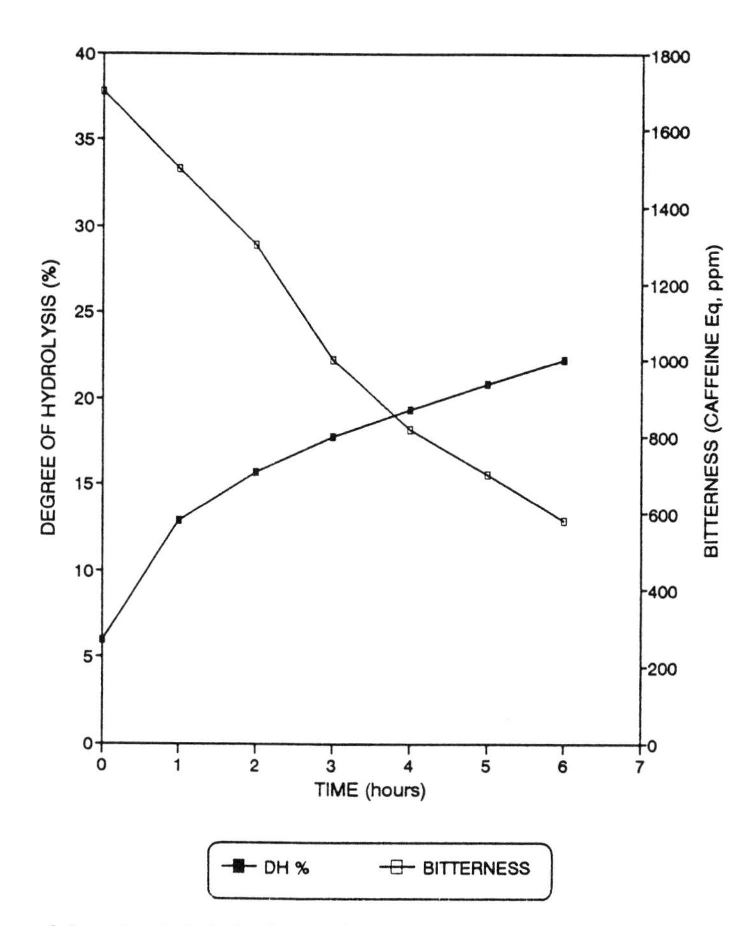

Figure 6 Secondary hydrolysis of a standard bitter casein hydrolyzate with leucine aminopeptidase from *Aspergillus oryzae* (dose rate, 100 LAP units/g protein, activity measured against leucine paranitroanilide). Bitterness threshold is 200 ppm caffeine equivalents.

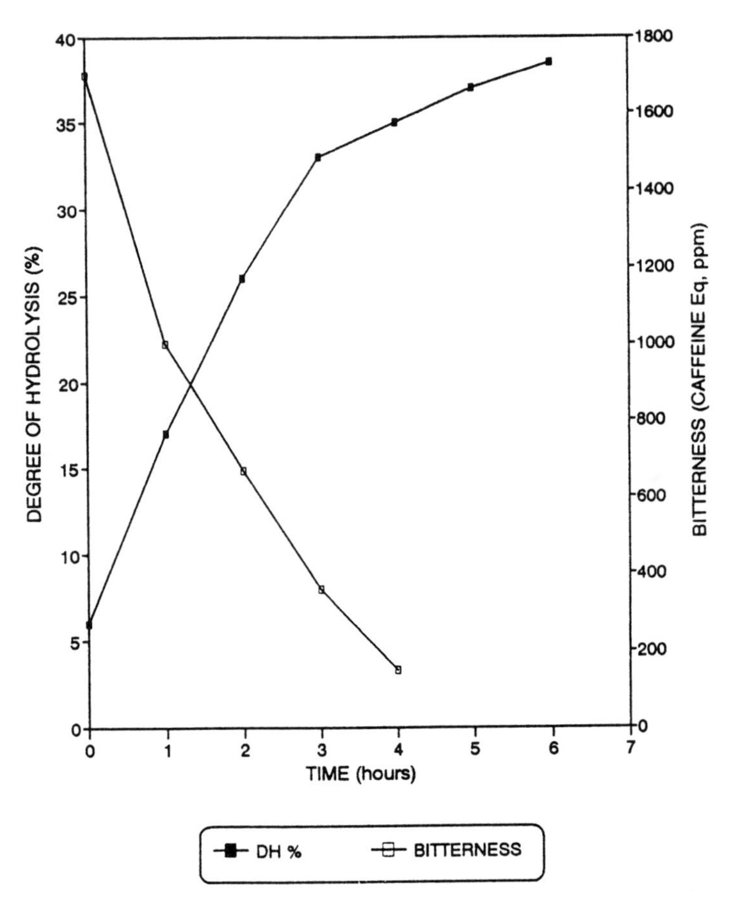

Figure 7 Hydrolysis of a standard bitter casein hydrolyzate to achieve maximum degree of hydrolysis with no bitterness (dose rate, *L. lactis* aminopeptidase 1.0 LAP units/g and *A. oryzae* aminopeptidase 40 LAP units/g protein). Bitterness threshold is 200 ppm caffeine equivalents.

in degree of hydrolysis (18%), and in this case, complete debittering was not achieved (Figures 5 and 6).

Where the aim is to maximize the level of free amino acids to provide a nonbitter base for use in flavorings, then it is beneficial to mix the peptidases from *Lactococcus* and *Aspergillus*. Using the same substrate and a ratio of 1:40 (based on activity against leucine pNA), a degree of hydrolysis of 37% can be achieved from the same bitter substrate. Clearly, by allowing the primary hydrolysis to proceed to completion, higher levels of hydrolysis can be obtained (Figure 7).

USE OF PEPTIDASES TO PREVENT BITTERNESS IN CHEESE

Bitterness is a well-recognized defect in matured hard cheeses such as cheddar and gouda and can also occur in mould-ripened cheeses such as stilton and camembert. Bitterness is a particular problem in all reduced-fat varieties. Bitterness in cheese results from the accumulation of bitter peptides derived from the enzymatic breakdown of casein as the cheese matures. Casein is the most hydrophobic food protein (Figure 1) and its breakdown, particularly of β-casein, can readily produce bitterness. The concentration of bitter peptides must exceed a certain threshold level before bitterness is detected and impairs the overall flavor profile. Below this threshold level bitter peptides should be seen as a normal constituent of cheese. Several bitter peptides have been isolated from cheese (Lemieux and Simard, 1992; Richardson and Creamer, 1973) with hydrophobicities (Q-values) up to 2300 cal/mol (Table 4).

THE FORMATION OF BITTER PEPTIDES IN CHEESE

There is no "one route" to bitterness, but a number of contributing factors that can make the cheese more likely to develop bitterness (Lemieux and Simard, 1991). All of these directly or indirectly affect the course of

TABLE 4. Bitter Peptides from Cheese.

α_{S-1} casein (26–32) Q 1930 kcal/mol *H-Ala-Pro-Phe-Pro-Glu-Val-Phe-OH*
β casein (61–67) Q 2300 kcal/mol *H-Pro-Phe-Pro-Gly-Pro-Ile-Pro-OH*
β casein (84–89) Q 1983 kcal/mol *H-Val-Pro-Pro-Leu-Gln-OH*
β casein (193–208) Q 1767 kcal/mol *H-Try-Gln-Gln-Pro-Val-Leu-Gly-Pro-Val-Argo-Gly-Pro-Phe-Pro-Ileu-Ileu-OH*

casein hydrolysis, leading to an accumulation of bitter peptides (see Figure 8). There are clearly economic implications for preventing bitterness, which can help to increase shelf life and produce a high-quality product.

Rennet and Endoprotease

High levels of calf rennet, crude preparations of rennet (high pepsin) and microbial rennets can all lead to bitterness. The pH at curd formation will affect the amount of calf rennet retained in the curd—and therefore can also impact bitterness (Lowrie and Lawrence, 1972).

Pure chymosin will hydrolyze both α- and β-caseins to produce bitter peptides, the rate of hydrolysis of a casein is faster and the rate of hydrolysis of both α- and β-caseins is reduced at high salt levels. The mild protease plasmin will also hydrolyze β-casein. Use of bacterial proteases to accelerate ripening will tend to produce bitter peptides unless used in conjunction with aminopeptidase.

Starter Culture

Certain strains of *Lactococcus lactis, Lactococcus cremoris* and other dairy starters can, as a result of their activity, produce bitterness in cheese. These strains have an *imbalance* of protease and peptidase activities with either an excess of cell wall proteases or insufficient levels of specific peptidase enzymes. Peptidase enzymes break-down bitter peptides as they are formed. Some strains appear to have a good balance of enzymes but have peptidase enzymes that are less effective at pH values typical in cheese (Crawford, 1977).

Make Procedure and Ripening

Given that starter strains are prone to bitterness, high levels of rennet or poor-quality rennet have been used, then a number of parameters, which can be controlled by the cheese make, will affect the final level of bitterness.

(1) Milk quality and presence of psychotrophic bacteria, can produce heat-stable proteases that survive pasteurization.
(2) Seasonal variations in milk quality. Milk which gives a poor set can give rise to bitter cheese due to casein breakdown prior to the cheese make.
(3) In preparation of bulk starter (mixed strains) use of external pH control has been shown to lead to dominance of “bitter” strains.

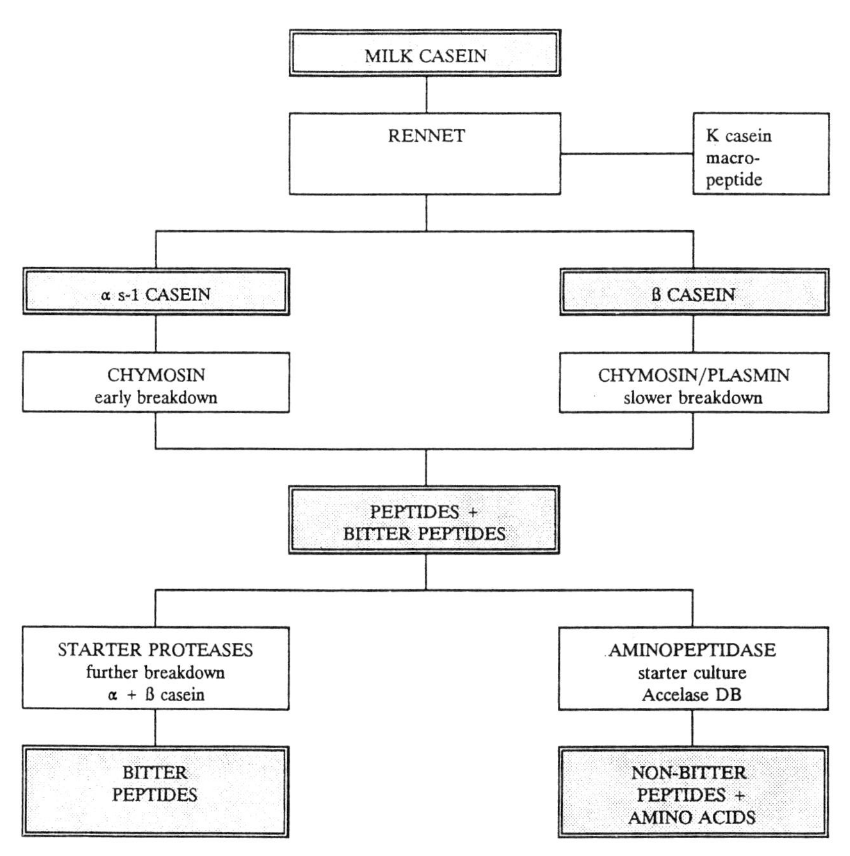

Figure 8 Formation and breakdown of bitter peptides in cheese.

(4) Several fast acid producers lead to bitter cheese but in general, "slow" starters, even when used at high levels, do not produce bitterness.
(5) Cook temperatures above 36–39°C are thought to inhibit growth of nonbitter starters.
(6) Low salt in moisture levels, low final pH of cheese (<pH 5), high-moisture cheese and high-ripening temperatures will all promote the development of bitter flavors.
(7) High salt in moisture levels slows the breakdown of bitter peptides by lactic acid bacteria but has no effect on the free aminopeptidase enzymes.

Activity of both proteases and aminopeptidases will increase with higher moisture contents or ripening temperatures—if endoprotease activities predominate, then bitterness will result.

Bitterness in Low-Fat Cheese

Bitterness is one of the three major defects in low-fat cheeses and along with rubbery texture and lack of flavor is one of the main criteria that limit sales of low-fat varieties. Bitterness in low-fat increases as the cheese ages and also during shelf storage. Bitterness is particularly a problem in low fat for several reasons:

(1) Higher protein content—up to 20% more
(2) Curd formation at higher pH increases retention of rennet.
(3) Increased proteolysis at higher moisture levels
(4) Increased proteolysis at low salt in moisture levels
(5) Fat helps to suppress or round off bitter and other harsh flavors—removing the fat reduces this effect.
(6) Casein fat matrix is different in low-fat cheese than full-fat cheese. This can affect the course of the hydrolysis.

Preventing Bitterness in Cheese

Bitterness develops when casein proteolysis is out of balance and bitter peptides accumulate. Bitterness can be minimized by careful selection of starter cultures and control of the other parameters of cheese making. However, other criteria are often of greater importance when selecting starters, such as fast and consistent acid production and phage resistance. The ability to add an additional source of aminopeptidase to restore the balance of proteolysis without having to interfere with the rest of the cheese make conditions is an attractive alternative.

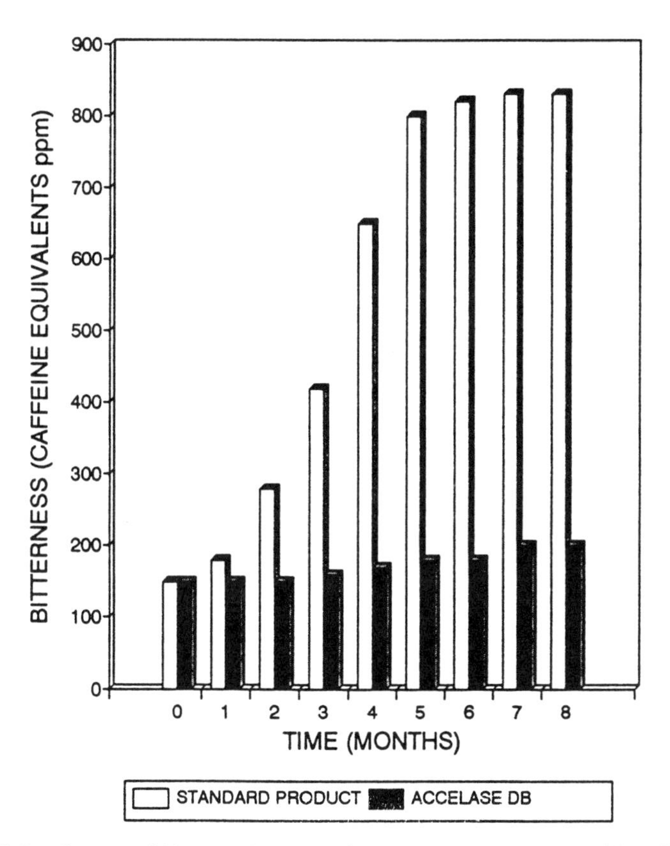

Figure 9 Development of bitterness in reduced fat cheddar and control by addition of *L. lactis* aminopeptidase (Accelase DB). Fifteen percent fat cheddar cheese, matured at 50°F, moisture 48%.

Aminopeptidase preparations derived from lactic acid bacteria can be added to cheese to restore the balance of aminopeptidase to endoprotease activity and therefore delay or prevent the onset of bitterness. Accelase® DB is a peptidase preparation containing predominantly leucine aminopeptidase (Pep N and Pep C) and prolyl dipeptidyl peptidase. The extent of peptidase action is influenced by the dose rate, pH, moisture content and storage temperature of the cheese. Ultimately, peptidase action is controlled by the level of polypeptides in the cheese, which have been produced by the action of endoproteases. These are the aminopeptidases' substrate and therefore rate limiting. Too little casein breakdown will produce bland-flavored cheese. Similarly, the accumulation of very high levels of bitter peptides will make the final flavor difficult to control without giving rise to savory notes.

Figure 9 illustrates the effect of adding additional peptidase (Accelase® DB) to control the bitterness in a reduced fat curd. Aminopeptidase can be added to cheese to accelerate the flavor development, here the aim was to add sufficient peptidase to "balance" the system and control bitterness. A

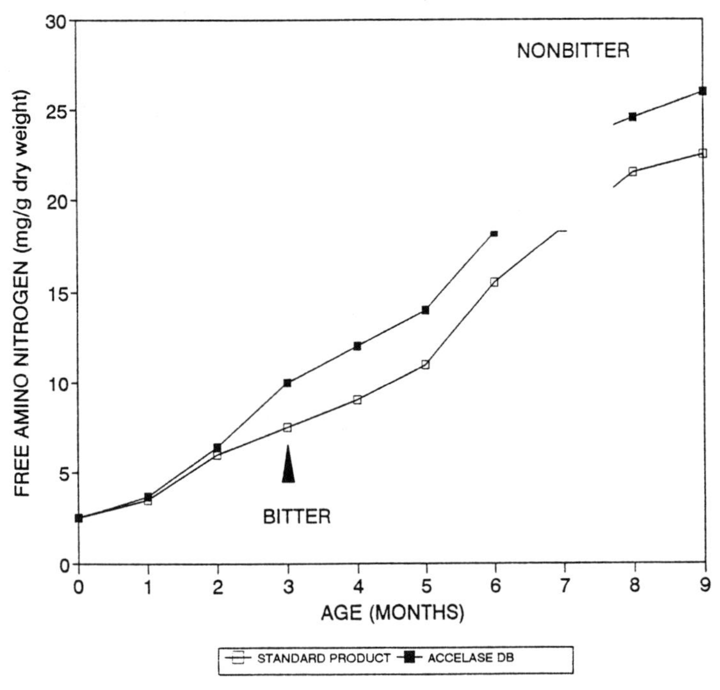

Figure 10 Free amino nitrogen levels in control (bitter) and peptidase treated (nonbitter) cheeses.

defined system for producing bitter cheese was utilized. Reduced-fat curd is manufactured using an increased level of microbial rennet (× 1-1/2 the standard dose of *Mucor miehei* protease), and a single strain of *L. lactis* with low peptidase levels; cheese was stored at 10°C (50°F) and graded at monthly intervals. To obtain a more qualitative indication of bitterness, a trained sensory panel was asked to compare water-soluble extracts of the cheeses to a series of caffeine standards. Protein breakdown was also monitored at monthly intervals (sulphosalicylic acid-soluble nitrogen). The results are presented in Figures 6 and 7 and show that in the control cheese bitterness was evident at three months, whereas in the peptidase-treated cheese no bitterness was present even at the end of the shelf life (eight months). At the end of the trial, the cheese with added peptidase had 15% higher levels of free amino nitrogen than the control cheese (Figure 10). The cheddar flavor intensity was judged to be similar in the control and trial cheeses.

SUMMARY

Bitterness remains a problem in protein hydrolyzates and is a too common defect in fermented products such as cheese. The understanding of the structure of bitter peptides and how they react with taste receptors has increased significantly and helps us interpret the action of exopeptidases in preventing bitterness in these products. Peptidases from lactic acid bacteria and *Aspergillus* are now commercially available and present the industry with enzymatic solutions to bitterness.

REFERENCES

Adler-Nissen, J. 1984. Control of the proteolytic reaction and the level of bitterness in protein hydrolysis processes. *J. Chem. Technol. Biotechnol., 34B,* pp. 215–222.

Adler-Nissen, J. 1986. Relationship and structure to taste of peptides and peptide mixtures. In *Protein Tailoring for Food and Medical Uses,* Feeney, R. E. and Whitaker, J. R., eds., Marcel-Dekker, New York, pp. 97–121.

Crawford, F. J. M. 1977. Introduction to discussion on bitterness in cheese. *International Dairy Federation Bulletin, 97,* pp. 1–10.

Cuthbertson, D. P. 1950. Amino-acids and protein hydrolysates in human and animal nutrition. *J. Sci. Food Agric., 1,* pp. 35–41.

Guiqoz, Y. and Solms, J. 1976. Bitter peptides, occurrence and structure. *Chem. Senses Flavour, 2,* pp. 71–84.

Ishibashi, N., Arita, Y., Kanehisa, H., Kouge, K., Okai, H. and Fukui, S. 1987. Bitterness of leucine-containing peptides. *Agric. Biol. Chem., 51,* pp. 2389–2394.

Ishibashi, N., Kubo, T., Chino, M., Fukui, H., Skinoda, I., Kikuchi, E., Okai, H.

and Fukus, S. 1988a. Taste of proline containing peptides. *Agric. Biol. Chem.*, *52*, pp. 95–98.

Ishibashi, N., Louge, K., Shinoda, I., Kanehisa, H. and Okai, H. 1988b. A mechanism for bitter taste sensibility in peptides. *Agric. Biol. Chem.*, *52*, pp. 819–827.

Lemieux, L. and Simard, R. E. 1991. Bitter flavour in dairy products. I. A review of the factors likely to influence its development, mainly in cheese manufacture. *Lait*, *71*, pp. 599–636.

Lemieux, L. and Simard, R. E. 1992. Bitter flavor in dairy products. II. A review of bitter peptides from caseins: their formation, isolation and identification, structure masking and inhibition. *Lait*, *72*, pp. 335–382.

Lowrie, R. J. and Lawrence, R. C. 1972. Cheddar cheese flavour. IV. A new hypothesis to account for the development of bitterness. *N.Z. J. Dairy Sci. Tech.*, *7*.

Matoba, T. and Hata, T. 1972. Relationship between bitterness of peptides and their chemical structures. *Agric. Biol. Chem.*, *36*, pp. 1423–1431.

Murray, T. K. and Baker, B. E. 1952. Studies on protein hydrolysis I—preliminary observations on the taste of enzymic protein hydrolysates. *J. Sci. Food. Agric.*, *3*, pp. 470–475.

Nakadai, T., Nasuno, S. and Iguchi, N. 1973. Purification and properties of leucine aminopeptidase from *Aspergillus oryzae*. *Agr. Biol. Chem.*, *37*(4), pp. 757–765, 767–774, 775–782.

Ney, K. H. 1971. Voraussage der Bitterheit von Peptidon aus deren Aminosoure-Zusammensetzung. *Z. Lebensm. Unters. Forsch.*, *147*, pp. 64–71.

Ney, K. H. 1979. Bitterness of peptides: amino acid composition and chain length. In *Food Taste Chemistry, ACS Series 115*, Boudreau, J. C., ed., Washington, pp. 149–173.

Ney, K. H. and Retzlaff, G. 1986. A computer program predicting the bitterness of peptides, esp. in protein hydrolysates. Based on amino acid composition and chain length (computer Q). In *The Shelf Life of Foods and Beverages, Proceedings of the 4th Int. Flavour Conference*, Charalambous, G., ed., Elsevier, Holland, pp. 543–550.

Pawlett, D. and Fullbrook, P. 1988. *Proc. Food Ingredients Europe, 3rd Int. Conf.*, London, Nov. 15–17.

Pawlett, D. and Bruce, G. 1996. The application of enzymes in industry. *Industrial Enzymology, 2*, Godfrey, T. and West, S., eds., Macmillan, UK (in press).

Pritchard, G. G. and Coolbear, T. 1993. The physiology and biochemistry of the proteolytic system in lactic acid bacteria. *FEMS Microbiology Reviews*, *12*, pp. 179–206.

Richardson, B. C. and Creamer, L. K. 1973. Casein proteolysis and bitter peptides in cheddar cheese. *N. Z. Journal Dairy Sci.*, pp. 46–51.

Shallenberger, R. S. 1993. Amino acids, peptides and proteins. In *Taste Chemistry*, Shallenberger, R. S., ed., Blackie Academics and Professional, Glasgow, UK, pp. 213–252.

Shinoda, I., Nosho, Y., Otagiri, K., Okai, H. and Fukue, S. 1986. Bitterness of diastereometers of a hexapeptide. *Agric. Biol. Chem.*, *50*, pp. 1785–1790.

Shiraishi, H., Okuda, K., Yamaoka, N. and Tuzimura, K. 1973. Taste of proline containing peptides. *Agric. Biol. Chem.*, *37*, pp. 2427–2428.

Tan, P. S. T., Poolman, B. and Konings, W. N. 1993. Proteolytic enzymes of *Lactococcus lactis. Journal of Dairy Research, 60,* pp. 269–286.

Tanford, C. 1962. Contribution of hydrophobic interactions to the stability of the globular conformation of proteins. *J. Am. Chem. Soc., 84,* pp. 4240–4244.

Taylor, C., Pawlett, D. and Brett, A. 1991. Natural meat flavours. In *FIE Conference Proceedings,* Expoconsult, Netherlands, pp. 207–210.

Zevaco, C., Monnet, V. and Gripon, J. C. 1990. Intracellular x-prolyl dipeptidyl peptidase from *L. lactis* ssp. *lactis*. Purification and Properties, *J. of App. Bacteriol., 68,* pp. 357–366.

CHAPTER 12

Specific Inhibitor for Bitter Taste

YOSHIHISA KATSURAGI[1]
KENZO KURIHARA[2]

INTRODUCTION

THERE are many drugs that elicit a bitter taste (Shallenberger and Acree, 1971). Several methods of masking bitterness such as coating of tablets with sugar or polymers (Fukomori et al., 1988; Fu Lu et al., 1991; Ueda et al., 1993) or chemical modification of drugs into insoluble derivatives (Bechtol et al., 1981) have been used, but are insufficient to mask the bitterness of many drugs. Hence the development of a method to mask bitterness is widely required in pharmaceutical sciences. Masking of bitterness is also required in food sciences.

To mask bitterness, a specific bitterness inhibitor would be most useful. Such an inhibitor would also be useful in elucidating the receptor mechanisms of bitter substances. No inhibitor has, however, been available. Bitter substances are commonly hydrophobic (Koyama and Kurihara, 1972; Kumazawa et al., 1986, 1988) and, hence, there may be hydrophobic substances that mask the target sites for bitter substances. Phospholipids, which are typical hydrophobic substances, may mask the target sites, but are not soluble in water. Recently, we found that lipoproteins, PA-LG made of phosphatidic acid (PA) and β-lactoglobulin (LG), which are easily dispersed in water, suppressed the taste responses to the bitter substances (Katsuragi and Kurihara, 1993).

The first part of this chapter deals with the effects of complexes made of various combinations of lipids and proteins on the frog taste nerve responses to various gustatory stimuli (Katsuragi et al., in press). We found

[1]Kao Corporation, Food Products Research Laboratories, Kashima 314-02, Japan.
[2]Faculty of Pharmaceutical Sciences, Hokkaido University, Sapporo 060, Japan.

that PA is an essential component of the bitterness inhibitor. Here we show that PA-LG inhibits the responses to all the bitter substances examined except for the hydrophilic bitter substance $MgCl_2$, without affecting the responses to other taste stimuli. In the second part of the chapter, we show that PA-LG among various complexes between lipids and proteins is mostly strongly adsorbed on the frog tongue surface and hydrophobic model membranes (Katsuragi et al., in press). This suggests that PA-LG is adsorbed on the hydrophobic region of gustatory receptor membranes and masks receptor sites for bitter substances without affecting those for other taste stimuli similar to the frog. The third part of this chapter shows the effects of PA-LG on taste sensation to various stimuli in humans (Katsuragi et al., 1995). The results show that PA-LG inhibits bitterness of various substances without affecting the taste sensation to other taste stimuli. The fourth part of this chapter presents basic studies for practical application of the inhibitor to drugs and foods. Here PA alone is used as an inhibitor. PA does not always inhibit the responses to all the bitter substances, but it is much cheaper than PA-LG and hence is a useful inhibitor for practical application. Both PA and LG originate from foods; PA and LG come from soybean and milk, respectively. Hence PA-LG and PA can be used to safely mask bitterness of foods and drugs.

INHIBITION OF FROG TASTE NERVE RESPONSES TO BITTER SUBSTANCES BY THE LIPOPROTEIN

PREPARATION OF LIPID-PROTEIN COMPLEXES

Several methods of preparing phospholipid-protein complexes have been published (Brown et al., 1983; Kanamoto et al., 1977; Mine et al., 1992; Mizutani and Nakamura, 1987; Ohtsuru et al., 1976) in which a mixture of phospholipid and protein was sonicated. We found that lyophilization instead of sonication also causes the formation of phospholipid-protein complexes. PA-LG was prepared as follows. Two grams of PA and 5 g of LG were suspended in 50 ml of water. The suspension was homogenized with a Polytron (Kinematica GmbH, Littau, Switzerland) and the resultant homogenate was freeze-dried. Extraction of the lyophilized powder with *n*-hexane, which is a good solvent for PA, transferred only 5% of the PA to the solvent, suggesting that the PA in the powder is mostly bound to LG. The powder was easily dispersed in water, which gave a translucent solution with only a slight turbidity. The pH of the solution was 6.5. Elution profile of PA-LG solution from a Sephacryl column showed a single peak at molecular weight of 500,000–1,000,000.

Complexes of PA with other proteins and those of phosphatidylcholine

(PC), triacylglycerol (TG) or diacylglycerol (DG) with LG were prepared by a method similar to that employed for preparing PA-LG.

MEASUREMENT OF FROG TASTE NERVE RESPONSES

The effects of PA-LG on the taste responses were examined by measuring the activities of the glossopharyngeal nerve (Kashiwagura et al., 1980; Yoshii et al., 1981). Briefly, adult bullfrogs, *Rana catesbeiana,* weighing 200–300 g were anesthetized with an intraperitoneal injection of urethane solution; then the glossopharyngeal nerve was dissected from the surrounding tissues and cut proximally. The nerve impulses were amplified with an AC amplifier and integrated with an electronic integrator. The peak height of the integrated response is taken as the magnitude of the response. The tongue was pretreated with PA-LG solution containing no bitter substances for 10 min and subsequently stimulated with bitter substances dissolved in water. That is, PA-LG was applied separately before each stimulus. In this method, the bitter substances do not directly contact with PA-LG, hence the possibility that suppression of the responses to bitter substances is brought about by binding of the substances to PA-LG in the medium can be eliminated.

EFFECTS OF PA-LG ON FROG TASTE NERVE RESPONSES

Figure 1 shows the effects of PA-LG on the integrated responses of the frog glossopharyngeal nerve to various stimuli. The tongue was pretreated with 0.3% PA-LG for 10 min. The PA-LG itself elicited no response in the nerve. Then the tongue was stimulated by various chemicals dissolved in water. The responses to bitter substances such as 0.1 mM quinine hydrochloride, 1 mM papaverine hydrochloride and 100 mM L-leucine were greatly suppressed by PA-LG. On the other hand, the responses to 200 mM NaCl, 200 mM L-alanine, 500 mM galactose and acetic acid (pH 3.0) were not practically affected.

The third record for quinine shows that the suppressive effect of PA-LG is completely reversible. That is, after the tongue was treated with PA-LG for 10 min, the tongue was washed with Ringer solution for 1 min and then subjected to stimulation by quinine. The stimulation brought about a similar magnitude of the response to that of the first record for quinine. Similar recovery was observed with the responses to other bitter substances.

Figure 2 shows the magnitudes of the responses to 0.1 mM quinine hydrochloride and 1 mM papaverine hydrochloride as a function of the period exposing the tongue with PA-LG where the magnitude of the response before treatment is taken as unity. After exposure to 0.3% PA-LG, the response to quinine hydrochloride is greatly reduced within 5 min and

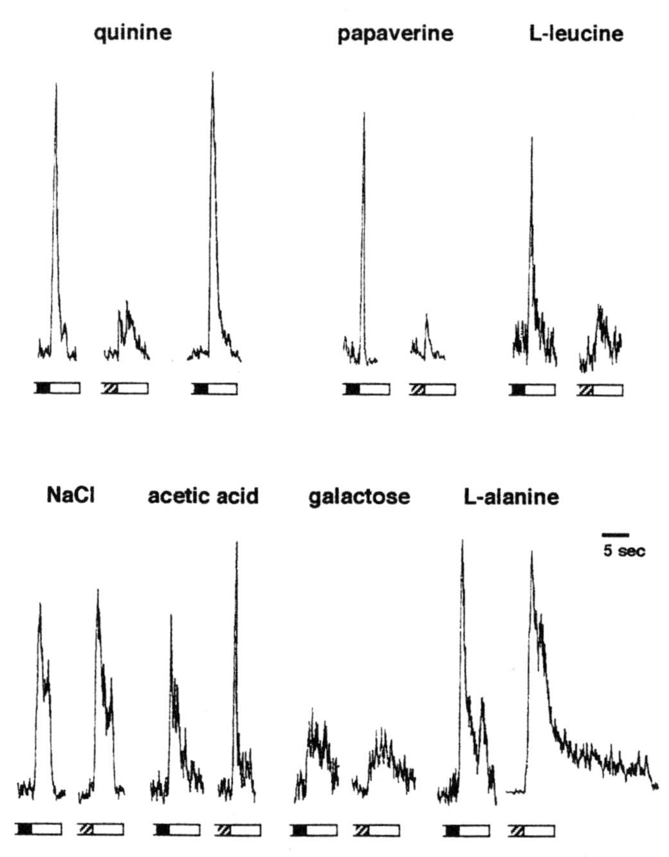

Figure 1 Typical integrated responses of the frog glossopharyngeal nerve to various stimuli before and after the tongue was pretreated with 0.3% PA-LG for 10 min. The tongue was adapted to water (closed bar) or 0.3% PA-LG (shaded bar) and stimulated with chemicals (open bar) dissolved in water. The third pattern for quinine hydrochloride was recorded as follows. The tongue was pretreated with 0.3% PA-LG for 10 min, washed with Ringer solution for 1 min and then subjected to stimulation by quinine hydrochloride. The concetrations used are 0.1 mM for quinine hydrochloride, 1 mM for papaverine hydrochloride, 100 mM for L-leucine, 200 mM for NaCl, pH 3.0 for acetic acid, 500 mM galactose and 200 mM L-alanine.

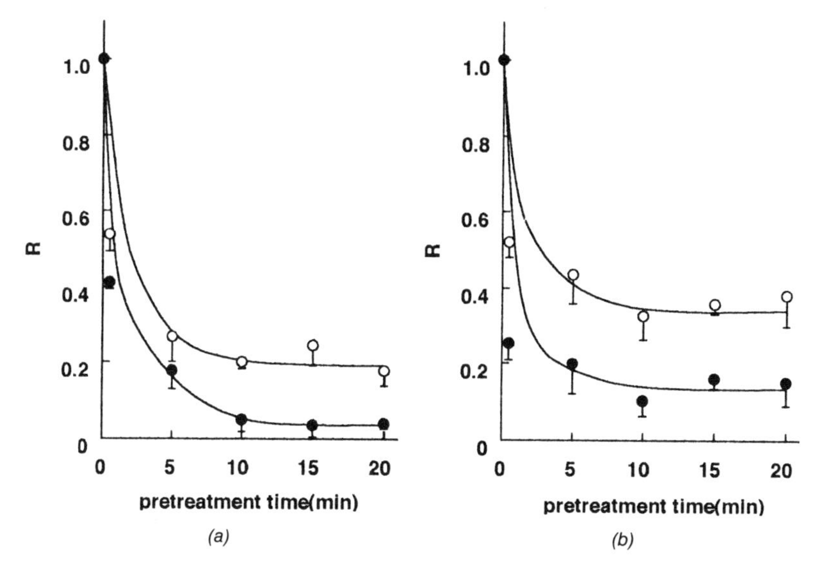

Figure 2 Relative magnitudes of the responses (R) to 0.1 mM quinine hydrochloride (a) and 1 mM papaverine hydrochloride (b) as a function of time for treatment with PA-LG. The tongue was adapted to 0.3% (○) or 1% (●) PA-LG solutions for different period and subsequently stimulated with a stimulus solution containing no PA-LG. The magnitude of the response to each substance before treatment is taken as unity.

reaches a steady level at 10 min. The pretreatment with 1% PA-LG reduces the response to a lower level than that for 0.3% PA-LG. Similar results were observed with the response to papaverine hydrochloride. That the magnitudes of the responses to the bitter substances decreased with an increase in the period of PA-LG exposure suggests that the binding of PA-LG to the target sites on the receptor membranes contributes to the suppression of the bitter responses.

In Table 1, the relative magnitude of the responses to various stimuli after treating the tongue with 0.3% PA-LG for 10 min is shown. Here the magnitude of the response to each stimulus before PA-LG exposure is taken as unity. As seen from the table, the responses to the bitter substances (quinine hydrochloride, strychnine nitrate, papaverine hydrochloride and caffeine) were greatly suppressed by PA-LG. On the other hand, the responses to the hydrophilic bitter substances such as CsCl, $MgCl_2$ and tetraethylammonium chloride (TEACl) were unchanged or increased by PA-LG, suggesting that the receptor mechanism of this type of bitter substances is different from that of the former bitter substances. It is known that there are multiple transduction pathways in reception for bitter substances, such as receptor-mediated pathways including the inositol-1,4,5-trisphosphate ($InsP_3$) pathway (Hwang et al.,

TABLE 1. Effects of Complexes of Various Lipids with LG and Components of PA-LG on the Frog Glossopharyngeal Nerve Responses to Various Stimuli.

	PA-LG	PS-LG	PA-LA	PA	LG
0.1 mM Quinine	0.204 ± 0.017	0.549 ± 0.096	0.252 ± 0.060	0.551 ± 0.069	1.000 ± 0.011
1 mM Strychnine	0.209 ± 0.030	0.803 ± 0.139	0.333 ± 0.055	0.534 ± 0.034	1.163 ± 0.089
1 mM Papaverine	0.328 ± 0.055	0.539 ± 0.114	0.539 ± 0.114	1.050 ± 0.041	0.906 ± 0.011
10 mM Caffeine	0.263 ± 0.034	0.263 ± 0.067	0.263 ± 0.067	0.624 ± 0.089	1.130 ± 0.106
200 mM NaCl	1.050 ± 0.122				
pH 3 Acetic acid	1.274 ± 0.213				
200 mM L-Alanine	0.895 ± 0.126				
500 mM Galactose	0.799 ± 0.180				
100 mM CsCl	1.216 ± 0.088				
100 mM $MgCl_2$	1.685 ± 0.202				
10 mM TEACl	1.003 ± 0.032				

The values in the table represent relative magnitudes of the responses to stimuli where the magnitude of a control response to each stimulus was taken as unity. The tongue was pretreated with the lipid-protein complexes, PA vesicle, or LG for 10 min and stimulated by a substance dissolved in water. The concentration of the complexes was 0.3%. The concentrations of PA vesicle and LG were 0.085 and 0.215%, respectively, which were used to prepare 0.3% PA-LG.

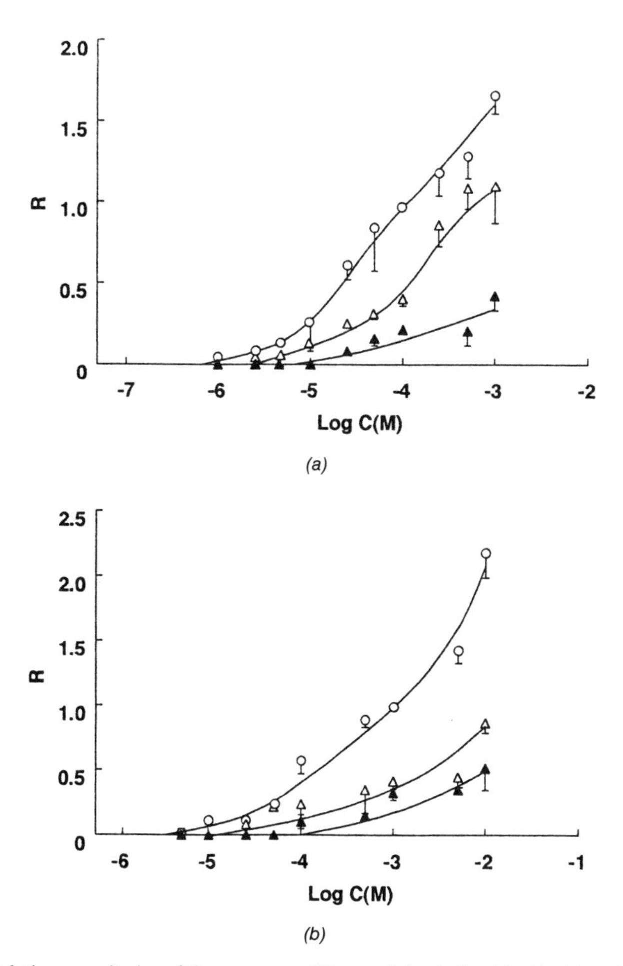

Figure 3 Relative magnitudes of the responses (R) to quinine hydrochloride (a) and papaverine hydrochloride (b) as a function of their concentrations before (○) and after the tongue was treated with 0.1% (△) and 0.3% (▲) of PA-LG for 10 min. The tongue was stimulated by the bitter substances dissolved in water. The magnitudes of the responses to 0.1 mM quinine hydrochloride and 1 mM papaverine hydrochloride are taken as unity, respectively.

1990), a pathway via phosphodiesterase (Kolesnikov and Margolskee, 1995; Kurihara, 1972; Margolskee, 1995; Price, 1973; Ruiz-Avila et al., 1995) and a nonreceptor-mediated pathway (Kumazawa et al., 1985, 1988), including direct activation of G protein (Naim et al., 1994) and direct block of a resting K-conductance (Cummings and Kinnamon, 1992; Spielman et al., 1989). The present results support the notion that the hydrophilic bitter substances induce the responses by the mechanism different from that for other bitter substances.

Figure 3 shows the magnitude of the responses to quinine hydrochloride and papaverine hydrochloride as a function of their concentrations before and after the tongue was treated with 0.1 and 0.3% PA-LG. If the inhibition of the responses to the bitter substances is of a competitive type, the dose-response curve should shift to the right with an increase in PA-LG concentration. The responses to the bitter substances do not, however, shift to the right, but are suppressed at all concentrations examined. The results suggest that the responses are inhibited by PA-LG in a noncompetitive manner.

Figure 4 shows the relative magnitudes of the responses to various bitter substances as a function of PA-LG concentration. The responses to quinine hydrochloride, L-leucine, caffeine and papaverine hydrochloride decreased with an increase in the concentration (%) of PA-LG and reached a zero

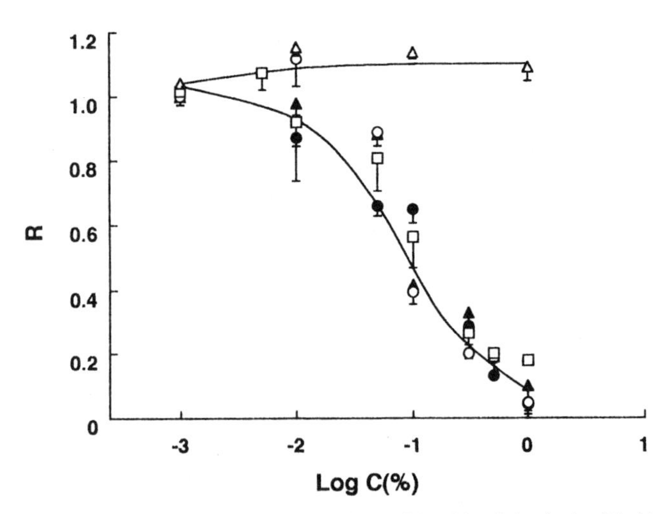

Figure 4 Relative magnitudes of the responses (R) to 0.1 mM quinine hydrochloride (○), 100 mM L-leucine (●), 1 mM papaverine hydrochloride (▲), 10 mM caffeine (□) and 200 mM NaCl (△) as a function of PA-LG concentrations. The tongue was treated with PA-LG of varying concentrations for 10 min and subsequently stimulated with a solution containing a bitter substance and no PA-LG. The magnitude of the response to each stimulus before the treatment is taken as unity.

level at 1%. Thus, PA-LG at high concentrations such as 1% completely suppressed the bitter responses by full occupation of the target sites with PA-LG. The data obtained with four bitter substances roughly follow a single curve. It is known that there are multiple receptor sites for bitter substances (Hall et al., 1975; Harder et al., 1984) and the results in Figure 4 do not always indicate that the receptor site for bitter substances is single. The results may show that the receptor sites for various bitter substances are hydrophobic in general, and adsorption of PA-LG to the hydrophobic region of taste receptor membranes uniformly masks all the receptor sites for hydrophobic bitter substances.

EFFECTS OF VARIOUS LIPID-PROTEIN COMPLEXES ON FROG TASTE NERVE RESPONSE

The effects of lipoprotein made of LG and other lipids such as PS, PC, TG and DG on the responses to the bitter substances were examined. It is noted that all the lipoproteins themselves elicited no response in the glossopharyngeal nerve. PS-LG shows suppressive effects on the response to the bitter substances, although its effects on the responses to quinine hydrochloride, strychnine nitrate and papaverine hydrochloride are much less than those of PA-LG (Table 1). Other complexes (PC-LG, TG-LG and DG-LG) do not practically suppress the responses to the bitter substances (data not shown). In order to examine the suppressive effect of lipoprotein composed of other proteins than LG, we prepared the lipoprotein composed of α-lactalbumin (LA) and PA. As shown in Table 1, PA-LA had a suppressive effect on the responses to the bitter substances similarly to that of PA-LG, suggesting that LG is not essential for suppression of bitter responses.

Next, we examined the effects of each component of PA-LG on the responses to the bitter substances. As shown in Table 1, LG itself had no suppressive effect on the responses to the bitter substances. PA vesicle suppressed the response to quinine hydrochloride, strychnine nitrate and caffeine, although its suppressive effect is much weaker than PA-LG. PA vesicle had no effect on the response to papaverine hydrochloride.

HIGH ADSORPTION ABILITY OF PHOSPHATIDIC ACID-CONTAINING LIPOPROTEINS TO FROG TONGUE SURFACE AND HYDROPHOBIC MODEL MEMBRANES

In a previous section, we showed that the lipoproteins containing PA were most effective at inhibiting bitterness and that those containing PS were moderately effective. The lipoproteins composed of other lipids showed practically no inhibitory activity. On the other hand, species of proteins was not essential for the inhibition; lipoproteins composed of PA and any protein examined showed high inhibitory activity similar to PA-LG.

In general, bitter substances are hydrophobic (Koyama and Kurihara, 1972; Kumazawa et al., 1986, 1988); hence their receptor sites seem to be hydrophobic. It is possible that lipoproteins such as PA-LG have a unique property to bind to hydrophobic sites on the taste receptor membranes. To test this possibility, we carried out two types of experiments. In the first experiment, we found that the lipoproteins having the inhibitory activity are adsorbed on the frog tongue surface while those having no inhibitory activity are not absorbed. In the second experiment, we examined adsorption of the lipoproteins on model lipid membranes coated on quartz crystal microbalance (QCM) by measuring changes in its frequency. It was demonstrated that the lipoproteins having inhibitory activity are strongly adsorbed on the hydrohobic lipid membranes, while those having no inhibitory activity are not practically adsorbed on the membranes. It was concluded that PA-containing lipoproteins selectively inhibit bitterness due to their unique affinity to the hydrophobic sites on the taste receptor membranes.

ADSORPTION OF LIPOPROTEINS ON FROG TONGUE SURFACE

A fluorescent derivative of PA (NBD-PA) and of PC (NBD-PC) were used for preparation of fluorescent-labeled lipoproteins. Twelve-AS was also used for preparation of fluorescent-labeled lipoproteins containing PS, DG and TG. The frog tongue was treated with protein-lipid complexes composed of LG and fluorescent-labeled lipids for 10 min and washed with Ringer solution three times for 10 sec. The tongue was fixed with formaldehyde and sectioned into thin slices.

A slice labeled with NBD-PA-LG was observed under a fluorescent microscope. The surface of the fungiform papillae, which bears the taste bud, is coated with the fluorescent lipoprotein. The surface of the papillae is rather densely coated with fluorescence, but the surface of the other area is also coated. We also used a lipoprotein composed of LG and PA containing 12-AS and confirmed that the tongue surface is coated with the lipoprotein similarly to the case of a lipoprotein composed of NBD-PA and LG. The tongue surface was also coated with PS-LG, but the fluorescence on the tongue surface is much weaker than with PA-LG.

No fluorescence on the tongue surface was observed with the complex composed of LG and PC. Nor was fluorescence observed with the complexes composed of LG and DG or TG. As shown in the previous section, PA-LG showed strong inhibitory activity of bitterness and PS-LG showed moderate activity. Other protein-lipid complexes showed practically no inhibitory activity. Hence the adsorption ability of the protein-lipid complexes on the tongue surface was closely related to their inhibitory activities.

PA-LG is rather densely adsorbed on the tongue surface. There is a possibility that a bitter substance is bound to PA-LG adsorbed on the tongue surface and consequently a concentration of the stimulus is decreased. We tested the possibility using quinine hydrochloride, a typical bitter substance. Ten milliliters of 0.1 mM quinine hydrochloride solution were applied to the tongue surface heated with 0.3% PA-LG solution for 10 min, which is the same condition as for usual chemical solution. Then the quinine hydrochloride concentration in the stimulating solution flowing out from the tongue surface was measured. The concentration was 99.0% (mean value of the data obtained with three measurements) when quinine hydrochloride concentration before application to the tongue was taken as 100%. This result shows that quinine hydrochloride does not bind to PA-LG bound to the tongue surface at all or a decrease in quinine hydrochloride concentration is negligibly small even if quinine hydrochloride is adsorbed to PA-LG bound to the tongue surface.

ADSORPTION OF LIPOPROTEINS ON THE LIPID-COATED QUARTZ CRYSTAL MICROBALANCE

It is known that the frequency changes of QCM are related to changes in mass. Hence the amount of a substance adsorbed on QCM can be estimated by measuring the frequency changes (Okahata et al., 1990, 1991). Figure 5(a) shows the frequency changes in DSPC-coated QCM in response to addition of 0.005% PA-LG at an open arrow. The frequency of the QCM decreased immediately and reached the equilibrium state within 5 min. Addition of deionized water at a closed arrow returned the frequency to the original level, suggesting that the adsorption of PA-LG on the DSPC membrane is completely reversible. Figure 5(b) shows the frequency changes of DSPC-coated QCM in response to addition of 0.005% PC-LG. Addition of PC-LG induced little change in the frequency.

In Figure 6, which shows the changes in frequency by the addition of various protein-lipid complexes, PA and LG are compared with their inhibitory activities of the responses to bitter substances (papaverine hydrochloride, quinine hydrochloride and strychnine nitrate). As seen from the figure, there is a close relationship between the inhibitory activities and the frequency changes.

Since PA vesicle itself induced a relatively large change in the frequency, the high affinity of PA-LG to the DSPC membrane seems to come mainly from the property of PA. Both PA and PS have a negative charge, hence one may consider that this negative charge contributes to adsorption of the membranes. On the other hand, DSPC has a positive charge of choline residue and a negative charge of phosphoric acid and a net charge of DSPC at neutral pH is zero. Nevertheless, PA-LG was strongly adsorbed on DSPC, suggesting that electrostatic interaction between PA-LG and the membranes does not mainly contribute to the adsorption. Probably hydrophobic interaction between DSPC and PA-LG contributes mainly to the adsorption of PA-LG on DSPC. To test this possibility, we used QCM coated with polystyrene, which has no charge and was hydrophobic. Among various protein-lipid complexes, PA-LG induced the largest frequency change. This implies that hydrophobic adsorption was largest in PA-LG.

One may consider that the adsorption ability of the lipid-LG complexes or PA vesicle to the tongue surface and the model membranes is related to their particle sizes. PA-LG, which has a high adsorption ability, is easily dispersed in water and gives a transparent solution, while PC-LG, having only a weak adsorption ability, gives turbid suspension. The average particle size of PC-LG was 3.5 mm. That of PA vesicles, which has a much higher adsorption ability than PC-LG, was 14.1 mm. These results suggest that the particle size is not a primary factor for the adsorption ability,

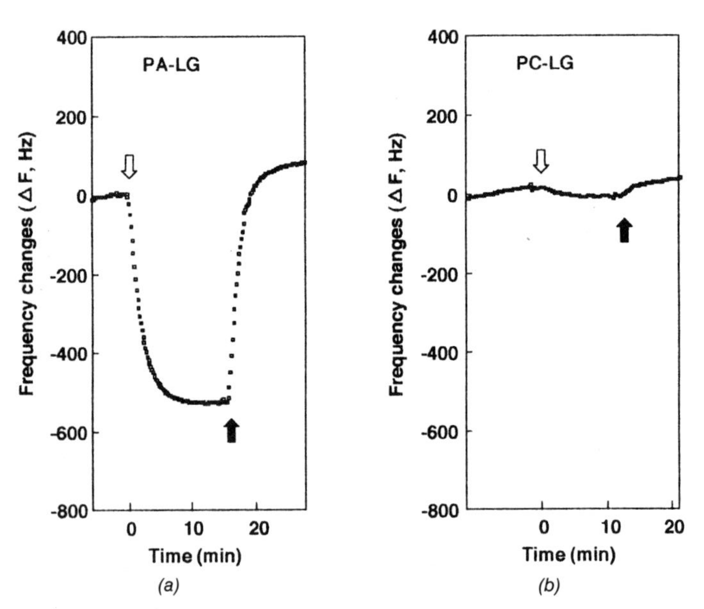

Figure 5 Frequency changes of DSPC-coated QCM by addition of 0.005% PA-LG (a) and 0.005% PC-LG (b). The lipoproteins were added at open arrow and deionized water was added at closed arrow.

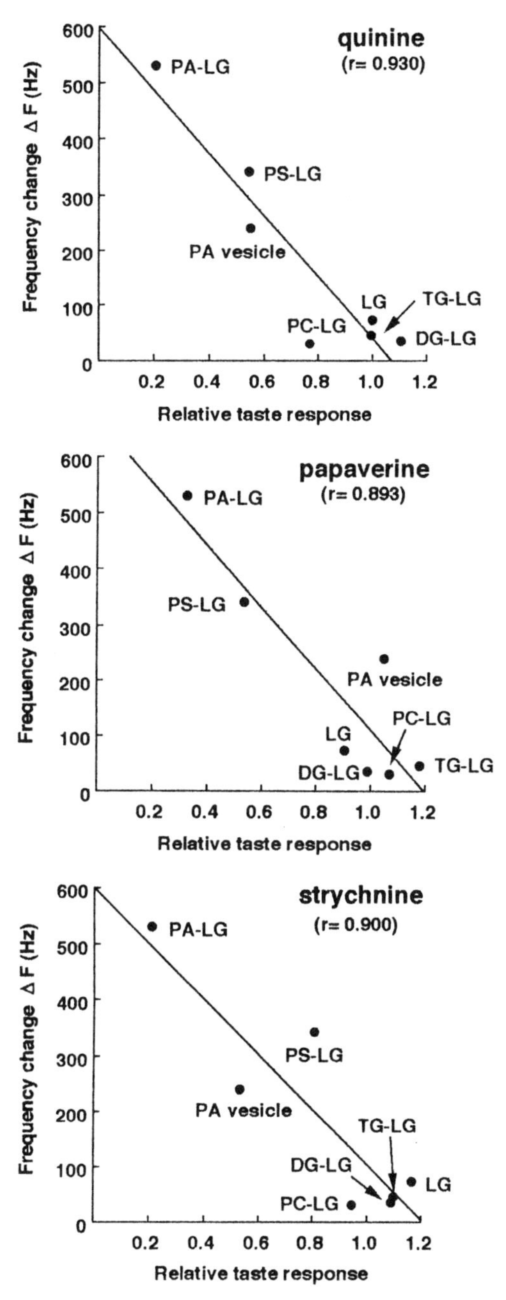

Figure 6 Relation between frequency changes of DSPC-coated QCM by adsorption of various lipid-LG complexes and PA vesicle and relative magnitude of frog taste nerve responses to quinine hydrochloride, papaverine hydrochloride and strychnine nitrate after the tongue was treated with various lipid-LG complexes and PA vesicle. The magnitudes of the responses to the bitter substances are relative when a magnitude of the control response to each bitter substance is taken as unity.

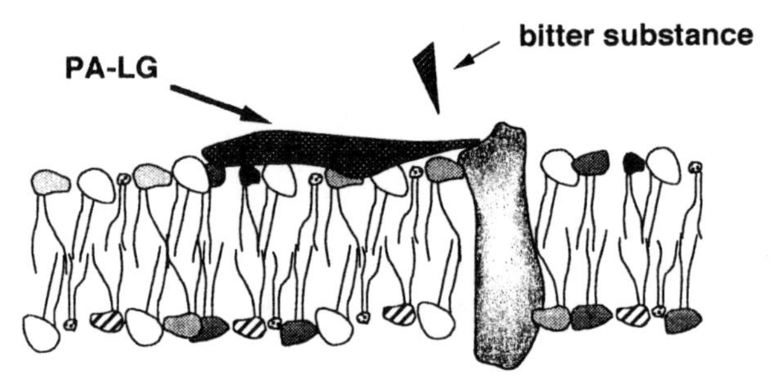

Figure 7 Schematic illustration for a mechanism of the response to a bitter substance.

although the difference in the adsorption ability between PA-LG and PA vesicle may come from their particle size.

Not only PA, but also PC, TG and DG are hydrophobic. But the complexes of the latter lipids with LG did not greatly adsorb on DSPC. Hence the hydrophobicity of the lipid is an essential factor for the adsorption, but not a sufficient factor. It seems that a unique unknown property of PA, in addition to its hydrophobicity, contributes to the high affinity for hydrophobic membranes such as DSPC.

MECHANISM OF INHIBITION OF BITTERNESS OF PA-LG

The finding that PA-LG was densely bound to both the tongue surface and DSPC suggests that PA-LG is adsorbed on hydrophobic sites of the tongue surface. It seems that the receptor sites for bitter substances are hydrophobic, hence PA-LG masks the receptor sites (Figure 7). Other tastants such as salts, acids and sugars are hydrophilic, hence their receptor sites also seem to be hydrophilic. Adsorption of PA-LG on the hydrophobic sites of the tongue surface will not mask the receptor sites for other stimuli than bitter substances. This mechanism explains the fact that the responses to the hydrophilic bitter substances were not suppressed by PA-LG. We emphasize that PA-LG nonselectivity binds to the hydrophobic region of the taste receptor membranes, which consequently brings about selective inhibition of the responses to the hydrophobic bitter substances.

SELECTIVE INHIBITION OF BITTER TASTE IN HUMANS BY PHOSPHATIDIC ACID-CONTAINING LIPOPROTEIN

In the present section, we show the effects of lipoproteins composed of various species of proteins and lipids on human taste sensation. The results show that lipoproteins composed of PA and the proteins inhibit bitter tastes of various substances without inhibiting the saltiness of NaCl or sweetness of sucrose and PA-containing lipoproteins can be used to inhibit bitterness in humans.

METHOD FOR EVALUATION OF EFFECTS OF LIPOPROTEINS ON HUMAN TASTE SENSATION

The effects of the lipoproteins on human taste sensation were psychophysically evaluated according to the method employed previously (Ugawa et al., 1992). Six to 12 subjects participated in each experiment.

The test solutions were prepared by dissolving chemicals in deionized water or solutions containing PA-LG of various concentrations. Standard solutions were prepared by dissolving quinine sulfate, NaCl and sucrose of different concentrations in deionized water.

About 5 ml of each solution was added to test tubes and presented at room temperature. The subjects were required to compare the taste intensity of a test solution with that of the standard solutions and to select a standard solution with a taste intensity equivalent to that of the given test solution. The subjects were instructed to scoop a teaspoonful of solution, to place the solution on the tongue, to taste it and to rinse their mouths throughly with deionized water after tasting each solution. The effects of lipoproteins on taste sensation were expressed by R (normalized concentration of standard solution, the taste intensity of which is equivalent to that of the test solution), defined as follows:

$$R = SR_i n_j / Sn_j \quad (1)$$

Here, R_i is the concentration of a normalized standard solution; $R_i = C_i / C_0$ where C_i is the concentration of the standard solution whose taste intensity is equivalent to that of the test solution and C_0 is the concentration of standard solution, the taste intensity of which is equivalent to that of the test solution containing no lipoprotein. n_j is the number of subjects.

EFFECTS OF LIPOPROTEINS ON TASTE SENSATION TO VARIOUS TASTE STIMULI

Figure 8 shows the effects of various lipoproteins on taste sensation to quinine hydrochloride as a function of their concentration. Relative bitterness is expressed by the normalized concentration (R) of the standard bitter solution, the bitterness of which is equivalent to the test solution. As the figure shows, the bitterness was greatly decreased with an increase in PA-LG concentration. On the other hand, lipoproteins of LG with other

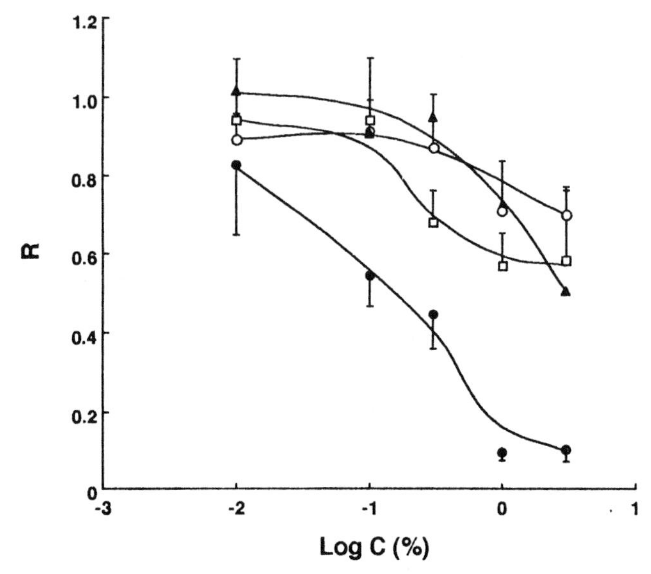

Figure 8 Effects of various lipoproteins on the bitter taste of 0.5 mM quinine hydrochloride as a function of their concentration. The relative taste sensation (R) was defined in Equation (1); ○, TG-LG; ▲, DG-LG; □, PC-LG; ●, PA-LG.

lipids only partly suppressed the bitterness. Thus, PA-LG was the most effective in suppressing bitterness of quinine hydrochloride among the lipoproteins examined. This is consistent with the results observed with the frog tongue.

Figure 9 shows the effects of PA-LG on the bitter taste of 5 mM promethazine hydrochloride, 10 mM propranolol hydrochloride and 50 mM caffeine, salty taste of 400 mM NaCl and sweet taste of 600 mM sucrose as a function of PA-LG concentrations. The bitterness of promethazine hydrochloride and propranolol hydrochloride was decreased to nearly zero at 3.0%, while that of caffeine was decreased to the 0.42 level at 3.0%. The effects of PA-LG were completely reversible since washing of the tongue with water after tasting solutions containing PA-LG and the bitter substances immediately recovered the function to taste their bitterness. On the other hand, saltiness and sweetness were practically unchanged by PA-LG below 1.0% and slightly increased with further increases in PA-LG concentration.

The effects of lipoproteins composed of PA and various proteins on the bitter taste of 0.5 mM quinine hydrochloride and 50 mM caffeine were examined (Table 2). PA-OVA had a suppressive effect similar to that of PA-LG, PA-BSA, PA-LA and PA-casein were less effective than PA-LG and PA-OVA, but significantly inhibited the bitterness of quinine hydro-

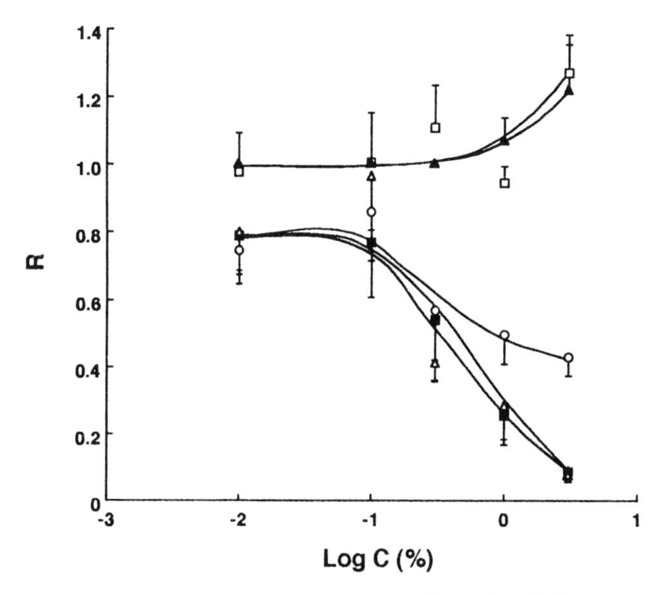

Figure 9 Effects of PA-LG of various concentrations on 400 mM NaCl (▲), sweetness of 600 mM sucrose (□) and bitterness of 10 mM propranolol hydrochloride (△), 5 mM promethazine hydrochloride (■) and 50 mM caffeine (○). The relative taste sensation (*R*) was defined in Equation (1).

chloride. These lipoproteins also suppressed the bitterness of caffeine, but not as strongly as they did the bitterness of quinine hydrochloride. There was no essential difference in the effectiveness of the inhibition between the different lipoproteins. These findings indicate that differences in the proteins in the lipoproteins did not greatly affect the ability of the lipoprotein to inhibit the bitter taste.

TABLE 2. Effects of Various Lipoproteins Composed of PA and Different Proteins on Bitter Taste of 0.5 mM Quinine Hydrochloride and 50 mM Caffeine.

Lipoprotein	Quinine	Caffeine
PA-LG	0.091 ± 0.013	0.428 ± 0.074
PA-BSA	0.204 ± 0.069	0.506 ± 0.067
PA-OVA	0.064 ± 0.011	0.415 ± 0.110
PA-LA	0.166 ± 0.043	0.466 ± 0.078
PA-casein	0.154 ± 0.031	0.452 ± 0.087

R was defined in Equation (1). Stimulating solutions were prepared by dissolving 0.5 mM quinine hydrochloride and 50 mM caffeine in 3.0% PA-LG solution.

TABLE 3. Effects of Each Component of PA-LG on Bitter Taste of 0.5 mM Quinine Hydrochloride and 50 mM Caffeine.

	Quinine	Caffeine
PA-LG	0.091 ± 0.013	0.428 ± 0.057
LG	0.792 ± 0.168	1.045 ± 0.107
PA vesicle	0.142 ± 0.035	1.093 ± 0.153

R was defined in Equation (1). The concentration of PA-LG was 3.0%. The concentrations for PA vesicle and LG were 0.85% and 2.15%, respectively, which were used to prepare 3.0% PA-LG.

Table 3 shows the effects of each component of PA-LG on the bitter taste of quinine hydrochloride and caffeine. LG itself had only a small suppressive effect on the bitterness of quinine hydrochloride and no effect on that of caffeine. The PA vesicle greatly suppressed the bitterness of quinine hydrochloride, hence PA can be used for the inhibition of the bitter taste

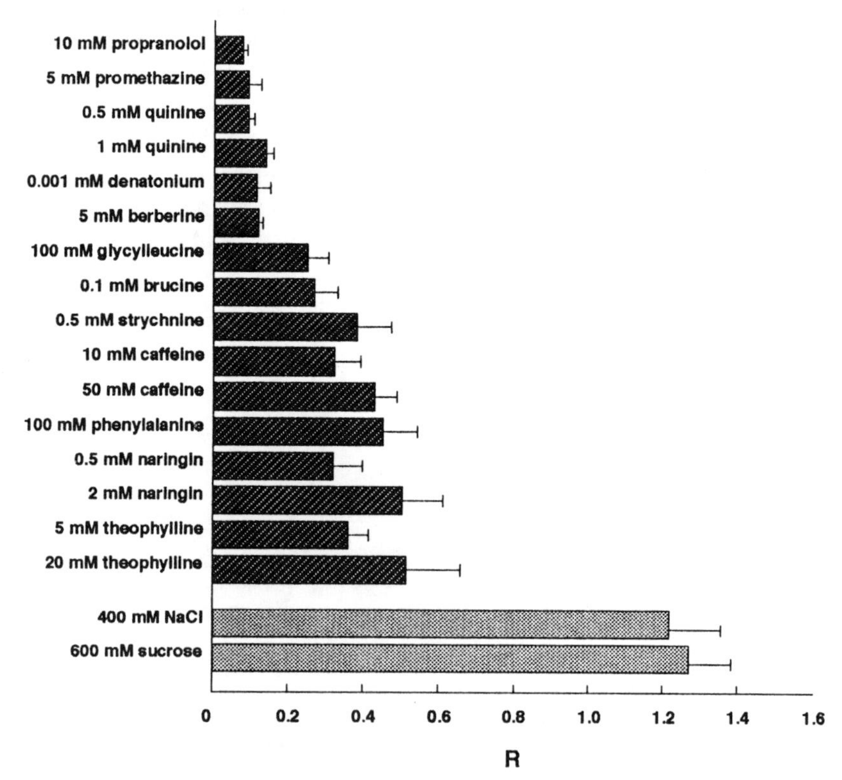

Figure 10 Effects of 3.0% of PA-LG on bitterness of various bitter substances, saltiness of NaCl and sweetness of sucrose. The relative taste sensation (*R*) was defined in Equation (1).

of substances such as quinine hydrochloride. However, PA vesicle had no effect on the bitter taste of caffeine. Thus, PA vesicle is not a complete inhibitor for bitterness, but is useful for practical use to inhibit bitterness of drugs and foods, as will be described in the next section.

Figure 10 shows the effects of 3.0% PA-LG on the bitterness of various substances; *R* values varied among the bitter substances and ranged from 0.075 to 0.51. The bitter tastes of basic substances such as propranolol hydrochloride, promethazine hydrochloride, quinine hydrochloride and denatonium benzoate were greatly suppressed by PA-LG. The bitter tastes of brucine and strychnine nitrate, which are also basic, were less effectively suppressed than those of the other basic substances. The bitter taste of glycyl-L-leucine, caffeine, L-phenylalanine, naringin and theophylline was also suppressed by PA-LG. The effects of PA-LG were examined with two different concentrations of quinine hydrochloride, caffeine, naringin and theophylline. The bitter tastes of these substances at lower concentrations were more effectively suppressed than those at higher concentrations, but the extent of the inhibition is not so appreciable.

EFFECTS OF PRETREATMENT OF THE TONGUE WITH PA-LG

In all the above experiments, the subjects tasted the bitter substances dissolved in solutions containing the lipoproteins. Hence the bitter substances may directly interact with the lipoproteins in the medium. To exclude this possibility, the tongue was first treated with PA-LG solution for

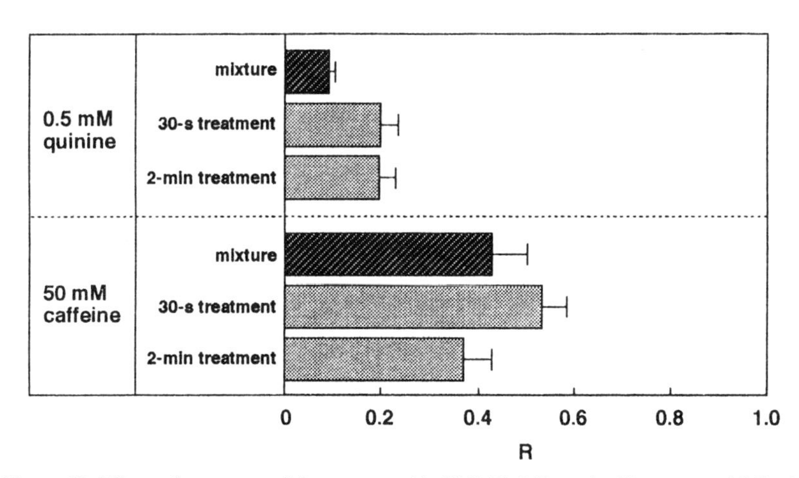

Figure 11 Effects of treatment of the tongue with 3.0% PA-LG on the bitter taste of 0.5 mM quinine hydrochloride and 50 mM caffeine. The tongue was treated with PA-LG solution for 30 sec and 2 min, then the bitter substances dissolved in deionized water were tasted. "Mixture" represents data when the bitter substances dissolved in 3.0% PA-LG solution were tested without treatment of the tongue with PA-LG. The relative taste sensation (R) was defined in Equation (1).

30 sec or 2 min, and then the subjects tasted the bitter substances dissolved in deionized water. As shown on Figure 11, inhibition of the bitter taste of 0.5 mM quinine hydrochloride had already reached a steady level at 30 sec and further inhibition was not brought about by a 2-min treatment. The extent of suppression brought about by this method was slightly less than that observed when quinine hydrochloride was tasted in the presence of PA-LG. The bitter taste of caffeine was suppressed to a greater extent by a 2-min than by a 30-sec treatment. This extent of suppression by 2-min treatment was close to that observed when caffeine was tasted in the presence of PA-LG.

BINDING OF BITTER SUBSTANCES TO PA-LG

The data shown in Figure 11 suggest that the suppression is brought about by masking of the target sites for bitter substances on the receptor membranes with PA-LG, but there is a possibility that the suppression was partly brought about by direct interaction of the bitter substances with PA-LG in the medium when tasted in its presence. To test this possibility, binding of the bitter substances to PA-LG was measured. Bitter substances were dissolved in 3.0% PA-LG solution and the resultant solutions were filtered with M. W. 5,000 cut-off filters by centrifugation. The quantity of the bitter substance in the filtrate was determined by reverse phase chromatography.

Table 4 shows the percentage of the quanity of a bitter substance in the fitrate to total quantity of the substance. The percentage was closely related to log *P*, which represents the hydrophobic parameter of the molecule. That is, the binding of the substances to PA-LG was due mainly to hydrophobic interaction. Only less than 5% of propranolol hydrochloride, promethazine hydrochloride, quinine hydrochloride, brucine and strychnine nitrate remained in the filtrate, suggesting that these substances are mostly bound to PA-LG in the medium. These findings suggest that the suppression of the bitter taste of these substances is mostly due to the binding of the substances to PA-LG in the medium when dissolved in PA-LG solution for tasting. This is not inconsistent with the finding that the bitter tastes of these substances were mostly suppressed by masking of their target sites with PA-LG when the tongue was treated with PA-LG and the bitter substances were tasted in the absence of PA-LG (see Figure 11). Brucine and strychnine nitrate were strongly bound to PA-LG in the medium (Table 4), but the extent of the suppression was lower than that of the other hydrophobic substances (Figure 10). No bitter taste was observed at the moment of tasting of a solution containing PA-LG and brucine, but gradually a bitter taste was noted. This suggests that the complex between PA-LG and brucine is gradually dissociated into its components in the saliva.

TABLE 4. Percentage of Unbound Bitter Substance to PA-LG.

Bitter Substance	Unbound (%)	Log *P*
10 mM Propanolol	4.60	2.57
5 mM Promethazine	0.30	4.65
0.5 mM Quinine	0.54	3.44
100 mM Glycyl-L-leucine	97.8	−1.96
0.1 mM Brucine	0.60	0.98
0.5 mM Strychnine	4.74	1.93
50 mM Caffeine	100	−0.07
100 mM Phenylalanine	97.1	−1.70
100 mM Naringin	81.9	−2.13
20 mM Theophylline	97.0	−0.02

Bitter substances were dissolved in 3.0% PA-LG solution, and the resultant solutions were filtered with MW 5000 cut filters by centrifugation. The quanity of the bitter substance in the filtrate was determined by reverse phase chromatography. The values in the table represent percentage of the quantity of the bitter substance in the filtrate to total quantity of the substance. Log *P* in the table represents hydrophobic parameter of the molecule, which was taken from the Pomona database. The parameters for denatonium benzoate and berberin were not found in the database.

In contrast to the hydrophobic substances, glycyl-L-leucine, caffeine, L-phenylalanine, naringin and theophylline are mostly unbound to PA-LG in the medium, which suggests that suppression of the bitter taste of these substances is brought about by masking of the target sites for the bitter substances with PA-LG. The percentage of the quantity of the unbound substance to the total quality of the substance is not related to the extent of suppression of the bitter taste by PA-LG. For example, glycyl-L-leucine was not bound to PA-LG in the medium, but its bitterness was greatly suppressed by PA-LG (see Figure 10). The bitter taste of propranolol hydrochloride was most effectively suppressed by PA-LG among the bitter substances tested, but the affinity of propranolol hydrochloride to PA-LG was much lower than those of promethazine hydrochloride, quinine hydrochloride and brucine. This is because masking of the target sites for bitter substances with PA-LG contributes to the suppression of the bitter taste.

MECHANISM OF INHIBITION OF BITTERNESS IN HUMANS BY PA-LG

In the first section, we found that PA-LG inhibits frog taste nerve responses to bitter substances without affecting the responses to NaCl, galactose, acetic acid or L-alanine. Here, the tongue was treated with a PA-LG solution containing no stimulus for 10 min and then a stimulus solution containing no PA-LG was applied to the tongue. Hence, inhibition of the responses to the bitter substances by PA-LG under these conditions is due to masking of the target sites for bitter substances with PA-LG.

In most experiments discussed in the present section, the tongue was not treated with PA-LG for practical reasons, but subjects tasted bitter substances dissolved in solutions containing PA-LG. In this method, both masking of the target sites for bitter substances with PA-LG and binding of bitter substances to PA-LG in medium contribute to inhibition of bitterness. Basic and hydrophobic substances such as quinine hydrochloride and denatonium benzoate have low taste thresholds and elicit a very bitter taste (Shallenberger and Acree, 1971). There are many such types of drugs. These bitter substances have a high affinity of PA-LG, hence binding effect to PA-LG greatly contributes to the inhibition of their bitterness. Hence, PA-LG is very effective in suppressing the bitterness of this type of substance.

PA and LG are not connected by covalent bonds, but by hydrophobic interactions and hydrogen bonding. Hence, PA-LG should be easily hydrolyzed in the digestive system. PA-LG forms a complex with certain bitter substances, but the bitter substances seem to be easily released from PA-LG since LG is hydrolyzed in the digestive tract.

INHIBITION OF BITTER TASTE IN HUMANS BY PHOSPHATIDIC ACID

In a previous section, we showed that PA vesicle itself has an inhibitory action on the bitter taste of certain substances. Although PA does not always inhibit the taste of all the bitter substances, it is a useful bitterness inhibitor for practical use because PA is much cheaper and much more stable than PA-LG. In this section, we deal with inhibition of bitterness by PA vesicle.

INHIBITION OF BITTERNESS OF VARIOUS SUBSTANCES IN SOLUTION

PA was added to solutions of bitter substances and its effects were examined. Table 5 shows the effects of 1.0% PA on the bitter taste of various substances dissolved in water. Here ten standard solutions (No. 1 to No. 10) of quinine sulfate in a range of 2.9×10^{-6} to 3.28×10^{-4} M were prepared and the intensity of bitterness of a test solution was expressed by the number of the standard solution. The solutions with points below 3.5 had no bitter taste. The solutions without PA vary from 9.5 to 5.0. The intensities of bitterness of basic bitter substances such as propranolol hydrochloride, quinine hydrochloride, papaverine hydrochloride, promethazine hydrochloride and berberine chloride were 7.8–9.0, which im-

TABLE 5. Effects of 1.0% PA Added to Aqueous Solutions Containing Bitter Substances.

	Intensity of Bitterness	
Bitter Substances	Without PA	With PA
10 mM Propranolol	9.0	4.2
0.5 mM Quinine	8.6	3.0
5 mM Papaverine	7.8	3.6
5 mM Promethazine	8.6	3.4
0.5 mM Berberine	8.4	4.2
5% Casein hydrolyzate	5.0	3.0
5% Soy hydrolyzate	5.0	3.0
5% Corn hydrolyzate	6.8	5.4
		3.4 (3% PA)

The values in the table represent relative taste intensity. Solutions below 3.5 elicit no bitter taste. The values are mean of data obtained from four to eight subjects.

plies that these substances elicit a strong bitter taste. Addition of 1% PA to the solutions containing these substances decreased the bitterness to 3.0–4.2, which elicits no or a very weak bitter taste. Hydrophobic amino acids such as leucine, phenylalanine, tyrosine and tryptophan elicit a bitter taste. Protein hydrolyzate, which contains free amino acids and peptides, usually elicits a bitter taste. Intensities of bitterness of casein hydrolyzate and soy hydrolyzate were decreased to 3.0. Corn hydrolyzate was decreased to 5.4 by 1% PA and to 3.4 by 3% PA. Thus, solutions of protein hydrolyzate containing 1–3% PA elicited no or weak bitterness.

INHIBITION OF BITTERNESS OF DRUGS IN GRANULES

Core granules were made of 1% quinine hydrochloride, hydroxypropylmethylcellulose (HPMC) and corn starch. Bitterness of the core granules was 8.14. The bitterness of granules containing 1% and 5% PA was 6.14 and 3.71, respectively. Thus, PA greatly inhibits the bitterness of the quinine in the granules.

The core granules were coated with 1% PA. The bitterness of the PA-coated granules was 3.3. Thus, the PA coating effectively inhibits the bitterness of the drug. It seems that suppression of bitterness is brought about both by preventing direct contact of the drug with the receptor sites on the tongue and by masking the receptor sites with PA. In the film coating, over 10% film is usually coated on the surface of the granules and in sugar coating, a much thicker coating is needed to mask bitterness of drugs. Thus, there is an advantage that the PA coating can suppress the bitterness with a much thinner layer than film or sugar coating.

SUMMARY AND CONCLUDING REMARKS

The development of a specific inhibitor for bitter substances is widely required in the fields of taste physiology, pharmaceutical sciences and food sciences, but no inhibitor has been available. We found that lipoproteins, PA-LG composed of phosphatidic acid (PA) and β-lactoglobulin (LG) and PA-LA composed of PA and α-lactalbumin (LA) reversibly suppressed the responses of the frog glossopharyngeal nerve to the bitter substances. The frog tongue was treated with PA-LG solution for 10 min and then stimulated by a stimulus dissolved in water. In these conditions, there is no chance that the stimuli contact the lipoprotein in the medium. The responses to the bitter substances such as quinine hydrochloride, papaverine hydrochloride, caffeine and L-leucine were completely suppressed by PA-LG, while those to the hydrophilic bitter substances such as CsCl, $MgCl_2$ and tetraethylammonium chloride were not suppressed. The responses to NaCl, galactose, acetic acid and L-alanine were unchanged or only slightly increased.

It was found that the lipoproteins having the inhibitory activity are adsorbed on the frog tongue surface while those having no inhibitory activity are not adsorbed. We also examined adsorption of the lipoproteins on model lipid membranes coated on quartz crystal microbalance by measuring changes in its frequency. The lipoproteins having inhibitory activity were well adsorbed on the hydrophobic lipid membranes, while the proteins having no inhibitory activity were adsorbed very little on the membranes. It seems that receptor sites for bitter substances on the taste cell membranes are hydrophobic whereas those for other taste stimuli such as salts, acids and sugars are hydrophilic. Hence the binding of PA-LG to hydrophobic sites of the receptor membranes leads to selective inhibition of bitterness.

The effects of various lipoproteins on the taste sensation to various stimuli in humans were examined by a psychophysical method. Using PA-LG, the effects on taste sensation to various stimuli were examined. The bitter taste of all twelve substances examined was inhibited, while saltiness of NaCl and sweetness of sucrose were not inhibited. The inhibition of the bitter taste was completely reversible. In these studies, subjects tasted a mixture of taste stimuli and PA-LG. Hence there is a chance that taste stimuli form a complex with PA-LG in the medium. It was shown that basic hydrophobic substances are bound to PA-LG in the medium. Such binding contributes to the inhibition of the bitterness. On the other hand, substances such as glycyl-L-leucine, caffeine, L-phenylalanine, naringin and theophylline were unbound to PA-LG in the medium, but their bitterness was suppressed by PA-LG. This suggests that masking of the target sites for bitter substances on the taste receptor membranes with PA-LG also contributes to the inhibition of the bitter taste.

PA does not always inhibit the responses to all the bitter substances, but PA has the advantage that it is much cheaper than PA-LG. This is an advantage in practical use for suppression of bitterness. PA alone inhibited bitterness of many drugs and the protein hydrolyzates. Among various drugs, basic and hydrophobic substances such as quinine hydrochloride and propranolol hydrochloride have low taste thresholds and are said to be the most bitter. PA most effectively suppressed the bitter taste of such substances. PA incorporated into granules containing quinine hydrochloride effectively inhibited its bitterness. Coating of the granules containing quinine hydrochloride with a small amount of PA completely inhibited its bitterness.

It was concluded that PA-LG is an ideal bitterness inhibitor as a physiological tool to study the receptor mechanism. PA is a good bitterness inhibitor for practical use to suppress the bitterness of drugs and foods. PA originates from soybeans and the proteins used, except for bovine serum albumin, originate from milk or eggs. Hence PA-LG or PA alone can be safely used to mask the bitter taste of drugs and foods.

REFERENCES

Bechtol, L. D., DeSante, K. A., Foglesong, M. A., Spradlin, C. T. and Winely, C. L. 1981. The bioavailability of pediatric suspensions of two erythromycin esters, *Curr. Ther. Res.*, 29: 52–59.

Brown, E. M., Carroll, R. J., Pfeffer, P. E. and Sampugna, J. 1983. Complex formation in sonicated mixtures of β-lactoglobulin and phosphatidylcholine, *Lipids,* 18: 111–118.

Cummings, T. A. and Kinnamon, S. C. 1992. Apical K^+ channels in Necturus taste cells; modulation by intracellular factors and tates stimuli, *J. Gen. Physiol.*, 99: 591–613.

Fu Lu, M.-Y., Borodkin, S., Woodward, K., Li, P., Diesner, C., Hernandez, L. and Vadnere, M. 1991. A polymer carrier system for taste masking of macrolide antibiotics, *Pharm. Res.*, 8: 706–712.

Fukumori, Y., Yamaoka, Y., Ichikawa, H., Fukuda, T., Takeuchi, Y. and Osako, Y. 1988. Coating of pharmaceutical powders by fluidized bed process II, *Chem. Pharm. Bull.*, 36: 1491–1500.

Hall, M. J., Bartoshuk, L. M., Cain, W. S. and Stevens, S. J. C. 1975. PTC (phenylthiourea or phenylcarbamide) taste blindness and the taste of caffeine, *Nature,* 253: 442–443.

Harder, D. B., Whitney, G., Frye, P., Smith, J. C. and Rashotte, M. E. 1984. Strain differences among mice in taste psychophysics of sucrose octaacetate, *Chem. Senses,* 9: 311–323.

Hwang, P. M., Verma, A., Bredt, D. S. and Snyder, S. H. 1990. Localization of phosphatidylinositol signaling components in rat taste cells: role in bitter taste transduction, *Proc. Natl. Acad. Sci. USA,* 87: 7395–7399.

Kanamoto, R., Ohtsuru, M. and Kito, M. 1977. Diversity of the soybean protein-phosphatidylcholine complex, *Agri. Biol. Chem.*, 41: 2021–2026.

Kashiwagura, T., Kamo, K., Kurihara, K. and Kobatake, Y. 1980. Nature of dynamic and steady components in the frog gustatory nerve response and interpretation by a theoretical model, *Am. J. Physiol.*, 238: 445–452.

Katsuragi, Y. and Kurihara, K. 1993. Specific inhibitor for bitter taste, *Nature,* 365: 213–214.

Katsuragi, Y., Sugiura, Y., Cao, L., Otsuji, K. and Kurihara, K. 1995. Selective inhibition of bitter taste of various drugs by lipoprotein, *Pharm. Res.*, 12: 658–662.

Katsuragi, Y., Sugiura, Y., Otsuji, K. and Kurihara, K. Characteristics of phosphatidic acid-containing lipoproteins which selectively inhibit bitter taste; high affinity to frog tongue surface and hydrophobic model membranes, *Biochim. Biophys. Acta,* in press.

Katsuragi, Y., Yasumasu, T. and Kurihara, K. Lipoprotein that selectively inhibits taste nerve responses to bitter substances, *Brain Res.*, in press.

Kolesnikov, S. S. and Margolskee, R. F. 1995. A cyclic-nucleotide-suppressible conductance activated by transducin in taste cells, *Nature,* 376: 85–88.

Koyama, N. and Kurihara, K. 1972. Mechanism of bitter taste reception: interaction of lipid monolayers from bovine circumvallate papillae with bitter compounds, *Biochim. Biophys. Acta,* 288: 22–26.

Kumazwa, T. Kashiwayanagi, M. and Kurihara, K. 1985. Neuroblastoma cell as a model for a taste cell: mechanism of depolarization in response to various bitter substances, *Brain Res.*, 333: 27–33.

Kumazawa, T., Kashiwayanagi, M. and Kurihara, K. 1986. Contribution of electrostatic and hydrophobic interactions of bitter substances with taste receptor membranes to generation of receptor potentials, *Biochim. Biophys. Acta,* 888: 62–69.

Kumazawa, T., Nomura, T. and Kurihara, K. 1988. Liposomes as model for taste cells: receptor sites for bitter substances including N–C=S substances and mechanisms of membrane potential, *Biochemistry,* 27: 1239–1244.

Kurihara, K. 1972. Inhibition of cyclic 3′,5′nucleotide phosphodiesterase in bovine taste papillae by bitter taste stimuli, *FEBS Lett.*, 27: 279–281.

Margolskee, R. F. 1995. Receptor mechanism in gustation. In R. L. Doty (Ed.), *Handbook of Olfaction and Gustation,* Marcel Dekker, Inc., New York, pp. 575–595.

Mine, Y., Kobayashi, H., Chiba, K. and Tada, M. 1992. ^{31}P NMR study on the interfacial adsorptivity of ovalbumin promoted by lysophosphatidylcholine and free fatty acids, *J. Agric. Food Chem.*, 40: 1111–1115.

Mizutani, R. and Nakamura, R. 1987. Emulsifying properties of a complex between apoprotein from hen's egg yolk low density lipoprotein and egg yolk lecithin, *Agri. Biol. Chem.*, 51: 1115–1119.

Naim, M., Seifert, R., Nurnberg, B., Grunbaum, L. and Schultz, G. 1994. Some taste substances are direct activators of G-proteins, *Biochem. J.*, 297: 451–454.

Ohtsuru, M., Kito, M., Takeuchi, Y. and Ohnishi, S. 1976. Association of phosphatidylcholine with soybean proteins, *Agri. Biol. Chem.*, 40: 2261–2226.

Okahata, Y., Enn-na, G. and Ebato, H. 1990. Synthetic chemoreceptive membranes. Sensing bitter or odorous substances on a synthetic lipid multibilayer film by using quartz-crystal microbalances and electric responses, *Anal. Chem.*, 62: 1431–1438.

Okahata, Y. and Ebato, H. 1991. Adsorption behaviors of surfactant molecules on a lipid-coated quartz-crystal microbalance. An alternative eye-irritant test, *Anal. Chem.*, 63: 203–207.

Price, S. 1973. Phosphodiesterase in tongue epithelium: activation by bitter taste simuli, *Nature,* 241: 54–55.

Ruiz-Avila, L., McLaughlin, S. K., Wildman, D., McKinnon, P. J., Robichon, A., Spickofsky, N. and Margolskee, R. F. 1995. Coupling of bitter to phosphodiesterase through transducin in taste receptor cells, *Nature,* 376: 80–85.

Shallenberger, R. S. and Acree, T. E. 1971. Chemical stucture of compounds and their sweet and bitter taste. In L. M. Beidler (Ed.), *Handbook of Sensory Physiology IV-2,* Springer-Verlag, Berlin, New York, pp. 221–277.

Spielman, A. I., Mody, I., Brand, J. G., Whitney, G., MacDonald, J. F. and Salter, M. W. 1989. A method for isolating and patch-clamping single mammalian taste receptor cells, *Brain Res.,* 503: 326–329.

Ueda, M., Nakamura, Y., Makita, H.and Kawashima, Y. 1993. Preparation of microcapsules masking the bitter taste of enoxacin by using one continuous process technique of agglomeration and microencapsulation, *J. Microencapsulation,* 10: 461–473.

Ugawa, T., Konosu, S. and Kurihara, K. 1992. Enhancing effects of NaCl and Na phosphate on human gustatory responses to amino acids, *Chem. Senses,* 17: 811–815.

Yoshii, K., Kobatake, Y. and Kurihara, K. 1981. Selective enhancement and suppression of frog gustatory responses to amino acids, *J. Gen. Physiol.,* 77: 373–385.

SECTION III

APPLICATIONS IN ORAL PHARMACEUTICALS

General Ingredient or Process Approaches to Bitterness Inhibition and Reduction in Oral Pharmaceuticals

GLENN ROY[1]

INTRODUCTION

THE previous chapter introduced a new ingredient pair for the treatment of bitterness perception in oral pharmaceuticals. Oral pharmaceuticals often impart unpleasant taste, primarily a bitter perception. The desire for improved palatability in those products has prompted numerous formulations of nonbitter oral pharmaceuticals. The formulations presented may be the commercial processes in practice.

Bitter taste modulation in oral pharmaceuticals appears to be most important and commands a market differentiation that leads to a large profit. However, the field has languished in the use of sweetness to mask bitter. Now superior encapsulating and coating compositions have revolutionized the marketplace. Many applications of bitter-reduced medications have been reviewed (Roy, 1994). Additionally, a compilation of 829 over-the-counter (OTC) pharmaceutical formulations is available and organized according to their therapeutic effect. The formulation listings include concentrations of the active and inactive ingredients and, in most cases, a recommended procedure for mixing the formulation (Braun, 1994).

In the last couple of years a marked increase in literature articles and patents on taste-masking in oral pharmaceuticals was compiled. These references generally include techniques of bitterness inhibition and reduction. Many of these contributions have been successfully commercialized in oral pharmaceutical preparations available over-the-counter or by prescription. Certainly, the ideal solution to bitterness inhibition or reduction

[1]Pepsi-Cola Co., 100 E. Stevens Ave., Valhalla, NY 10595, U.S.A.

is the discovery of a universal inhibitor of all bitter-tasting substances. Taste sensors may allow high through-put screening of new chemical entities prepared for oral medical administration. Two comprehensive reviews of recent efforts to control bitter taste in foods and beverages were presented along with reflection on the discovery of a universal bitterness inhibitor (Roy, 1990, 1992).

This chapter will inform the reader about the most recent referenced applications of taste masking in oral pharmaceuticals affecting primarily bitterness. The applications include ingredient or processing approaches in formulations to improve oral administration by masking bitter taste. The ingredients discussed are flavors, sweeteners, acidic amino acids, lipids or lecithin-like substances and surfactants. Processing approaches include coating and inclusion complexes with proteins (gelatin, zein), gelatinized starch, gums, cyclodextrins (β-CD), chitosan and liposomes with adsorption by polymeric membranes, ion exchange resins, zeolites. Chemical modifications and salt preparations are briefly mentioned. An effort has been made to organize the text by ingredient or process formulation approaches. Individual medicinal substances are indexed at the end of the book. There are, in some cases, many ways to mask the same oral drug.

The classes of oral pharmaceuticals covered include:

- antibiotics, antibacterials
- antiasthma
- antidiarrheal
- analgesics, cold and allergy
- vitamins and minerals
- sedatives and adrenergics (stimulants), antidepressants
- antihistamines
- antiulceratives
- vasodialators

Sensory developments in bitterness reduction and inhibition are an important mainstay of product evaluation in oral pharmaceutical formulations. The consumer does not care for the taste modality of bitter in the broadest sense and will continue to be displeased with products whose bitter taste is perceived. The unusual flavanone for oral anesthesia, eriodictyol, imparts a bitter taste. Oral pharmaceuticals with perceived bitter taste may include chewable and nonchewable tablets, capsules, syrups, suspensions, concentrates, lozenges, dentifrices and mouthwashes, atomized solid or liquid inhalants and ingestible ointments. However, improved palatability has resulted from proven methods for bitterness reduction and inhibition. Orally available treatments are preferred by the American public. Oral dosages of normally injectable peptides such as long-term renin inhibitor treatments are approaching commercialization as potentially useful classes of medications (Kleinert et al., 1993).

For those plagued with taste disorders (dysgeusia), the perception of unpleasant tastes may be quite prevalent when sweetness is not perceived. Drug-induced dysgeusia, its frequency, mechanisms, effectors and treatment of the disease have been reviewed in Japanese (Tanaka et al., 1994). The Taste and Smell Clinic in Washington, D.C., has published a very comprehensive, 682-reference review describing impairment of taste and smell sensory function at a molecular level causing two major behavioral changes: loss of acuity (hypoguesia and hyposmia) and/or distortion of function (dysgeusia and dysosmia). Loss of acuity occurs primarily by drug inactivation of receptor function through inhibition of tastant/odorant receptor. Those inactivations occur by binding; Gs protein function; inositol triphosphate function; Ca^{+2}, Na^{+} channel activity; other receptor-inhibiting effects; or some combination of these effects. Distortions occur primarily by a drug inducing abnormal persistence of receptor activity (i.e., normal receptor inactivation does not cocur) or through failure to activate: various receptor kinases; Gi protein function; cytochrome P450 enzymes; or other effects that usually turn off receptor function, inactivate tastant/odorant receptor binding, or some combination of these effects (Henkin, 1994). Henkin emphasizes that dysgeusia is more common than presently appreciated. The taste changes can impair appetite, food intake, cause significant lifestyle changes and may require discontinuation of drug administration.

SWEETENERS, FLAVORS AND AMINO ACIDS

Numerous dentifrices and mouthwashes applied to the oral cavity elicit unpleasant taste perceptions. An Italian review discusses the side effects of drugs on taste (Sapone et al., 1992). Mouthwashes or cough drop formulations with medicinal or bitter taste, such as those containing up to 0.35% eucalyptus oil, can be masked with 0.0025% of fenchone, borneol or isoborneol. The taste-masking agents significantly suppress the perception of unpleasant organoleptic sensations such as bitterness or medicinal off-taste of the volatile oil (Hussein and Barcelon, 1991). Certainly, the cooling effect of the taste-masking agents must be aiding the reduced perception of bitterness. Other mentholated compositions containing di-D-fructofuranose 1,2′:2,3′-dianhydride are useful for dentifrices, mouthwashes and foods. The bitter taste is reduced with about 0.02 wt.% menthol and the low-calorie formulations show beneficial anticaries effect (Kondo and Nishimura, 1991). Dentifrices with anethole and menthofuran to mask bitterness also have improved low-temperature stability (Ueki et al., 1993). A formulation for dentifrices with masked bitterness and improved low-temperature stability contains glycerin 20, calcium monophosphate 40, propylene glycol 3, sodium saccharin 0.1, sodium laurylsarcosinate 0.1, sodium lauryl sulfate 1.5, sodium carboxymethyl cellulose 0.8, carrageenan 0.2, tranexamic acid 0.05, dipotassium glycyrrhetinate 0.2, methyl paraben 0.2, butyl paraben 0.02, sodium benzoate 0.5, spearmint flavor 1, anethole 0.15, menthofuran 0.01 and water to 100 mL.

A therapeutically effective amount of an analgesic and antitussive can be prepared in a solid, swallowable form for temporary relief of coughs and a sore throat. A menthol-flavored film coating is applied to the core in an amount sufficient to provide sensory stimulation of the throat and nasal passageways when the dosage form is placed in the oral cavity and swallowed. A regular-strength caplet containing acetaminophen and dextromethorphan-HCl was coated with a solution containing hydroxypropyl methylcellulose 9.6, peppermint flavor 2.4, mint flavor 1.7, and purified water 86.3% (Sonley and Turnbull, 1995).

An improved liquid oral-hygiene composition for dentifrices or mouthwashes contains peppermint oil to achieve a clean, less bitter taste as well as lower the need for the presence of menthol and surfactants. The composition contains glycerin, sodium saccharin, citric acid, sodium citrate, sodium lauryl sulfate, colors and Poloxamer 407 dissolved in water with spearmint, methyl salicylate, menthol and peppermint dissolved in ethanol (Hussein et al., 1994; Carlin et al., 1994). A plant extract containing polygodial provides improved mint flavoring for beverages, pharmaceuticals and toiletries by a cooling sensation, suppressing bitterness and prolonging flavor. A mint-flavor base was manufactured by adding 15 g l-menthol and 10 g edible oil to 75 g spearmint oil with 0.05 g polygodial. A gum with the formulation had a longer-lasting, fresher taste than gum with mint flavoring not containing the polygodial (Ishikawa et al., 1995).

Bronchodilators (antiasthmatics) and peripheral dilators give a bitter taste of atomized theophylline, the most common oral bronchodilator. Oral liquid compositions containing theophylline salts are formulated with sorbitol to give a solution that was less bitter than aqueous theophylline solution (Maegaki et al., 1993). A formulation for bronchodilator mists with reduced bitterness contains theophylline salts 0.8, sodium bicarbonate and/or carbonate 1.0, D-sorbitol 20, sodium saccharin 0.04, sodium glutamate 0.05, hydroxymethyl cellulose 0.017 g, vanilla essence and water to 100 mL.

Dentifrice compositions containing betain-type amphoteric surfactants with saccharin or in conjunction with other sweetening agents such as thaumatin or stevioside remove the bitter after-taste of the betain-type amphoteric surfactants. Certainly the silicas, sugar alcohol and gum are assisting in taste masking (Hammond, 1995).

A dentifrice as toothpaste contains 70% sorbitol 45.0, thickening silica 10.0, abrasive silica 10.0, xanthan gum 1.0, PEG 5.0, titanium dioxide 1.0, sodium monofluorophosphate 0.8, cocamidopropyl betaine 2.0, saccharin 0.2, flavor 1.2, and water to 100 g.

Oral compositions in the form of microcapsules that reduce oral bacteria and provide long-lasting breath protection contain bitter substances. For example, cetyl pyridinium chloride and domiphen bromide are mixed with sweeteners to reduce bitterness (Peterson et al., 1994). As expected, lingering sweeteners offer the masking effect for the breath-freshening microcapsule containing gelatin 9.84, 70% sorbitol solution 3.616, saccharin 0.418, acetosulfam 0.495, aspartame 0.495, monoammonium glycyrrhizin 0.027, neohesperidine dihydrochalcone 0.020, FD&C Blue No. 1 0.010, FD&C Yellow No. 5 0.005, Captex-300 8.352, flavor 7.158, cetyl pyridinium chloride 0.675, domiphen bromide 0.075, propylene glycol 2.017, glycerin 0.270, PEG 29.522, sucrose acetate isobutyrate 33.408 and water 3.397%.

Lozenges used for the common cold possess a bitter or astringent taste due to zinc acetate dihydrate. Taste masking is accomplished with saccharin, anethol-β-cyclodextrin complex and magnesium stearate followed by tableting with compressible polyethylene glycol and fructose (Eby, 1992). A formula of 25% acetaminphen granules was prepared with 0.8% aspartame (NutraSweet®) to effectively reduce bitterness. Starch, lactose and mannitol offer additional taste-masking properties of the 1.7% caffeine ingredient (Matsubara et al., 1990). Aspartame, starch, lactose and mannitol are added to the analgesic acetaminophen to reduce its bitter taste (Motola et al., 1991; Matsubara et al., 1990). A taste-masked oral ibuprofen composition contains pyridoxine and ibuprofen in a ratio of 1:8 to 1:2 and formulated as a syrup along with flavors and sweeteners (Depalmo, 1993). The oral composition of ibuprofen contains ibuprofen 4.0, pyridoxine hydrochloride 1.0, potasium hydroxide 1.4, glycerol 3.0, sodium EDTA 0.01, sodium sulfate 0.2, methyl paraben 0.15, propyl paraben 0.03, sodium citrate dihydrate 3.0, sodium saccharin 0.5, refined sugar 20.0, flavors 2.2 and water to 100 mL.

Acetaminophen bitterness is effectively masked by suspending the drug (2.5 g) in a medium containing guar gum 0.08 and sweetening agents as sucrose. The mixture is treated until wetted and homogeneous, then admixed with 34.7 g of a syrup containing 65% sucrose and 0.08% sodium benzoate (Yiv and Tustian, 1995).

The taste-enhancing properties of a sweet-lingering artificial sweeteners, neohesperidine dihydrochalcone and hesperidine dihydrochalcone 4′-β-D-glucoside, have the ability to reduce the perception of bitterness in pharmaceuticals by virtue of the lingering sweetness. The ability to mask bitterness and saltiness has been reviewed (Baer et al., 1990). Similarly, MacAndrews & Forbes Company, Mafco (Camden, NJ) offers a line of mono-ammonium glycyrrhizinate (licorice) products, Magnasweet®. Low levels of use are said to mask bitter, harsh and astringent taste in chewable multivitamins, cough/cold syrups, oral antibiotics, chewable analgesics and alcohol-based oral antiseptics. A lingering sweetness provides taste masking primarily because the taste profile of a bitter substance appears later than normal sugar sweetness generally lasts. An oral suspension of an antihistamine is said to mask bitter taste and offensive odor (Blank et al., 1996). The oral suspension contains the antihistamine nizatidine 1.5, sugar 25, colloidal silica 1, aspartame 1, Mafco Magnasweet-135 1 and Captex (C8–C10 glycerides) to 100%.

Some sweetness inhibitors possess a lingering, faint sweetness. Lactisole®, a sweetness inhibitor recently formulated in Domino sugar for a "non-sweet" sugar, possesses taste-masking potential in pharmaceuticals. Several tasteless sweetness inhibitors are being actively pursued as bitter inhibitors (Kurtz and Fuller, 1993). Nonbitter dentifrices are prepared by sweetening a quaternary ammonium salt-type bactericide (Yokoo and Hirohata, 1993). The mouthwash showed 100% bactericidal activity against *E. coli* and contained benzethonium chloride 0.02, mono- and disodium phosphate buffers 0.66, ethanol 10.0, tranexamic acid and/or ϵ-caproic acid 0.025, Stevia-based sweetener extract 0.01, glycerin (85%) 5.0 and water to 100 mL.

Bitterness of oral pharmaceuticals is alleviated by spraying crystalline cellulose granules with a solution containing vinpocetine 10 (a cerebral vasodilator), glycyrrhizic acid di-K salt 10, aspartame 12, mannitol 28, talc 40, low-substituted hydroxypropylcellulose 40, hydroxypropyl cellulose 100 and water 2160 g. After drying for two minutes, the dispersion did not show bitterness. A control preparation formulated with additional 22 g mannitol instead of glycyrrizhinic acid salt and aspartame showed bitter taste (Ito et al., 1994). Bitter dentrifices may be masked with sodium sulfate (1.0), silicon dioxide (25.0) and di-potassium glycyrrhizic acid salts (0.05 w%) (Ishii and Uno, 1995).

Anticholesterolemic saponin-containing foods, beverages and pharmaceuticals are supplemented with amino acids for bitterness control (Watabe et al., 1992). A formulation to give a bitterness-free beverage contains tea saponin 0.02, quillaja saponin 0.05, glycine 2.0, alanine 1.0, sucrose 10.0, citric acid 0.5, flavor 0.1 and water to 100 mL.

Protein-like compositions, useful for improvement of liver disorders, severe burn, trauma, etc., contain branched amino acid-modified proteins. Whey powders (100 g) were treated with papain in the presence of the amino acids ethyl L-leucine (16.1 parts), ethyl L-iso-leucine (7.4 parts), ethyl L-valine (10.2 parts), cysteine hydrochloride (1.5 parts) and sodium carbonate (26 g) in water at 40°C for 20 minutes to manufacture tasteless and odorless powders containing 10% free amino acids and 43 wt% branched amino acids. Whey powders not treated in that manner and containing L-leucine, L-iso-leucine and L-valine tasted bitter. The branched amino acid-modified powders were mixed with fats, dextrins, salts, vitamins, etc., to make a nutrient for highly invasive liver diseases (Ishibashi and Shinoda, 1993). Oral preparations with bitter taste due to L-Ile, L-Leu and L-Val as active ingredients may be taste masked with D-Trp. For example, an aqueous solution (1 L) with DL-Trp 0.075 g, L-Ile 0.95 g, L-Leu, 1.15 g and L-Val 0.90 g tasted sweet and not bitter. In the same formulation, reducing the DL-Trp concentration to 0.038 g with DL-Ala 1.70 g reduced bitterness as well (Akyama, 1995a, 1995b). Trehalose masks bitterness and unpleasant taste of amino acids and peptides and does not permit Maillard reaction with them. An example cites granules containing Leu, Ile and Val coated with a mixture of trehalose, hydroxypropyl cellulose and aspartame (Hatsuda and Takeda, 1994). Oral preparations for prevention of or recovery from fatigue contain L-Ile, L-Leu and L-Val as active ingredients and D-isomers of Trp and Val for masking the bitterness of the L-amino acids. Bitterness of an aqueous solution (1 L) containing L-Ile 7, L-Leu 11, DL-Val 20, DL-Ala 8, Gly 6, malic acid 6, maltose 20 and aspartame 0.1 g/L was masked by the addition of the DL-Val and was given to adult men after they had played tennis to show recovery from fatigue in 30 min (Iwata, 1995).

Bitterness-free Vitamin B oral solutions contain sugars, amino acids and apple flavors. The formulation was dissolved in water at 80°C, cooled to room temperature and mixed with 30 mg of Coix seed apple flavor composition to give 30 mL of bitter-free oral solution (Kobayashi et al., 1993). A formulation for bitter-free Vitamin B oral solutions contains sodium riboflavin phosphate 19 g, pyridoxine hydrochloride 50 g, sucrose 1500 g, maltitol 3000 g, sodium L-asparate 60 g, sodium benzoate 21 g, ethyl para-hydroxybenzoate 1.5 g, L-malic acid 90 mg, Coix seed extract 2 mL. Oral compositions consisting of Vitamin B, sodium 5′-ribonucleoside (inosinate), citrus (orange) flavor or fruit flavor also have improved taste (Kobayashi et al., 1992).

Adrenergics or stimulants impart a strong bitter taste. Imagine a gum with enough caffeine to equal a gulp of strong coffee. The "chewing-not-brewing" trend has the demand for high-caffeine chewing gums and lozenges growing. The Morishita Jintan Co. market for such a masticated product tops $76.3 MM with Strong Man. Cleverly marketed as a gum or lozenge that "charges the energy inside you," what makes it palatable? The gum is black and contains caffeine, vitamins, essence of garlic, spices, carrot concentrate and four types of sugar for taste masking. Could this be a "nutrachewtical?" The Japanese unit of Warner-Lambert also jumped into the market with a white version of the gum with caffeine crystals, Sting, "the gum that stings drowsiness." Caffeine may be coated with an inner coating containing Eudragit RL30D, followed by an outer coating solution containing Eudragit RS30D to give coated beads. The caffeine beads were mixed with compressible sugar and magnesium stearate to obtain chewable tablets with no bitter taste and good bioavailability (Bhardwaj and Hayward, 1993).

"Brush, floss, chew tannin" read a highlighted article by Mike Snider in *USA Today* on January 4, 1991. All those bitter substances in tea, coffee, chocolate that are masked by sugar may actually help fight the tooth decay that the sugar promotes. Researchers have proof that tannins, natural bitter components in food and beverages, make plaque harder to stick to teeth. Nigerians have a habit of gnawing on sticks containing tannins along with the lowest incidence worldwide of cavities and gum disease. In an example of periodontal research on saliva-coated tooth enamel, it was found that a tannin-spiked, plaque-producing bacterial culture was 85% dead in one hour. The journalist warns us that indulging in chocolate, tea and coffee is not the best way to avert cavity-producing bacteria. The tannin-rich compositions will probably be formulated in new dental rinses, toothpastes and chewing gums. But "the problem is they have fairly bitter tastes, so you have to hide that." The use of flavored taste-masking agents such as fenchone, borneol and isoborneol, commonly used in the manufacture of mouthwashes and cough drops, may be a useful approach (Kurasumi et al., 1991). Dental anesthetics are bitter for the moment just prior to desensitization when oral injections leak in the oral cavity. Topical anesthetic swabbing of the gingiva accomplishes reduced bitter taste perception. The addition of an extremely small quantity of an anesthetizing agent such as sodium phenolate to an aspirin-medicated floss serves to numb the taste buds sufficiently within 4–5 seconds. The bitter aspirin taste is not perceived compared to a control without the phenolate. For example, white beeswax and Chloroseptic lozenge (Norwich Eaton Pharmaceuticals, Norwich, NY) containing 32.5 mg of phenol and sodium phenolate and inactive ingredients were blended into a fine powder. The powder was mixed with Crisco vegetable oil, "lime floss sugar" and aspirin. The mixture was then spun using a floss-spinning device (Fuisz, 1991).

Gelatin and flavoring materials mask bitter taste of tannic acid presumably by viscosity effects when made into a jelly by cooling (Aoi and Murata, 1992). A gelatin gum-like formula to mask bitter contains tannic acid 0.05 g, gelatins 1.25 g, chocolate flavor 5.22 g and water to 25 mL. Also, a 50 mL aqueous solution of tannic acid (0.1 g) and sodium alginate (0.4 g) had reduced bitterness compared to a control of tannic acid alone in water (Kikuta et al., 1992).

Cacao powder is said to be key to the masking potential in a chewable tablet and a tablet-compression agent (Martani et al., 1995). The tablet contains calcium carbonate 1250, cacao powder 250, sorbitol 200, povidone 30, aspartame 2, flavors 14, talc 10 and magnesium stearate 4 mg.

A lure for oral administration of veterinary pharmaceutical compositions contains medicaments, flavoring agents, masking agents and excipients (Kerouedan, 1994). The excipient and lure contain flour 70, starch 15, sorbitol 15% as excipient with vanilla flavor 0.4–0.8, pork flavor 0.4–0.8 and citrus flavor 0.2 g/kg.

LIPIDS

Oils, surfactants, and polyalcohols can effectively increase viscosity in the mouth and coat taste buds. A taste-masking carrier for acetaminophen pharmaceutical compositions comprises aliphatic or fatty acid esters. Stearyl stearate was melted at 75°C and the molten solution was sprayed into a fluidized bed of acetaminophen granules. The resultant granules were mixed with suitable excipients and incorporated into a chewable tablet formulation (Gowan and Bruce, 1993). Waxes of higher alcohols, fatty acids and esters and higher hydrocarbons are heated and spray-coated on pharmaceutical particles to mask the taste. This process is simpler than conventional methods employing organic solvents (Nishii et al., 1993).

The efficacy and safety of chloroquines as an antimalarial has contributed to the survival of millions in the past 50 years. Chloroquine is widely available, cheap, well tolerated and orally well absorbed. An oral administration, especially to children, imparts an unpleasant taste. Multiple W/O/W emulsions of chloroquine phosphate showed faster release rates due to the inherent smaller internal aqueous globules and, therefore, an increased interfacial area (Vaziri and Warburton, 1994). The use of multiple emulsions of paraffin with chloroquine indicated that emulsions of O/W/O could mask the taste of chloroquine to some extent. Paraffin was superior to the use of vegetable oil due to differences in the stability of the preparations (Rao and Bader, 1993).

Mono-, di-, and tri-esters of hexoses such as glucose (lipid sugars) are useful as foaming agents for dentifrices and as emulsifiers for food. Their preparation in a nonbitter form requires sequential solvent separation by chromatography on silica gel to isolate the monoesters free of bitter taste (Endo et al., 1991). A bitterness-free syrup may be prepared and remain stable at 60°C for longer than a month (Miura et al., 1992).

The bitter-reduced liquid or syrup of carbetapentane citrate contains butyl 4-hydroxybenzoate 0.04, sucrose 180.0, acetaminophen 3.0, diphenhydramine-HCl 0.25, carbetapentane citrate 0.16, noscapine-HCl 0.16, dl-methylephedrin-HCl 0.2 and flavor 0.036, polyglycerin fatty acid ester 3.6, glycerin 5.4, medium chain triglyceride 18.0, water to 360 mL.

Similarly, the experimental drug for seizures gabapentin (a cyclic amino acid) has improved taste when coated with gelatin and then with a mixture of partially hydrogenated soybean oil and glycerol monostearate

(Cherukuri and Chau, 1991). The antidepressant indeloxazine-HCl is coated with a mixture of hydrogenated rape oil 10, polyoxyethylene polyoxypropylene glycol 1.4, sucrose fatty acid ester 0.1 wt.% in methylene chloride to give 120% coated granules. This provides the rapid-release serotonin uptake inhibitor with taste-masked properties (Shiozawa et al., 1992). Oral pharmaceutical preparations with bitter taste is controlled by adding >0.01-w% glycerol diester of C6–C22 fatty acid or diglyceride or sucrose fatty acid ester. As an example, an aqueous solution (1 L) containing quinine sulfate (0.036 g) with diglycerides from rapeseed oil (9 g) and sucrose ester (9 g) did not taste bitter (Suguira et al., 1995).

Hardened oil 15 and sugar ester 10 wt. parts were mixed and melted at approximately 80°, and to this was added ticlopidine-HCl 10 and low-substitution hydroxypropyl cellulose 5 wt. parts. The mixture was granulated to give a product with a high solution rate and little unpleasant taste and odor (Nakamura et al., 1995).

Affinity Biotech (Aston, PA) specializes in microemulsion technology for taste-masking powders, chewable tablets and liquid suspensions. Numerous generic drugs and proprietary compounds have been taste-masked with their lipids and polymers. A short review discusses drug and polymer coatings, dosage forms and in vivo studies (Bakan et al., 1992).

A substantially bitter-free ranitidine composition comprises a dispersion of molten glyceryl tristearate and glycerol trilaurate with ranitidine hydrochloride. Ethyl cellulose core particles were added to give a homogeneous lipid-coated suspension and the mixture spray-dried. Chewable tablets were prepared from the mixture (Douglas and Evans, 1994). Good palatability and good dissolution characteristics are achieved in chewable tablets also by granulating cimetidine with mono-, di- or tri-esters of sucrose or glycerol (polyhydroxy compounds having 2–10 hydroxyl groups) and stearic and/or palmitic acid used at about 20% w/w on cimetidine (Gottwald et al., 1989).

A particulate form of a drug with an unpalatable taste may be masked by mixing with a lipid, followed by an emulsifying agent, a polymer solution and a sweetening dilution solution to provide the final stable drug composition. Cimetidine is mixed with a lipid and to this drug/lipid mixture are added an emulsifying agent, a polymer solution and a sweetening dilution solution to provide a final taste-masked drug composition. Thus, Panodan 205 70 mg, Lipo GMS-470 50 mg and Pureco 92 400 mg were heated to 90°F until a clear homogeneous liquid phase was formed to which was added 400 mg of cimetidine. To the above mixture was added 788 mg polymer solution containing 8% w/v carboxymethylcellulose Aqualon 7LF PH and 20% w/v glycerin, followed by addition of 10.56 g sucrose solution containing 65% sucrose and 0.050 w/v methylparaben and flavors comprising 1 mg peppermint oil and 2.8 mg spearmint oil. The composition had pH 7.7 and contained 400 mg cimetidine/10 mL liquid (Yiv and Tustian, 1994a, 1994b).

Time-intensity measures were used to assess the effects of mouthcoatings of sunflower or coconut oils vs. water on subsequent perception of intensity of sucrose, salt, citric acid and quinine sulfate in gelatin gels. Panel mean curves for each stimulus were produced and compared by means of a new averaging method. In general, both oils reduced maximum intensity and suppressed responses on a number of other time and intensity measures. Compared to sunflower oil, coconut oil tended to have more consistent and suppressive effects on response measures. The possible mechanism has been discussed (Lynch et al., 1993).

LECITHIN-LIKE SUBSTANCES

Most recent research has found a specific inhibitor for bitter taste that is universal to several test substances and may be useful for masking bitter tastes of drugs and foods. Dr. Katsuragi at the Kao Corporation in Japan along with Dr. Kurihara at the Faculty of Pharmaceutical Sciences of the Hokkaido Unviersity have found homogenated suspensions of phosphatidic acid and β-lactoglobulin from soybean and milk, respectively, completely suppress bitter stimulants such as quinine, L-leucine and isoleucine, caffeine and papaverine hydrochloride in frogs and humans. The suspension ingredients do not suppress sweet, sour or salt taste. Other lipids such as phosphatidylcholine, triglycerides or diglycerides with β-lactoglobulin did not have as marked an effect. The potentially commercial product is said to be prepared by homogenizing a suspension of PA and LG in water and dehydrating the homogenate, yielding a powder that is easily dissolved in water (Katsuragi and Kurihara, 1993).

The frog taste selectivity and reversibility of nerve responses to bitter substances with lipoproteins including PA-LG led to human taste testing. Lipoproteins including bovine serum albumin, phosphatidic acid, lactoglobulin, ovalbumin, α-lactoalbumin or casein similarly suppressed bitter taste. The bitter taste of 12 drugs was inhibited, while saltiness of NaCl and sweetness of sucrose were not inhibited. The inhibition of bitter taste was completely reversible. Masking of the target sites for bitter substances on the taste receptor membranes with PA-LG in the medium seems to contribute to the inhibition of bitter taste of certain substances. Basic and hydrophobic substances such as quinine, denatonium and propranolol have low taste thresholds and are said to be the most bitter. PA-LG most effectively suppressed the bitter taste of such substances. PA originates from soybeans and the proteins used, except for bovine serum albumin, originate from milk and eggs; hence the lipoproteins can be safely used to mask the bitter taste of drugs. Phospholipids from soybean (1 part) can be emulsified with D-mannitol (2 parts) and water (7 parts) and spray-dried to prepare the product. The bitter of one part of quinine is nearly eliminated in 50 parts of the solid (Katsuragi et al., 1995a, 1995b); Yamazawa and Katsuragi, 1995). The technology of acidic phospholipids or an acidic lysophosolipids has taken the forefront in suppressing the bitter taste of foods, beverages and pharmaceuticals (Katsuragi et al., 1995a). Interestingly, the effective use of PA alone was shown to have inhibition qualities. PA-LG may be too expensive for widespread commercial use in foods and beverages except only in high value products such as pharmaceuticals (cf. Chapter 12). However, formulations with a large excess of lecithin or lecithin-like substances are claimed to control bitter taste in pharmaceuticals (Kinoshita and Mutsumi, 1987).

SURFACTANTS

A pleasant-tasting toothpaste base is prepared with an aqueous paste of 25–75% Texapon N70, 100 g (alkyl ether sulfates, Henkel Corp., Düsseldorf, Germany) mixed with an excess by weight of glycerin, 400 g. The water content is distilled off at 80°/100 Torr and then 110°/20 Torr. Superheated steam at 120°/5 Torr is passed through the mixture for 30 minutes with intense agitation to remove by-products with an unpleasant flavor and aftertaste (Ansmann et al., 1993).

COATINGS AND COMPLEXES WITH CARBOHYDRATES, RESINS, PROTEINS AND ZEOLITES

Microencapsulated coatings, bonding to resins, and inclusion complexes comprise techniques of adsorption and/or inclusion to mask the bitter taste. Coating of chewable tablets provides excellent taste masking while still providing acceptable bioavailability. Particles of core materials successively coated with drugs and polymers mask bitter and astringent taste with smooth mouth-feel. In the formulation of spherical particles by extrusion/spheronization, the microcrystalline cellulose brands cannot be interchanged without also changing the water content of the formulations (Newton et al., 1993). Pleasant-tasting chewable tablets of cimetidine are prepared by including hygroscopic water-insoluble cellulose as excipient (France and Leonard, 1994). The tablets contain cimetidine 12.7, Eudragit E100 1.3, microcrystalline cellulose 9.5%, sorbitol 38.0, lactose 31.6, sodium croscarmellose 2.5, aspartame 0.6, anise seed flavor 1.3, butterscotch flavor 1.3 and magnesium stearate 1.3.

CARBOHYDRATES

Increasing viscosity with rheological modifiers such as gums or carbohydrates can lower diffusion of bitter substances in the saliva to the taste buds. Acetaminophen suspensions are formulated with xanthan gum (0.1–0.2%) and microcrystalline cellulose (0.6–1%) to reduce bitter taste (Blase and Shah, 1993). A syrup composition comprising phenobarbital or acetaminophen in a polyhydric alcohol such as polyethylene glycol or propylene glycol with polyvinylpyrrolidone, gum arabic or gelatin is said to have no bitter taste (Kawasaki and Suzuki, 1991).

The taste of orally administered drugs may be masked by coating the drug with polymeric carbohydrate membranes. Bitter solid drugs are formulated in an organoleptically acceptable manner by particle coating with a mixture of a water-insoluble film-forming polymer (cellulose or shellac) and a second film-forming polymer soluble at pH <5 [(dimethylamino)ethyl methacrylate-methacrylic acid ester copolymer]. With this process, a spasmolytic, pinaverium bromide has no bitter taste (Block et al., 1990). Antihistamines may be adsorbed onto Avicel PH 101 porous particles (FMC Corp., Philadelphia, PA). An aqueous solution containing 50 parts chlorpheniramine maleate was impregnated onto 3000 parts of the polymeric carbohydrate and spray-coated with an aqueous solution containing xylitol to seal the polymeric pores (Ogasawara and Ueda, 1992). Taste masking, compressible grade formulations of xylitol are available as Xylitab® from Xyrofin (Schaumburg, IL). Triprolidine-HCl, an oral form of antihistamine, was taste-masked with a dispersion coating of the water-soluble polymer hydroxypropyl cellulose, plasticizing agent, sweetening and flavoring agent (McCabe et al., 1992). Atenolol (an anti-arythmic) 250, Florite (Ca) 250, calcium lactate 250 and water 900 g were mixed and granulated to give bitterness-free granules (Sugawara et al., 1995).

Oriental drugs are masked with starch and/or celluloses containing carboxymethyl (CM) groups. Examples include: carboxymethyl cellulose (CMC), sodium CMC, calcium CMC, cross-linked sodium CMC and sodium CM starch (Hayashida and Hatayama, 1993). Core elements of drugs coated with water-insoluble polymer such as ethyl cellulose offer taste masking and reduced dissolution profiles for a variety of drugs including paracetamol, ranitidine hydrochloride, doxycycline hydrochloride, pseudoephedrine hydrochloride, sodium naproxen, theophylline and aspirin (Morella and Lukas, 1992). An oil-free, extrudable composition containing cholestyramine was homogenized and extruded at 49.4 bar and 174° to release steam and expand the product into a taste-masked porous structure (Heckenmueller and Friess, 1994). The base formula contained cholestyramine 400, wheat starch 860, wheat bran 70, flavoring 0.4 and water 332.6 kg.

Bitter-masked oral preparations of ecabet sodium-containing cores comprise coating film layers, and overcoating film layers containing menthol or sodium glutamate (Nakajima et al., 1994). The process uses granules 70, D-mannitol 5, hydroxypropylcellulose 11 and hydroxypropyl methyl cellulose 4 weight parts, which were coated with an aqueous solution containing hydroxypropyl methyl cellulose 8, Macrogol 6000 2 and talc 1 weight part, and overcoated with 0.03 weight part menthol and 0.02 parts silica.

A suspension of beeswax and ethyl cellulose was used to coat pseudoephedrine hydrochloride to form a tasteless particle (Cuca et al., 1994). Pharmaceutical granules with bitter taste are coated with water-soluble polymers of hydroxypropyl cellulose and sugars such as sucrose and lactose to decrease the bitter perception at the time of oral administration (Moroi et al., 1993). Ethyl cellulose (ETHOCEL, Dow Chemical Co., Midland, MI) coating systems are commercially applied with acetaminophen and are available as SURELEASE from Colorcon (West Point, PA). Bitter basic pharmaceutical salts may be bitter reduced with weakly alkaline compounds of good bioavailability. Cefcanel daloxate hydrochloride (90 g), lactose (360 g) and corn starch (90 g) were mixed and granulated with 15 g PVP (a weakly alkaline substance) containing ethanol solution to give granules. The granules (400 g) were first coated with an ethyl cellulose containing methylene chloride-methanol mixed solution. They were further coated with a similar solution of 50 g trisodium citrate and 10 g ethyl cellulose and mixed with lemon oil and granules prepared from sodium saccharin (4 g) sucrose (300 g) and 2% hydroxypropyl cellulose aqueous solution. The final product contained 100 mg cefcanel daloxate hydrochloride per gram. There was no bitterness and good bioavailability to volunteers (Tabata and Yoshimi, 1992a, 1992b). Sparfloxacin, a new quinolone antibacterial, is optimally taste-masked by preparing film-coated granules. Higher levels of ethyl cellulose reduce bitter most effectively. Optimal bioavailability is also achieved with 20% drug, 52% low-substituted hydroxypropyl cellulose in the cores and 10% ethyl cellulose (4 parts)/hydroxymethyl cellulose (2 parts) in the coating film. Increasing the core carbohydrate content causes the granules to burst prematurely as they hydrate (Shirai et al., 1993).

Further research by the authors documented the dissolution and bioavailability claims. Coated fine granules with a water-insoluble film composed primarily of ethyl cellulose, containing 20% sparfloxacin and various amounts of low-substituted hydroxypropyl cellulose in the cores, were orally administered to fasting rats to determine the effect of the core cellulose on bioavailability. The release of sparfloxacin in water from four kinds of coated fine granules containing 0, 25, 40 and 52% hydroxypropyl cellulose followed pseudo first-order kinetics, followed by the second phase, with refractive points between 0.25 and 0.5 h. The rate constant

(K_1) up to 0.25 h increased with an increase in the amount of cellulose in the core, and rate constant (K_2) in subsequent release (the second phase) was lower than K_1 in each fine granule. The areas under plasma-concentration-time curves of the drug and the peak plasma levels (C_{max}) after oral administration of coated fine granules lacking the cellulose were suppressed to 1/8 to 1/9, respectively, of those obtained from the core granules that rapidly released the drug. However, the plasma-concentration-time curves and (C_{max}) from the coated fine granules increased linearly with an increase in the amount of the cellulose in the cores, and nearly equaled those from the core fine granules when the content of the core cellulose was 52%. These results confirmed that the addition of the cellulose to the cores increases not only the dissolution rate but also the bioavailability of the drug (Shirai et al., 1994). Heat treatments of granules containing sucrose fatty acid esters cause changes in the film properties. It is hypothesized that tensile strength and wettability are altered and result in an increased dissolution level (Shirai et al., 1996).

The antidiarrheal loperamide hydrochloride or Imodium® (Janssen Pharmaceutica, Piscataway, NJ; McNeil Consumer Products, Co., Fort Washington, PA) in tableted form contains loperamide or its salts as sugar- or film-coated to give bitterness-free tablets (Sakakibara et al., 1993). The tablet formulation for antidiarrheals contains loperamide hydrochloride 1.0, water-soluble sugars or starches (lactose) 70.0, crystalline cellulose 15.0, hydroxypropyl cellulose 13.0, water-insoluble metal salts or their corresponding acids and magnesium stearate 1.0 mg.

An analgesic cold-medicine mixture contains acetaminophen 150, *d*-chlorpheniramine maleate 0.583, Medicon P salt (dextromethorphan hydrobromide phenolphthalein salt) 12, noscapine 8, D-mannitol 187.417, hydrogenated castor oil 90, low-substituted hydroxypropyl cellulose 80, partial gelatinized starch 60 and HPC-SL 12 parts. The contents were kneaded with ethanol and granulated, then heated at 90° for 20 minutes to give granules with an in vivo dissolution rate of 22% in 30 s versus a 90–100% dissolution rate at 30 s without the heat melding process. The early-prevented dissolution rate is key to masking the bitter taste of drugs. At 10 minutes, both forms of granules have 100% dissolution in vivo (Nagafuji et al., 1995).

Geriatric and pediatric applications of a common cold formula (Ratnaraj and Sunshine, 1994) contain acetaminophen powder 3.2, pseudoephedrine HCl 0.3, chlorpheniramine maleate 0.02, HFCS 73.0, water 20.0, sorbitol solution 20.0, glycerin 10.0, xanthan gum 0.14, CM cellulose 0.56, Na CM cellulose 0.03, butylparaben 0.025, Na benzoate 0.2, propylene glycol 0.25, malic acid 0.076, citric acid 0.038, coloring 0.002 and grape flavoring 0.2 g/100 mL.

A nonrupturable, fast-dissolving, taste-masking composition provides immediate release of pharmaceutically acceptable active ingredients (Brideau, 1995). The formula contains magnesium stearate 0.5, sorbitol 44.17, silica 0.1, citric acid 6.25, $NaHCO_3$ 18.75, chlorpheniramine maleate 0.8, phenylpropanolamine HCl 5.0, xanthan gum 24, orange flavor 0.4 and cream flavor 0.3%.

Sulpyrin, an aryl-alkyl pyrazolone analgesic, may be prepared with no bitter taste by mixing 100 g of the drug in 100 g water and 200 g partially gelatinized starch, crushed and dried to give pellets (Kamata and Hirano, 1994).

In order to mask the bitter taste of drugs, a novel microencapsulation process combined with the wet spherical agglomeration (WSA) technique was developed by using a modified phase separation method. The spherical agglomerates of enoxacin (a fluorinated quinolone antibacterial) with various additives, including disintegrants, were succesfully produced in the system of acetone-*n*-hexane-ammonia water or acetone-*n*-hexane-distilled water by the WSA, using the flocculation phenomena of particles in liquids. The resultant agglomerates could be microencapsulated continuously with Eudragit RS utilizing the phase separation method in the same system as agglomeration under stirring. Explosible microcapsules which were free from the bitter taste could be produced in formulating finer particle sizes of enoxacin and 50% of Primojel in core agglomerates, using distilled water as a bridging liquid, and treating with 20% polymer coating level. These microcapsules were bioequivalent to the commercial enoxacin 100 mg tablets in beagle dogs. Thus, one continuous process technique of agglomeration and microencapsulation was useful for the design of enoxacin powders that masked the bitter taste and controlled the drug release rate (Ueda, 1993). Wet spherical agglomeration of improved core particles and microencapsulation using the coacervation method in a continuous process is reviewed with 32 references (Ueda, 1995).

Oral pharmaceutical tablets are manufactured by coating the active ingredient amino acid or peptides with aminoalkyl methacrylate copolymer Eudragit RS and then with a layer of rape oil and insoluble (inorganic) talc and silica to cover the unpleasant odor and taste and to promote active ingredient release (Mito et al., 1994).

Ibuprofen may also be formulated as Formula 23, rotogranulated and coated with a solution containing hydroxyethyl cellulose (400 parts) and hydroxypropyl methyl cellulose (400 parts) in water (9200 parts) to obtain coated granules where the coating was 18% of the total weight of the granules. The granules were compressed into chewable tablets (Roche and Reo, 1992) for rotogranulation containing ibuprofen 5000, PVP 210.5 and sodium lauryl sulfate 52.6.

Ibuprofen encapsulated using cellulose acetate phthalate and gelatin as the microencapsulating wall material provided high bioavailability at high payload (>83%) and taste masking for chewable tablets and liquid aqueous suspensions for medicinal use (Ghanta and Guisinger, 1995). Acetaminophen was coated with a blend of cellulose acetate and PVP and compressed with other conventional ingredients to form a wafer. The wafer was placed on the tongue and found to disintegrate in <30 sec. without a bitter aftertaste (Gowan, 1995).

A novel neomorphic form of ibuprofen is characterized by a distinctly less bitter taste and caused less of a burning sensation upon swallowing. The form exhibits no birefringence. Tests also indicate less irritation to the gastrointestinal tract of animals upon administration. The preparation involves conventional ibuprofen heated to its molten state at 77–80° and cooled to 0° in a pliable plastic container; the vessel is struck repeatedly by a hammer to induce the supercooled ibuprofen to resolidify. This product showed improved properties compared to conventional ibuprofen (Geyer and Tuliani, 1995).

Aspirin tablets may be taste-masked with a plasticized thin film of cellulose acetate latex and triacetin (glycerin triacetate) at not more than 1% of the coated medicament (Wheatley and Erkoboni, 1992). The advent of gel caps and liquid-filled capsules has revolutionized taste masking for oral ingestion. A coating composition for acetaminophen tableting contains cellulose acetate (39.8% acetylation) 15, Eudragit E100 50 and PVP (avg. MW 40,000) 35% (Hoy and Roche, 1993).

The antiulcerative propantheline bromide coated on low-substitution spherical hydroxypropyl cellulose, and further coated with ethyl cellulose, masks the unpleasant taste while readily releasing the active ingredient from the preparation (Block et al., 1990). The chewable tablet may be prepared from the formulation of famotidine 5.2, hydroxypropyl methyl cellulose 2.4 and lactose 32.4 kg, rotogranulated and coated with a 10% solution of cellulose acetate and hydroxypropyl cellulose (70:30) in a mixture of 80:20 acetone and methanol (Roche et al., 1993). Famotidine (Pepcid; Merck Sharp & Dohme, West Point, PA) has been rotogranulated with PVP and lactose and coated with cellulose acetate and PVP (80:20), then compressed with binders and other excipients to prepare a chewable tablet (Roche et al., 1992).

PROTEINS

Notably, Tylenol® Geltabs (McNeil Consumer Products Co., Fort Washington, PA), employing the Gelkote™ gelatin-coating process, offer ease of swallowing and taste masking. The singular disadvantage arises in hot, humid climates where product stability may be compromised. Hydrolyzed gelatin improves taste and mouth-feel when incorporated in small amounts into chewable tablets containing ingredients requiring taste masking. Medicinals and nutritional supplements may now be prepared as chewable tablets for those children or adults who find regular tablets hard to swallow, or for those desiring the convenience. In particular, Riopan (magnesium sulfate/aluminum hydroxide) or calcium carbonate 500, starch 120, sweeteners 360, flavors 8, lubricants 15 and hydrolyzed gelatins 3 mg possessed a large improvement in taste and mouth-feel (Alexander et al., 1995).

A tamper-resistant, caplet-like medicament that masks the taste of poor-tasting drugs and possesses improved lubricity for easier swallowing is produced by shrinking a gelatin-based capsule about a caplet-shaped tablet of the medicament at specific temperatures, pressure and relative humidity. Thus, two halves of gelatin capsules were placed about the caplet core either manually or mechanically at 65–72° and relative humidity of 70–75% for 120–180 minutes with tablet defects of 3–17% (Lebrun and Massey, 1995).

Essential oils in a gelatin capsule film are said to enhance efficacy of the active ingredients. Examples include capsules for rhinitis, common cold, antipyretic analgesic and antitussive expectorants containing chlorpheniramine maleate 4.0, phenylpropanolamine hydrochloride 24.0, belladona alkaloids 0.13, caffeine 40.0, corn oil 175.0 and monoglycerides 10.0 mg. The mixture was encapsulated with a capsule base containing gelatin 140.56, concentrated glycerin 39.36, preservatives 0.08, menthol 0.40 mg and colorant to give a soft capsule. The capsule was orally administered to 10 patients with rhinitis to ameliorate snivel, nasal congestion and heaviness of head at effective rates of 90, 88 and 93%, respectively, versus 75, 72 and 80%, respectively, for a control capsule containing no menthol (Takahashi et al., 1995).

Mint-flavored chewing gum exhibits reduced bitterness by virtue of a reduction in the amount of 1-menthol in the flavoring agent (Record et al. 1994). The optimum performance of chewing gum may be achieved from mathematical predictions of flavor release at any time during mastication leading to reformulation of flavor compositions (de Roos and Wolswinkel, 1994).

However, gums that have a high level of mint flavor (>4–7% of flavor to gum base) also achieve high flavor impact upon chewing. For potential mint-flavored oral pharmaceutical gums, late chew bitterness and harshness of the flavor may be reduced by incorporating a prolamine/cellulose ingredient. A comparative and inventive example using sugar in the formulas from Wm. Wrigley, Jr. Co. is shown in Figure 1. Bitterness escalates as flavor decreases over time. The granulated ingredient containing zein (an alcohol-soluble protein of the prolamine class obtained from corn) and a hydroxypropylmethylcellulose (or other cellulose derivatives) at 24%

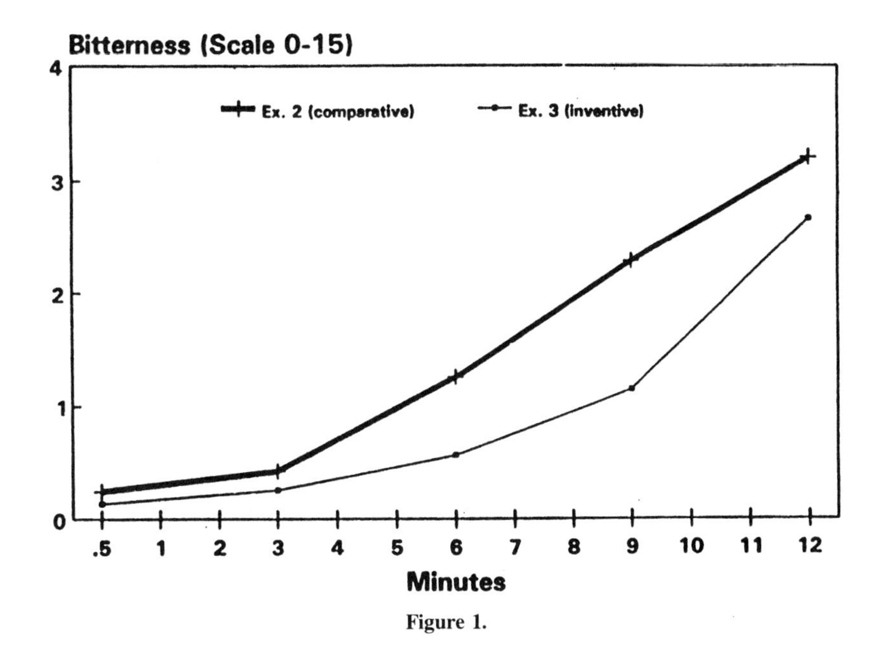

Figure 1.

zein, 71% HPMC, 0.5% sodium hydroxide and 4% water. A high pH aqueous zein solution used to coat hydroxypropyl cellulose is particularly effective in combating the bitterness of spearmint flavor. Peppermint flavor can be used at higher flavor levels before becoming objectionable. It is believed the zein ingredient opens up ionically or covalently with select flavor molecules (trace aldehydes and phenolic compounds). The HPMC and/or zein is believed to coat mucous membranes and prevent the perception of bitterness or harshness (Patel et al., 1993). Macrolide antibiotics are coated with a mixture of prolamine (zein and/or gliadin) and plasticizer, preferably vegetable oil or wax. The coating prevents dissolution of the drug in the mouth and acts as a taste masker. Cores of clarithromycin-PVP mixtures (9:1) were coated with a mixture of 93% zein and 7% medium-chain triglycerides to accomplish bitterness reduction (Meyer and Mazer, 1993).

A rokitamycin monohydrate crystal process is said to hide the bitterness of the substance. Wettability is also said to be improved, thus allowing the preparation of bitter-free syrups. The process dissolves rokitamycin in aqueous 3% acetic acid cooled to 10°; the solution is adjusted to pH 5 with 10% aqueous sodium hydroxide and left to stand at room temperature for 24 h. The precipitated crystals are collected by filtration and dried in vacuo to give the monohydrate (Kinoshita, 1994). Various microencapsulated dosage forms were prepared to limit release of an antibiotic into solution for up to three days and in the oral cavity following administration.

INCLUSION COMPLEXES

The strong bitter taste of liquid or syrup carbetapentane citrate is reduced 50% by preparing a 1:1 complex with cyclodextrin (Kurasumi et al., 1991). Palatable ibuprofen solutions are prepared by forming a 1:11 to 1:15 inclusion complex with ibuprofen and hydroxypropyl-β-cyclodextrin, respectively. The complex removes the bitter component but creates a sour taste that is masked by sweeteners (Motola et al., 1991). An ibuprofen-β-cyclodextrin complex is prepared for taste masking of ibuprofen by dissolving β-cyclodextrin in water at 100° and adding ibuprofen, then drying the resulting solution at 60° for 16 h to give a powder with about 7% ibuprofen content. Diffierent formulations are also given to prepare hot water-soluble and pleasant-tasting solutions (Grattan, 1993).

The preparation and design of bitterness-suppressed granules of benexate hydrochloride-cyclodextrin (Betadex; Eurand Intl., S.p.A.) is effective in the oral treatment of gastric ulcer. The inclusion compounding also protects benexate's instability in water and organic solvents. The powder-coating granulation is accomplished by hot melt methods using PEG 6000 or 4000 and stearyl alcohol as binders. A centrifugal fluidizing granulator can coat the drug successively on swollen microcrystalline cellulose spheres, then with bitterness-suppressive layers of hydrophobic compounds such as cyclodextrins (Susuki et al., 1993b). The molecular structure of the inclusion complex of CD with quinine in aqueous solution was determined by FTIR (Wang and Chen, 1994).

Pharmaceuticals, foods or feed additives containing *Gymnema sylvestre*, a bitter and astringent-tasting sweetener for diabetes control, can have their unpleasant tastes masked by the aminopolysaccharide chitosan (Ikezuki, 1990). Chitosan is a major component from the hydrolytic treatment of insect and crustacean skeletons, besides calcium carbonate. A therapeutic tea bag containing *Gymnema sylvestre* with 5% added chitosan is said to mask bitter taste in the *Gymnema sylvestre* tea (Kaneko, 1991). The leaf extract of *Gymnema sylvestre* is less bitter when mixed with cyclodextrin, which further enhances the blood sugar-lowering effect of the extract containing gymnemic acids (Ueno, 1992).

The bitter and astringent taste of antidiabetic *Gymnema sylvestre* extract (containing gymnemic acid glycosides) is masked by hydrolysis of the extract under alkaline (pH 8-14) and anaerobic conditions. The treatment has no adverse effect on blood sugar-lowering activity as determined in Wister rats (Yumoto et al., 1994).

The bitter flavor of ursodeoxycholic acid in formulations for treatment of cholestasis in children is masked by dispersing finely crystalline ursodeoxycholic acid in an a aqueous medium containing a thickening or swelling agent so that only a small portion of the acid is dissolved. Residual bitterness is removed by addition of β-cyclodextrin or flavor-masking agents. For example, a suspension as a nonbitter liquid dosage (Widauer, 1995) contained ursodeoxycholic acid 0.05 g, β-cyclodextrin 0.1, Avicel RC591 0.01, sucrose 0.3, methylparaben 0.0013, propylparaben 0.0002, propylene glycol 0.05, flavoring 0.0013 and demineralized water to 1 mL. Remarkably, cyclodextrins have been known since 1891 and hopefully will be gifted with an "Additive" approval in the U.S. in the very near future.

RESINS

The peripheral vasodilator buflomedil may be taste-masked by bonding to a cation exchange resin such as Amberlite IRP 69 (Rohm and Haas, Philadelphia, PA) at 60% resinate powder (Honeysett et al., 1992). The Amberlite IRP 64 resin is also recommended with isopropyl alcohol as solvent. The dried resin powder was incorporated into oral formulations (Lafon, 1992). A chlorotheophylline salt of the antihistamine, diphenhydramine, is called dimenhydrinate and indicated for treatment of motion sickness (Dramamine®; Richardson-Vicks Health Care, Wilton, CT; G. D. Searle & Co., Skokie, IL). The formulation is available as chewable tablets with methacrylic acid copolymer (Eudragits; Roehm GmbH, Darmstadt, Germany) or carboxymethyl cellulose (CMC) and starch for taste masking. Alternatively, a delayed-release matrix in an effervescent form of disintegrating tablets of instant granulates is available. The drug-containing matrix is applied to a carrier with fatty acid esters to contribute taste masking. The formulations release about 65% of the drug within 2 minutes in aqueous solution or >80% of the drug in 0.1 N hydrochloric acid at body temperature (Gergely et al., 1993).

The base pharmaceutical formulation to be added to effervescent formulation contained dimenhydrinate 2, Eudragit L 3, Eudragit S 3, xylitol 1-3, blended in: ethyl alcohol 35 parts, isopropyl alcohol 10 parts and sorbitol 200 parts, evaporated and added to effervescent formulations containing fatty acid esters to contribute taste masking. Similarly, dimenhydrinate (10 g) in chewable formulations comprises Eudragit L 10 g dissolved in 100 mL isopropyl alcohol and then 200 mL hexane added to precipitate the product. This complex of basic and acidic components, respectively, was then granulated with conventional tableting agents (Francese et al., 1995).

Sagebrush extract and absinthe oil have found application in the preparation of hard candies and chewing gum for mouth-odor control and beverages for use to improve skin smoothness. Absinthe oil is a bitter, green, licorice-flavor distilled liqueur principally flavored with an aromatic oil obtained from wormwood *(Artemisia absinthium)*, also known as absinthe. Its alleged toxicity led to earlier prohibition in the U.S. and Canada, when in fact only large doses cause GI problems, nervousness, stupor, convulsions and finally death. Vermouth is a blend of white wines containing traces of absinthium and other flavors. The bitter components, mainly azulenoids, may be removed by gel permeation chromatography on cross-linked dextran oxypropyl ether gel such as Sephadex LH-20 (Kodama et al., 1992a, 1992b, 1992c).

The pH-sensitive polymethyacrylate (Eudragit resin) microcapsules were suitable for taste masking of antibiotics (Friend, 1992a, 1992b). A polymer carrier system was developed to reduce the bitterness of erythromycin and its 6-*O*-methyl derivative, clarithromycin, by adsorption to Carbopol (R.I.T.A. Corp., Woodstock, IL). The mechanism involves ionic bonding of the amine macrolide to the high molecular weight polyacrylic acid, thereby removing the drug from the solution phase in an ion-free suspension. After ingestion, endogenous cations displace the drug from the polymer into the gastrointestinal tract to achieve bioavailability (Lu et al., 1991).

Poly(vinyl acetal) (diethylamino)acetate is soluble in gastric juice and organic solvents. Hence, it is used as a coating polymer to prevent water entry into tablets and to mask drug bitterness. In aqueous 10% solution with increasing temperature, the solution coagulates hydrophobically to form a hydrogel. The gel formation impacts the thermo shrinking and drug-dissolution properties of clarithromycin, and a large study identified critical ratios of clarithromycin in the gel to optimize the degree of microencapsulation giving efficacy and taste masking (Shimano et al., 1994). Clarithromycin was also dispersed in cacao fat at 35–50°C and atomized to give fine granules, which were suspended in 7 wt% of poly (vinyl acetal) (diethylamino)acetate at 0°C and spray-dried. The preparation has no bitter taste and good bioavailability (Koyama et al., 1990a).

Oral preparations are manufactured by soaking particles containing water-soluble biologically active compounds encapsulated in polymers in buffer solutions containing carboxylic acids and metal hydroxides. An aqueous suspension contained water 94, poly(vinyl acetal) (diethylamino) acetate 5, sodium lauryl sulfate 0.02, cetyl alcohol 0.005 and clarithromycin 1 wt. part. The mixture was sprayed through a nozzle to give aqueous particles, which were soaked in glycine buffer containing sodium hydroxide at 52° for 14 h and dried at 7° for 30 min. to give a nonbitter oral preparation (Koyama and Shimano, 1993a). Similar formulations are given to prepare an aqueous clarithromycin gel that is not bitter (Koyama and Shimano, 1993b).

β-Lactam antibiotics having basic groups may have their unpleasant odor and bitter taste masked while still retaining excellent dissolution properties and absorbability of the active ingredient. Water-soluble enteric polymers and ion-exchange resins provide masking potential in granules containing milled granulated sugar 282 parts, a penem antibiotic 15, hydroxypropylcellulose 3, ethanol 147 and coated with a solution comprising hydroxypropylmethylcellulose phthalate (95), PEG 6000 (5), acetone (1142.9), ethanol (285.7) to obtain enteric coated granules which were sieved to provide granules having a mean particle diameter of 400 mm. The above granules were mixed with a styrenic strongly acidic cation-exchange resin (Diaion SK110) to provide the invention (Matoba et al., 1994; Maruyama, 1993).

Ranitidine (Zantac; Glaxo Pharmaceuticals, Research Triangle Park, NC) possesses a bitter taste that can be masked by adsorbing on a cation-anion exchange resin (Douglas and Bird, 1991). Cross-linked gelatin via glutaraldehyde and Eudragit resin L100, S100 and E100 coated capsules were effective in preventing release of clarithromycin under simulated conditions of storage (Friend, 1992a). Erythromycin is also prepared with acceptable taste (Koyama et al., 1990b). Stearyl alcohol (700 g) was melted at 100°C and Eudragit E (100 g) dispersed in the melt followed by dispersing clarithromycin (200 g). The resultant dispersion was spray-cooled and granulated. Sorbitol 90 g, magnesium oxide 0.2 g and crystalline cellulose 9.8 g were mixed with 100 g of the granules for an oral administration containing 10% clarithromycin (Yajima et al., 1993a). Alternatively, the granules (600 g), sugar (450 g) and hydroxypropylcellulose (50 g) were mixed to give a "dry syrup" (Yajima et al., 1993b). The taste of oral pharmaceuticals in aqueous suspensions is also controlled by liposome-associated flavorants (Popescu and Mertz, 1991).

A method of taste masking and spray-congealing technique for clarithromycin (CAM) product does not dissolve in the mouth, yet is bioavailable in the gastrointestinal tract with bioequivalence to the conventional dosage form. Glyceryl monostearate (GM) and aminoalkyl methacrylate copolymer E (AMCE) were selected as ingredients since GM, a substance with a low melting point, is able to be decomposed by an enzymic reaction in the intestinal tract where the solubility of AMCE is very low. On the other hand, AMCE was selected becaue it is freely soluble at low pH (as in stomach), but insoluble in higher pH levels (as in mouth). Spherical particles of the matrix and discs of the matrix with various ratios were prepared and the optimum ratios of CAM, GM and AMCE determined (by the simple method of experimental design) as 3:6:1. The release delivery criteria were:

- 100 mg/L of CAM in the spherical matrix should be completely released within 20 minutes in pH 4.0.
- Less than 14 mg/L of CAM in spherical matrix should be released at 60 minutes in pH 6.5 (Yajima et al., 1996).

The antibacterial roxithromycin (Rulid; Hoechst-Roussel, Somerville,

NJ) and polyethylene glycol were mixed with water, granulated, dried and coated with a mixture of Eudragit L100-55, sodium hydroxide, talc, triethyl citrate, licorice flavor and water and incorporated into tablets (Mapelli et al., 1991). Melt granulation with water-soluble substances successfully reduces bitterness for cetraxate and ofloxacin. Granules consisting of cetraxate hydrochloride, corn starch and Macrogol-6000 (PEG-6000) were coated with a mixture of Eudragit S100, talc and silica to mask bitter taste (Haramiishi, 1993a, 1993b). Antibacterial quinolonecarboxylic acid derivatives, such as ciprofloxacin, are adminstered to animals in their food loaded on weak cation exchangers such as the H^+ forms of Lewatit S100 or Lewatit CNP. The taste is improved as judged by the animals accepting the material more readily (Lange et al., 1992). Ciprofloxacin (betaine form) may be microencapsulated into a mixture of esters or quaternary ammonium salts of Eudragit NE 30D and hydroxypropyl cellulose (Poellinger et al., 1993). Remoxipride, a D2-dopamine receptor antagonist, is well tolerated and completely absorbed after oral administration. Because of its extremely bitter taste, an oral palatable suspension was developed by using microencapsulation providing complete bioavailability, but the rate of absorption is delayed 3 h. In comparison, an aqueous 0.5% sodium lauryl sulfate oral form delayed absorption only 1 h and a capsule 1.6 h (Sjoeqvist et al., 1993).

Ibuprofen is encapsulated with a chewable methacrylic acid copolymer to reduce bitterness. A fluidized bed of ibuprofen crystals was spray-coated with an aqueous dispersion of Eudragit L 30D coating polymer, propylene glycol as plasticizer and talc. The encapsulated ibuprofen was mixed with mannitol and flavor and compressed into tablets (Shen, 1991). Acetaminophen granules were coated with a solution of cellulose acetate (39.8% acetylation) 15, cellulose acetate butyrate or hydroxypropyl cellulose (Roche, 1991) or cellulose acetate (39.8% acetylation), Eudragit E100 50, and PVP (avg. MW 40,000) 35 (Hoy and Roche, 1993).

Morphine hydrochloride coated on spherical cellulose was further coated with an aqueous solution containing Eudragit NE30D and talc. The particles were then overcoated with Avicel RC-591NF (FMC Corp., Philadelphia, PA), sucrose, D-sorbitol, sodium saccharin, methylparaben and vanilla essence to give powders with good sustained release properties in water suspensions at pH 1.2–6.8 with no bitter taste (Mori et al., 1993).

"Dry syrups" are formulated, kneaded, granulated, coated with glycerin monostearate and overcoated with Eudragit L 30D55, polyethylene glycol and talc and sprayed with sucrose to afford a nonbitter dry syrup (Nomura and Izumida, 1993).

One formulation for "dry" syrups contains bifemelane hydrochloride salt 1, sodium citrate 1, mannitol 1.38, corn starch 0.58 and hydroxypropyl cellulose 0.04 kg. Potato starch (1800 g) was treated with 100 mL ethanol containing a diazepine, and the composition then dried at 60°C for 8 h. The powder was mixed with magnesium stearate and spray-coated with an ethanol solution of amino alkyl methacrylate copolymer to give coated particles. A mixture containing the coated particles 300 g, lactose 480 g, sucrose 240 g, mannitol 240 g and corn starch 2150 g was granulated while spraying an aqueous hydroxypropyl cellulose solution (Nitsuta et al., 1992).

ZEOLITES

Bactericidal feeds for domestic animals generally impart bitter tastes and may create feeding aversion among the animals in treatment. Tiamulin fumarate may be dissolved in methanol, supported on mordenite-type zeolite or starch, dried and further premixed with the supports to produce sustained-release bitterness-free granules. The results were determined by a stronger bactericidal effect on *Mycoplasma, Staphylococcus* and *Corynebacterium* than when tiamulin fumarate was in feed by itself, and not overly averted by pigs (Ryu, 1991).

SALT PREPARATION

The unpleasant taste of water-soluble ibuprofen salts in aqueous solution is masked by adding alkali metal bicarbonate, e.g., sodium bicarbonate (Gregory et al., 1990). Carbon dioxide-generating agents without fatty acid esters may mask bitter taste (Nishikawa and Hayashi, 1992). They contain caffeine 1, sodium bicarbonate 5, ascorbic acid 3, citric acid 3 and tartaric acid 3.

Some interesting evaluations appear in The Merck Index. Penicillin prepared as the *N,N'*-dibenzylethylenediamine diacetate salt or *N,N'*-bis (dehydroabietyl)ethylenediamine disalt is a tasteless material whose safety may need more study. Magnesium aspirin is almost tasteless. Quinine salicylate is bitter. Formulated, bitterness-reduced antitussive and expectorant compositions contain magnesium salts, sweeteners, starch and cellulose (Nishikawa and Hayashi, 1993).

The formulation for antitussives contains dihydrocodeine phosphate 1, *dl*-methylephedrine hydrochloride 2.5, *d*-chlorpheniramine maleate 0.2, magnesium silicate 5, magnesium oxide 2, mannitol 25.3, lactose 45, corn starch 10, crystalline cellulose 8 and magneisium stearate 1.

Oral preparations of ecabet sodium as an ulcer inhibitor use alkali metal salts as bitterness-masking agents containing Ecabet 700, D-mannitol 252.7, NaCl 20, aspartame 5 and magnesium stearate 20 g. The ingredients were mixed and the mixture was granulated and mixed with 0.3 g *l*-menthol and 2 g SiO_2 to give a granule with no bitter taste while a control granule without NaCl tasted bitter (Hirakawa et al., 1995). The bitter taste of microbicidal pyridonecarboxylic acids is decreased in aluminum or magnesium salt preparations of aqueous solutions (40 mL) containing enoxacin (400), alum 400, aspartame 200 and citric acid 400 mg (Hachiman et al., 1994).

Bitter-tasting decongestants, antihistamines, antitussives and expectorants are effectively taste-masked using a magnesium trisilicate/fumed silica adsorbate that is undetectable during dissolution in the mouth, yet provides a high degree of drug bioavailability when it reaches the acidic conditions of the stomach (Lech et al., 1995).

Gastrointestinal treatments can be manufactured containing Veegum N 1.0, Mg trisilicate 6.75, granulated $CaCO_3$ 42.87, Polysorbate 80 0.1, citric acid 0.55, sucrose 12.0, mannitol 10.0, flavor 1.25, color 0.005 and water to 100 (Carella and Opiola, 1994). A multisymptom relief formula is prepared using a wet granulation process onto silicon dioxide (50–85% as carrier). A chewable tablet was formulated containing diphenhydramine-HCl 1.25, pseudoephedrine-HCl 3.0, Cab-O-Sil 20.75, tartaric acid 2.0, monoammonium glycyrrhizinate 0.25, aspartame 1.20, purple lake color 0.25, Emdex (corn syrup) 69.35, magnesium stearate 0.75 and grape flavor 1.25% (Lech et al., 1994). Azithromycin 600, and magnesium oxide 20.65 mg in 50 mL of water was prepared as a taste-masked composition (Catania and Johnson, 1994).

Formulation of alpiropride (analgesic for migraine headache) as a solid dispersion is said to be comparable in dissolution to that of a salt and much less bitter (Duclos et al., 1992).

FUNCTIONAL GROUP ALTERATION

The alkyloxyalkyl carbonates of the clarithromycin 2′ position have remarkably alleviated bitterness and improved bioabsorbability when administered orally (Table 1). Mice administered 100 mg/kg by mouth had several mg/mL detectable serum concentration of the modified drug active against *Micrococcus luteus* (Asaka et al., 1992).

TABLE 1. Serum Concentration of Modified Drug in Tested Mice in mg/mL.

Modified drug position	T = 1 h	2 h	3 h	4 h
2′-(2-Ethyloxy carbonate)	11.2	7.37	5.46	1.96
2′-Ethyl carbonate	0.31	0.15	0.15	0.02

And lastly, derivatization of the single alcoholic functional group of quinine can change the taste perception. The bitterness of quinine is removed by preparing quinine tannate, quinine carbonate or quinine ethyl carbonate.

REFERENCES

Akyama, Y. 1995a. Bitterness-free oral preparations containing branched amino acids, in JP 07 69,877.

Akyama, Y. 1995b. Oral pharmaceutical preparations containing branched amino acids, in JP 07,118,150.

Alexander, T. A., Daher, L. J., Gold, G., Hancock, C. L. and Peterson, D. L. 1995. Hydrolyzed gelatin as a flavor enhancer in a chewable tablet, in PCT Int. Appl. WO 95 05,165 to Miles, Inc.

Ansmann, A., Breitzke, W. and Gantke, K. H. 1993. Preparation of flavor-neutral pastes of alkyl ether sulfates in glycerin, in DE 4,131,118 to Henkel K.-G.a.A.

Aoi, M. and Murata, K. 1992. Control of bitter taste of pharmaceuticals and food, in JP 04,346,937 to Kibun Shokuhin K. K.; Kibun Food Chemifa Co., Ltd.

Asaka, T., Misawa, Y., Kashimura, M., Morimoto, S., Watanabe, Y. and Hatayama, K. 1992. Preparation of 2-O-modified erythromycin or derivatives thereof as antibacterial agents, in PCT Int. Appl. WO 92 06,991 to Taisho Pharmaceutical Co. Ltd.

Baer, A., Borrego, F., Benavente, O., Castillo, O. and Del Rio, J. A. 1990. Neohesperidin dihydrochalcone: properties and applications, *Lebensm.-Wiss. Technol.*, *23*(5), 371–6.

Bakan, J. A., Powell, T. C. and Szotak, P. S. 1992. Recent advances using microencapsulation for taste-masking of bitter drugs. *Microcapsules Nanopart. Med. Pharm. 1992*, 149–56. Edited by Donbrow, M. CRC Press, Boca Raton, FL.

Bhardwaj, S. and Hayward, M. 1993. Palatable pharmaceutical compositions, in PCT Intl. Appl. WO 93 24,109 to SmithKline Beecham Corp.

Blank, R. G., Mogavero, A. and Seabrook, A. 1996. Lipid-based liquid medicinal compositions for H_2-antihistamines, in U.S. Pat. No. 5,484,800 to American Home Products Corp.

Blase, C. M. and Shah, M. M. 1993. Taste-masked pharmaceutical suspensions containing xanthan gum and microcrystalline cellulose, in Eur. Pat. Appl. EP 0556057 to McNeil-PPC Inc.

Block, J., Cassiere, A. and Christen, M. O. 1990. Polymer coatings for masking the bitter taste of drugs, in Ger. Offen. DE 3,900,811 to Kali-Chemie A.-G.

Braun, D. B. 1994. *Over-the-counter pharmaceutical formulations.* Noyes Publ., Noyes Data Corp.

Brideau, M. E. 1995. Fast-dissolving dosage forms containing taste-masking agents, in PCT Int. Appl. WO 95 33,446 to Procter and Gamble Co.

Carella, A. M. and Opiola, E. J. 1994. Adsorbate compositions for taste masking, in PCT Int. Appl. WO 94 20,074 to Procter and Gamble Co.

Carlin, E., Chu, P., Dills, S. S. and Talwar, A. 1994. Antiplaque and antigingivitis mouth rinse containing essential oils and quaternary ammonium salts, in PCT Intl. Appl. WO 94 18,939 to Warner Lambert Co.

Catania, J. S. and Johnson, A. D. 1994. Taste-masking of bitter pharmaceutical agents with alkaline earth oxides and hydroxides, in Eur. Pat. Appl. EP 0582396 to Pfizer, Inc.

Cherukuri, S. R. and Chau, T. L. 1991. Coated delivery system for cyclic amino acids with improved taste, texture and compressibility, in Eur. Pat. Appl. EP 0458751 to Warner-Lambert Co.

Cuca, R. C., Harland, R. S., Riley, T. C., Jr., Lagoviyer, Y. and Levinson, R. S. 1994. Taste masked pharmaceutical materials, in PCT Int. Appl. WO 94 12,157 to KV Pharmaceutical Co.; *Chem. Abstr.*, 121: 117730w.

Depalmo, G. A. 1993. Taste-masked oral compositions containing ibuprofen, in Eur. Pat. Appl. EP 0560207 to Aziende Chimiche Riunite Angelini Francesco (ACRAF) S.p.A.

de Roos, K. B. and Wolswinkel, K. 1994. Non-equilibrium partition model for predicting flavor release in the mouth, *Dev. Food Sci., 35 (Trends in Flavor Research),* 15–32.

Douglas, S. J. and Bird, F. R. 1991. A taste-masking resin for ranitidine, in Can. Pat. Appl. CA 2,029,667 to Glaxo Group Ltd.

Douglas, S. J. and Evans, J. 1994. Taste-masking compositions of ranitidine, in PCT Int. Appl. WO 94 08,576 to Glaxo Group Ltd.

Duclos, R., Saiter, J. M., Grenet, J. and Orecchioni, A. M. 1992. *Ars. Pharm., 33*(1–4, Vol. 1), 167–74, in Spanish.

Eby III, G. A. 1992. Taste-masked zinc acetate compositions for oral absorption, in U.S. Pat. No. 5,095,035.

Endo, Y., Matsutani, N. and Yasumasu, T. 1991. Separation of mono-, di-, and triesters of hexoses with fatty acids, in JP 03,184,988 to Lion Corp.; *Chem. Abstr.*, 115: 280469c.

France, G. and Leonard, G. S. 1994. Taste-masked pharmaceutical chewable tablets containing H_2 receptor antagonists, in U.S. 5,275,823 to SmithKline and French Laboratories Ltd.

Francese, F., Maneschi, M. and Oldani, D. 1995. Taste-masked composition containing a drug/polymer complex, in PCT Intl. Appl. WO 95 15,155 to SmithKline Beecham Farmaceutici S.p.A.

Friend, D. R. 1992a. Polyacrylate resin microcapsules for taste-masking of antibiotics, *J. Microencapsulation, 9*(4), 469–80.

Friend, D. R. 1992b. Polyacrylate resin microcapsules for taste-masking antibiotics, in *Proc. Int. Symp. Controlled Release Bioact. Mater.*, 19th, 289–90. Edited by Kopecek, J. Controlled Release Society, Deerfield, IL.

Fuisz, R. C. 1991. Taste-masking of pharmaceutical floss with phenol, in U.S. Pat. No. 5,028,632 to Fuisz Pharmaceutical, Ltd.

Gergely, G., Gergely, T. and Gergely, I. 1993. Taste-masked pharmaceutical preparation in the form of an effervescent and/or disintegrating tablet or an instant granule, in PCT Int. Appl. WO 93 13,760.

Geyer, R. P. and Tuliani, V. V. 1995. Neomorphic ibuprofen and methods of using same, in PCT Int. Appl. WO 95 01,321 and U.S. Pat. No. 5,466,865 to Ibah, Inc.

Ghanta, S. R. and Guisinger, R. E. 1995. Pharmaceutical microcapsules containing non-steroidal anti-inflammatory drugs, in PCT Intl. Appl. WO 95 05,166 to Eurand America, Inc.

Gottwald, E. F., Osterwarld, H. P., Machoczek, H. M. and Mayron, D. 1989. Pharmaceutical compositions, in Eur. Pat. Appl. EP 0322048 to SmithKline Dauelsberg GmbH.

Gowan, Jr., W. G. and Bruce, R. D. 1993. Aliphatic and fatty esters as taste-masking agent for pharmaceuticals, in Can. Pat. Appl. CA 2,082,137 to McNeil-PPC, Inc.

Gowan, Jr., W. G. 1995. Rapidly disintegrating pharmaceutical dosage form and process for preparation thereof, in Eur. Pat. Appl. EP 0636364 to McNeil-PPC, Inc.

Grattan, T. J. 1993. Oral pharmaceutical composition containing ibuprofen-β-cyclodextrin complex, in PCT Int. Appl. WO 93 20,850 to SmithKline Beecham PLC.

Gregory, S. P., Jozsa, A. J. and Kaldawi, R. E. 1990. Non-effervescent ibuprofen compositions, in Eur. Pat. Appl. EP 0418043 to Nicholas Kiwi Pty. Ltd.

Hachiman, T., Tanimoto, M. and Takakura, I. 1994. Pharmaceuticals containing pyridonecarboxylic acids, metal compounds, and aspartame, in JP 06,340,554 to Toyama Chemical Co. Ltd.; *Chem. Abstr.*, 122: 142633v.

Hammond, K. 1995. Dentifrice compositions containing betaine-type amphoteric substances, in Eur. Pat. Appl. EP 0658340 to Unilever N.V.; Unilever plc.

Haramiishi, O. 1993a. Granule coating for masking bitter taste of pharmaceuticals, in JP 05 58,880 to Daiichi Seiyaku Co.

Haramiishi, O. 1993b. Granulation of drugs with polymers for taste-masking, in JP 05,194,193 to Daiichi Seiyaku Co.

Hatsuda, Y. and Takeda, D. 1994. Stable solid preparations containing amino acids and/or peptides, in JP 06,227,975 to Morishita Pharma; Ajinomoto Kk.

Hayashida, T. and Hatayama, M. 1993. Carbohydrates as taste-masking agents for oriental drugs, in JP 05,17,372 to Yoshitomi Pharmaceutical Industries Ltd.

Heckenmueller, H. and Friess, S. 1994. Cholestyramine-containing composition with improved organoleptic properties, in Ger. Offen. DE 4,314,583 to Astra Chemicals GmbH.

Henkin, R. I. 1994. Drug-induced taste and smell disorders. Incidence, mechanisms and management related primarily to treatment of sensory receptor dysfunction. *Drug Safety, 11*(5), 318–77.

Hirakawa, Y., Yao, T. and Kurachi, S. 1995. Bitterness-masked oral preparations of ecabet sodium, in JP 07,165,572 to Tanabe Seiyaku Co.

Honeysett, R. A., Feely, L. C., Hoadley, T. H. and E. E. Sims. 1992. Taste-masked buflomedil preparation, in Eur. Pat. Appl. EP 0501763 to Abbott Laboratories.

Hoy, M. R. and Roche, E. J. 1993. Taste-mask coating for preparation of chewable pharmaceutical tablets, in Eur. Pat. Appl. EP 0523847 to McNeil-PPC, Inc.

Hussein, M. M. and Barcelon, S. A. 1991. Taste-masking agents for bitterness of volatile oils, in U.S. Pat. No. 4,983,394 to Warner Lambert Co.

Hussein, M. M., Barcelon, S. A. and Carlin, E. 1994. Improved peppermint flavor for oral hygiene products, in PCT Intl. Appl. WO 94 27,566 to Warner Lambert Co.

Ikezuki, Y. 1990. *Gymnema sylvestre* extract in pharmaceuticals or foods for diabetes control, in JP 02,291,244 to Doitsu Yakuhin Kaisha, Ltd.

Ishibashi, N. and Shinoda, H. 1993. Bitterness-free protein-like compositions and nutrients containing them for highly invasive disease and liver diseases, in JP 05 15,339 to Terumo Corp.

Ishii, S. and Uno, D. 1995. Dentifrices containing sodium sulfate and glycyrrhizic acid salts, in JP 07,285,838 to Lion Corp.; *Chem. Abstr.*, 124: 126934e.

Ishikawa, H., Suzuki, Y., Sakai, A. and Ishizuka, S. 1995. Method for improving mint flavoring, in JP 07,145,398 to Lotte Co., Ltd.; Toyotama Perfumery.

Ito, S., Kurihara, M. and Kashiwabara, T. 1994. Oral pharmaceutical preparations containing glycyrrhizic acid salts and sweeteners, in JP 06,298,668 to Takeda Chemical Ind., Ltd.; *Chem. Abstr.*, 122: 89494x.

Iwata, N. 1995. Oral amino acid preparations for prevention or treatment of fatigue, in JP 07 25,838 to Yotsuba Yuka Kk.

Kamata, E. and Hirano, J. 1994. Starch compositions and their uses as masking agents for oral drugs, in JP 06,100,602 to Asahi Chemical Ind.; *Chem. Abstr.*, 121: 65577.

Kaneko, K. 1991. Chitosan in antidiabetic *Gymnema sylvestre* tea extract to mask bitter taste, in JP 04 09,335 to Aamu K.K.

Katsuragi, Y. and Kurihara, K. 1993. Specific inhibitor for bitter baste, *Nature, 365,* 213–214.

Katsuragi, Y., Sugiura, Y., Lee, C., Otsuji, K. and Kurihara, K. 1995a. Selective inhibition of bitter taste of various drugs by lipoprotein, *Pharm. Res.*, *12*(5), 658–62.

Katsuragi, Y., Yasumasu, T., Umeda, T., Yamazawa, S. and Mitsui, Y. 1995b. Method of suppressing bitter taste of oral preparation, in Eur. Pat. Appl. EP 0631787 to Kao Corp.

Kawasaki, Y. and Suzuki, Y. 1991. Syrup composition of acetaminophen and phenobarbital with reduced bitter taste, in Eur. Pat. Appl. EP 0441307 to Showa Pharmaceutical Chemical Industry Co.

Kerouedan, B. 1994. A lure for oral administration of veterinary pharmaceutical compositions, in Fr. Demande FR 2,702,960 to Bonnet, Christophe.

Kikuta, Y., Aoi, M. and Murata, K. 1992. Reducing bitterness with alginates, in JP 04,235,136 to Kibun Shokujin K. K.; Kibun Food Chemifa Co. Ltd.

Kinoshita, Y. and Mutsumi, S. 1987. Pharmaceuticals containing lecithins controlling bitter taste, in Japanese Patent JP 62,265,234 to Showa Pharmaceutical Chemical Industry Co., Ltd.

Kinoshita, Y. 1994. Rokitamycin monohydrate crystal and a process for producing the same, in PCT Int. Appl. WO 94 05,683 to Asahi Kasei Kogyo K.K.

Kleinert, H. D., Baker, W. R. and H. H. Stein. 1993. Orally bioavailable peptidelike molecules: a case history, *Pharm. Tech.* (March), 30–36.

Kobayashi, S., Nagatomi, Y., Yomoda, S., Hitomi, N. and Suzuki, A. 1992. Oral solutions containing vitamin B, in JP 04,247,024 to Kanebo, Ltd.

Kobayashi, S., Nagatomi, Y., Yomoda, S., Hitomi, N. and Suzuki, A. 1993. Vitamin B oral solutions containing sugars, amino acids, and flavors, in JP 05 04,291 to Kanebo, Ltd.

Kodama, M., Miyahara, R. and Tate, K. 1992a. Sagebrush extract free of bitterness for hard candies for mouth odor control, in JP 04,278,050 to Shiseido Co., Ltd.; *Chem. Abstr.*, 118: 21410d.

Kodama, M., Miyahara, R. and Tate, K. 1992b. Sagebrush extract for preparation of chewing gum for mouth odor control, in JP 04,278,052 to Shiseido Co., Ltd.; *Chem. Abstr.*, 118: 21411e.

Kodama, M., Miyahara, R. and Tate, K. 1992c. Removal of bitterness from absinthe oil for beverages by gel permeation chromatography, in JP 04,278,068 to Shiseido Co., Ltd.; *Chem. Abstr.*, 118: 21412f.

Koyama, I., Shimano, K., Makabe, E. and Ozawa, Y. 1990a. Oral pharmaceuticals containing poly(vinyl acetal) (diethylamino)acetate, in JP 02,279,622 to Taisho Pharmaceutical Co. Ltd.

Koyama, I., Shimano, K., Makabe, E. and Ozawza, Y. 1990b. Method of producing oral pharmaceuticals with modified taste, in PCT Int. Appl. WO 90 12,566 to Taisho Pharmaceutical Co., Ltd.

Kondo, T. and Nishimura, A. 1991. Sweetening compositions containing di-D-fructofuranose 1,2′:2,3′-dianhydride and menthol, in JP 03 67,560 to Mitsubishi Kasei Corp.

Koyama, I. and Shimano, K. 1993a. Oral preparations containing polymers, carboxylic acids, and metal hydroxides, in JP 05,255,073 to Taisho Pharma Co., Ltd.; *Chem. Abstr.*, 120: 62323n.

Koyama, I. and Shimano, K. 1993b. Bitterness-masked preparations containing magnesium oxide and/or hydroxide, in JP 05,255,120 to Taisho Pharma Co., Ltd.; *Chem. Abstr.*, 120: 38194a.

Kurasumi, T., Imamori, K. and Iwasa, A. 1991. Compositions containing carbetapentane citrate with less bitter taste, in JP 03,236,316 to S. S. Pharmaceutical Co., Ltd.

Kurihara, K., Katsuragi, Y., Matsuoka, I., Kashiwayanagi, M., Kumazawa, T. and Shoji, T. 1994. Receptor mechanisms of bitter substances, *Physiol. Behav.*, *56*(6), 1125–32.

Kurtz, R. J. and Fuller, W. D. 1993. Ingestibles containing substantially tasteless sweetness inhibitors as bitter taste reducers or substantially tasteless bitter inhibitors as sweet taste reducers, in U.S. Pat. No. 5,232,735 to Bioresearch Inc.

Lafon, L. 1992. Binding of aminobutanone drugs to ion exchange resins for taste-masking, in Fr. Demande FR 2,676,364.

Lange, P. M., Mitschker, A., Naik, A. H., Rast, H., Scheer, M. and Voege, H. 1992. Ion exchange resins loaded with quinolonecarboxylic acid derivatives for taste-masking in feed, in U.S. Pat. No. 5,152,986 to Bayer AG.

Lebrun, J. C. and Massey, J. L. 1995. Method for the preparation of an encapsulated caplet, in U.S. Pat. No. 5,460,824 to Warner Lambert Co.

Lech, S., Schobel, A. M. and Denick, J., Jr. 1994. Chewable preparations containing silicon dioxide to improve the taste thereof, in PCT Intl. Appl. WO 94 28,870 to Warner Lambert Co.

Lech, S., Denick, J., Jr. and Schobel, A. M. 1995. Pleasant tasting effervescent cold/allergy medications, in PCT Intl. Appl. Wo 95 03,785 to Warner Lambert Co.

Lu, M. F. Y., Borodkin, S., Woodward, L., Li, P., Diesner, C., Hernandez, L. and Vadnere, M. 1991. A polymer carrier system for taste-masking of macrolide antibiotics, *Pharm. Res.*, *8*(6), 706–712.

Lynch, J., Liu, Y. H., Mela, D. J. and MacFie, H. J. H. 1993. A time intensity study of the effect of oil mouth-coatings on taste perception, *Chemical Senses, 18*(2), 121–9.

Maegaki, H., Kawasaki, Y. and Suzuki, Y. 1993. Oral liquid preparations containing theophylline with less bitterness, in JP 05,124,963 to Showa Pharm. Chem. Ind.

Mapelli, L. G., Marconi, M. G. R. and Zema, M. 1991. Masking taste with polymers in pharmaceutical formulations, in PCT Intl. Appl. WO 91 16,043 to Eurand International S.p.A.

Martani, R., Demichelis, A. G., Barbero, M., Le Vu, D., Lecointre, B. and Rabot, P. 1995. Chewable tablets containing cocoa powder as flavor-masking agent, in PCT Intl. Appl. WO 95 24,890 to Hi Pharmtech.

Maruyama, N., Kokubo, H. and Muto, Y. 1993. Core-containing pharmaceutical preparations with unpleasant odor masking, in JP 05,163,163 to Shinetsu Chem. Ind. Co.

Matoba, H., Ohmori, S., Koyama, H. and Kashhara, S. 1994. Taste masked pharmaceutical preparations containing ion exchange resins, in Eur. Pat. Appl. EP 0622083 to Takeda Chemical Industries, Ltd.

Matsubara, Y., Kawajiri, A. and Ishiguro, F. 1990. Granules with controlled bitter taste, in JP 02 56,416, to Daikyo Yakukin Kogyo KK.

McCabe, T. T., Stagner, R. A. and Sutton, Jr., J. J. 1992. Flavored film-coated tablet for taste-masking, in U.S. Pat. No. 5,098,715 to Burroughs Wellcome Co.

Merck Encyclopedia of Chemical Drugs and Biologicals, The Merck Index 10th Ed. 1983. Rahway, NJ; Nos. 5471, 6948, 7970, 7973, 7981, 7983.

Merck Encyclopedia of Chemical Drugs and Biologicals, The Merck Index 12th Ed. 1996. Nos. 5689, 7221.

Meyer, G. A. and Mazer, T. B. 1993. Prolamine coating for taste-masking oral drugs, in PCT Int. Appl. WO 93 12,771 to Abbott Laboratories.

Mito, Y., Uchida, K., Kishida, S. and Nomura, S. 1994. Materials for coating solid pharmaceuticals to cover the unpleasant odor and taste and to promote active ingredient release, in JP 06,256,170 to Morishita Pharma; *Chem. Abstr.*, 122: 38836c.

Miura, S., Matsushita, M. and Fujinaga, T. 1992. Bitterness-free syrups containing oils, surfactants, and polyalcohols, in JP 04,187,629 to Sumitomo Pharmaceuticals Co. Ltd.

Mogensen, L. and Adler-Nissen, J. 1988. Evaluating bitterness masking principles by taste panel studies, in *Proc. 5th Int. Flavor Conf.*, July 1 to 3, 1987, in *Frontiers of flavor,* 79. Edited by Charalambous, G. Elsevier, Amsterdam.

Morella, A. M. and Lukas, S. 1992. Microcapsule compositions and process, in Can. Pat. Appl. CA 2,068,366 to Faulding (F.H.) and Co. Ltd.

Mori, M., Shimono, N., Kitamura, K., Tanaka, T. and Nakamura, Y. 1993. Sustained-release pharmaceutical powders, in JP 05,213,740 to Dainippon Pharmaceutical Co.

Moroi, M., Nakajima, Y., Imamori, K. and Iwasa, A. 1993. Masking of stimulants and bitter taste of pharmaceutical granules, in JP 05,201,855 to Ss Pharmaceutical Co.

Motola, S., Agisim, G. R. and Mogavero, A. 1991. Palatable ibuprofen solutions, in U.S. Pat. No. 5,024,997 to American Home Products Corp.

Nagafuji, N., Hatsushiro, S., Suzuki, Y., Ogura, T. and Takagishi, Y. 1995. Taste improved granules and the manufacture by wet-granulation and heating, in JP 07,188,058 to Shionogi Seiyaku Kk.

Nakajima, K., Hirakawa, Y., Koida, Y. and Matsubara, K. 1994. Bitterness-masked ecabet sodium oral preparations, in JP 06,279,275 to Tanabe Seiyaku Co.; *Chem. Abstr.*, 122: 38889x.

Nakamura, K., Kawamura, M., Oosawa, S. and Sugawara, S. 1995. Compositions for masking the unpleasant taste and odor of drugs, in JP 07,242,568 to Eisai Co., Ltd.

Newton, J. M., Chow, A. K. and Jeewa, K. B. 1993. The effect of excipient source on spherical granules made by extrusion/spheronization, *Pharm. Tech.* (March), 166–174.

Nishii, H., Kobayashi, M., Toya, K. and Uchama, N. 1993. Coating of pharmaceuticals with molten waxes for taste masking, in JP 05,309,314 to Sumitomo Pharma; *Chem. Abstr.*, 120: 200427c.

Nishikawa, M. and Hayashi, H. 1992. Bitterness-controlled oral solid preparations, in JP 04,327,526 to Lion Corp.

Nishikawa, M. and Hayashi, H. 1993. Antitussives and expectorants containing magnesium compounds, in JP 05,139,996 to Lion Corp.

Nitsuta, K., Aoki, S., Uesugi, K. and Ozawa, H. 1992. Pharmaceutical granules and tablets containing drug particles coated with polymers, in JP 04,282,312 to Eisai Co. Ltd.

Nomura, T. and Izumida, Y. 1993. Bitterness-free bifemelane hydrochloride dry syrups, in JP 05 97,664 to Mitsubishi Chem. Ind.

Notola, S., Agisim, G. R. and Mogavero, A. 1991. U.S. Patent 5,024,997.

Ogasawara, S. and Ueda, S. 1992. Bitterness- and foul odor-controlled oral solid preparations, in JP 04,327,528 to Lion Corp.

Patel, M. M., Broderick, K. B., Meyers, M. A., Schnell, P. G., Song, J. H., Yatka, R. J. and Zibell, S. M. 1993. Strongly mint-flavored chewing gums with reduced bitterness and harshness, in U.S. Pat. No. 5,192,563 to Wm. Wrigley, Jr. Co.

Peterson, L. G., Sanker, L. A. and Upson, J. G. 1994. Breath protection microcapsules containing antimicrobial agents and sweeteners, in U.S. Pat. No. 5,370,864 to Procter & Gamble Co.

Poellinger, N., Michaelis, J., Benke, K., Rupp, R. and Buecheler, M. 1993. Microencapsulated taste-masked pharmaceutical compositions, in Eur. Pat. Appl. 0551820 to Bayer AG.

Popescu, M. C. and Mertz, E. T. 1991. Taste moderating pharmaceuticals, in U.S. Pat. No. 5,009,819 to Liposome Co., Inc.

Rao, Y. M. and Bader, F. 1993. Masking the taste of chloroquine by preparing multiple emulsions, *East. Pharm.*, *36*(431), 123–4.

Ratnaraj, S. M. and Sunshine, W. L. 1994. Taste-masked aqueous pharmaceutical suspension and process for preparation thereof, in Eur. Pat. Appl. EP 0620001 to McNeil-PPC, Inc.

Record, D. et al. 1994. Mint flavored chewing gum having reduced bitterness and methods, in U.S. Pat. No. 5,372,824 to Wm. Wrigley, Jr. Co.

Roche, E. J. 1991. Taste-masking and sustained-release coatings containing cellulose derivatives for pharmaceuticals, in Eur. Pat. Appl. EP 0459695 to McNeil-PPC, Inc.

Roche, E. J., Papile, S. M. and Freeman, E. M. 1992. Rotogranulations and taste-masking coatings for preparation of chewable pharmaceutical tablets, in Eur. Pat. Appl. EP 0473431 to McNeil-PPC, Inc.

Roche, E. J. and Reo, J. P. 1992. Rotogranulation and taste-masking coatings for preparation of chewable pharmaceutical tablets, in CA 2,063,141 to McNeil-PPC, Inc.

Roche, E. J., Freeman, E. M. and Papile, S. M. 1993. Taste-mask coatings for preparing chewable pharmaceutical tablets, in Eur. Pat. Appl. EP 0538034 to McNeil-PPC, Inc.

Roy, G. 1990. The applications and future implications of bitterness reduction and inhibition in food products, *Crit. Rev. Food Sci. Nutr.*, *29*(2), 59–71.

Roy, G. 1992. Bitterness: reduction and inhibition, *Trends in Food Science & Technology, 3,* 85–91.

Roy, G. 1994. *Pharmaceutical Technology* (April), 84–99; (May), 36.

Ryu, S. 1991. Sustained-release bitterness-free granules containing tiamulins supported on zeolite as bactericides for domestic animals, in JP 03,101,619 to SDS Biotech K.K.

Sakakibara, T., Kobayashi, T., Saito, M., Fukushima, F. and Mizunoya, T. 1993. Bitterness-free loperamide tablets with sugar or film coating, in JP 05,117,149 to Ota Pharma.

Sapone, A., Basaglia, R. and Biagi, G. L. 1992. Drug-induced changes of the teeth and mouth. Note 1, *Clin. Ter. (Rome), 140*(5), 487–98, in Italian.

Schiffman, S. S. and Warwick, Z. S. 1989. Use of flavor-amplified foods to improve nutritional status in elderly patients, *Ann. NY Acad. Sci., 56,* 267–276.

Shen, R. W. W. 1991. Controlled-release chewable taste-masked tablets of ibuprofen, in PCT Intl. Appl. WO 91 15,194 to Upjohn Co.

Shimano, K., Kondo, O., Miwa, A., Higashi, Y., Koyama, I., Yoshida, T., Ito, Y., Hirose, J. and Goto, S. 1994. Evaluation of temperature-sensitive and drug dissolution properties of poly(vinyl acetal) (diethylamino)acetate, *Yakuzaigaku, 54*(2), 69–76.

Shiozawa, H., Sugao, H., Yamazaki, S. and Yano, K. 1992. Rapid release coated pharmaceuticals, in PCT Int. Appl. WO 92 09,275 to Yamanouchi Pharmaceutical Co. Ltd.

Shirai, Y., Sogo, K., Yamamoto, K., Kojima, K., Fujioka, H., Makita, H. and Nakamura, Y. 1993. A novel fine granule system for masking bitter taste, *Biol. Pharm. Bull., 16*(2), 172–7.

Shirai, Y., Sogo, K., Fujioka, H and Nakamura, Y. 1994. Role of low-substituted hydroxypropyl cellulose in dissolution and bioavailability of novel fine granule system for masking bitter taste, *Biol. Pharm. Bull., 17*(3), 427–31.

Shirai, Y., Sogo, K., Fujioka, H. and Nakamura, Y. 1996. Influence of heat treatment on dissolution and masking degree of bitter taste for a novel fine granule system, *Chem. Pharm. Bull., 44*(2), 399–402.

Sjoeqvist, R., Graffner, C., Ekman, I., Sinclair, W. and Woods, J. P. 1993. In vivo validation of the release rate and palatability of remoxipride-modified release suspensions, *Pharm. Res., 10*(7), 1020–6.

Sonley, C. R. and Turnbull, M. A. 1995. Solid pharmaceutical dosage form containing flavored film coating, in Eur. Pat. Appl. EP 0679391 to McNeil-PPC, Inc.

Sugawara, S., Kawamura, M. and Oosawa, S. 1995. Bitterness-masked pharmaceutical granules, in JP 07,126,188 to Eisai Co., Ltd.

Suguira, Y., So, R. and Ootsuji, K. 1995. Bitterness control with diglycerides and less bitter pharmaceutical and food compositions, in JP 07 53,410 to Kao Corp.

Susuki, Y., Tsukada, T., Nagafuji, N., Tomoda, Y., Hayashi, T., Tanaka, H., Fujimoto, A., Nakajima, C. and Shima, K. 1993a. Bitterness-suppressive formulation of benexate hydrochloride-Betadex prepared by melt granulation, *Yakuzaigaku, 53*(4), 201–9, in English; *Chem. Abstr.,* 120: 307218n.

Susuki, Y., Ogura, T. and Takagishi, Y. 1993b. Bitterness-suppressive formulation of benexate hydrochloride-Betadex prepared by melt granulation, *Pharm. Tech. Japan, 9*(9), 999–1008, in Japanese.

Tabata, T. and Yosimi, A. 1992a. Bitter taste-improved oral preparations prepared by coating basic pharmaceutical acid adduct salts with weakly alkaline compounds, in JP 04,327,529 to Kyoto Pharmaceutical Industries, Ltd.

Tabata, T. and Yoshimi, A. 1992b. Bitter taste-improved oral pharmaceutical compositions containing basic β-lactam antibiotics acid adduct salts and weakly alkaline compounds, in JP 04,327,531 to Kyoto Pharmaceutical Industries, Ltd.

Takahashi, M., Wada, K. and Mochizuki, H. 1995. Gelatin capsules containing essential oils in the film, in JP 07,242,536 to Toyo Capsel Kk.

Tanaka, M., Yoshikawa, T. and Tomita, H. 1994. Drug dysgeusia, *Nippon Yakuzaishikai Zasshi, 46*(12), 1765–1770, in Japanese; *Chem. Abstr.*, 122: 45426x.

Ueda, M., Nakamura, Y., Makita, H. and Kawashima, Y. 1993. Preparation of microcapsules masking the bitter taste of enoxacin by using one continuous process technique of agglomeration and microencapsulation, *J. Microencapsulation, 10*(4), 461–73.

Ueda, M. 1995. Recent pharmaceutical techniques and future scope for taste masking of granules, *Gifu Yakka Daigaku Kiyo, 44,* 18–31, in Japanese.

Ueki, T., Kameda, S., Uno, D. and Kaneko, Y. 1993. Dentifrices containing acylsarcosine salts, anethole, and menthofuran, in JP 05,155,744 to Lion Corp.

Ueno, M. 1992. *Gymnema sylvestre* leaf extract for food or feed additives, in JP 04 11,865 to Meiji Sugar Mgf. Co., Ltd.

Vaziri, A. and Warburton, B. 1994. Slow release of chloroquine phosphate from multiple taste-masked W/O/W multiple emulsions, *J. Microencapsulation, 11*(6), 641–8.

Wang, X. M. and Chen, H. Y. 1994. Characterization of the β-cyclodextrin/quinine complex, *Chin. Chem. Lett.*, *5*(5), 399–402; *Chem. Abstr.*, 121: 205825m.

Watabe, S., Kato, T. and Nagata, N. 1992. Bitterness-free saponin compositions containing amino acids, in JP 04,207,161 to Rohto Pharmaceutical Co. Ltd.

Wheatley, T. A. and Erkoboni, D. F. 1992. Taste-masked medicaments having a cellulose ester film, and their preparation, in PCT Int. Appl. WO 92 19,209 to FMC Corp.

Widauer, J. O. 1995. Pharmaceutical composition in liquid dosage form containing ursodeoxycholic acid with improved flavor, in Eur. Pat. Appl. EP 0640344 to Medichemie AG.

Yajima, T., Ishii, K., Umeki, N., Itai, S., Hayashi, H. and Shimano, K. 1993a. Taste-masked oral compositions containing low melting point polymers, sugar alcohols and basic oxides, in PCT Int. Appl. WO 93 17,667 to Taisho Pharmaceutical Co., Ltd.

Yajima, T., Ishii, K. and Itai, S. 1993b. Bitterness-free oral antibiotics, in JP 05,255,075 to Taisho Pharmaceutical Co., Ltd.

Yajima, T., Nogata, A., Demachi, M., Umeki, N., Itai, S., Yunoki, N. and Nemoto, M. 1996. Particle design for taste-masking using a spray-congealing technique, *Chem. Pharm. Bull.*, *44*(1), 187–91.

Yamazawa, S. and Katsuragi, Y. 1995. Bitterness reducing agents, the manufacture, bitterness reducing method, and compositions with reduced bitterness, in JP 07,300,429 to Kao Corp.

Yiv, S. H. and Tustian, A. K. 1994a. Taste-masking pharmaceutical compositions with lipids and emulsifiers and polymers, in PCT Int. Appl. WO 94 05,260 to Affinity Biotech, Inc.

Yiv, S. and Tustian, A. K. 1994b. Taste-masking pharmaceutical compositions comprising surfactants and lipid-coated particles, in PCT Intl. Appl. WO 94 25,006 to Affinity Biotech, Inc.

Yiv, S. H. and Tustian, A. 1995. Taste-masked acetaminophen suspensions and methods of making the same, in PCT Int. Appl. WO 95 00,133 to Ibah, Inc.

Yokoo, T. and Hirohata, H. 1993. Dentifrices containing quaternary ammonium salt-type bactericides, in JP 05 00,931 to Lion Corp.

Yumoto, T., Iida, S. and Gunji, Y. 1994. Method for masking bitter and astringent taste of antidiabetic *Gymnema sylvestre* extract, in JP 06,321,797 to Toyo Sugar Refining; *Chem. Abstr.*, 122: 114905s.

CHAPTER 14

Cautions and Prospects in Taste-Masking Formulations

GLENN ROY[1]

APPLICATIONS THAT NEED COMMERCIALIZATION OF NEW TECHNOLOGIES AND ADDITIVES

WE have read ingredient and process approaches modifying bitter taste in foods, beverages and oral pharmaceuticals. Unfortunately, no single or universal ingredient or process approach can yet be applied to all bitter taste formulations. Ideally, a potentially bitter formulation should have a safe, stable, legal, infallible and cost-effective means of masking bitter taste. Formulation chemists should not have the guesswork of tailoring a technique for a specific application. Eventually, research will identify a universally applicable and bitter-taste specific inhibitor for commercialization. After all, the human population is nearly homogeneous to bitter taste perception. Genetic engineering approaches to remove bitter tastants from particularly bitter food sources may remove a potentially healthful benefit. The more appropriate vision should be to discover an ingredient that possesses a specific bitter taste inhibition with maximal universal applicability. The pursuit of discovering a universally applicable and specific bitter taste-masking ingredient is only as important as the financial risk the producer is willing to take. Product development has become increasingly expensive for every industrial operation. Yet, do novelty and specificity for bitter taste masking offer a differentiation in the marketplace with significant financial return for the producer? The phosphatidic acid-lactoglobulin product we have discussed is derived from natural sources and may be ideally suited to generate the market-driven in-

[1]Pepsi-Cola Co., 100 E. Stevens Ave., Valhalla, NY 10595, U.S.A.

terest and differentiation for a bitter inhibitor with to-be-proven universal applicability if ingredient cost could be lowered. The use of encapsulated coatings for tablets is distinct evidence of market success in a very narrow field of applications. Though comfortable with sugar primarily as a means to reduce bitter taste, will the marketplace awaken to the age-old discomforting problem of dental caries specifically in oral liquid applications of medicinals? The trend is ever increasingly towards oral liquid medicaments. Artificial sweeteners have made progress in offering anticariogenic properties to some formulations. Caution should also be exercised with sugar alcohols as they have laxative effects at high doses. A new bitterness inhibition ingredient does not appear to have an insurmountable barrier against the competition of substitute technologies.

New applications for a bitter taste inhibitor are constantly growing. New herbal supplements with efficacious doses in ready-to-drink form have primarily a bitter taste. Goldenseal extract for body organ cleansing is especially bitter tasting. A parsley seed oil-containing beverage for breath freshening is particularly unpalatable. Sports drinks with excessive levels of minerals or vitamins present palatability challenges. Imagine the taste-improved, large assembly of bitter emulsifiers offering potentially improved functionality because the taste was more palatable for use at higher levels. The market surge of low-fat products, formulated merely with more water and less fat or carbohydrate fat mimetics, will create a burgeoning demand on better-performing emulsifiers. Imagine that immense caches of lower-quality wines, whose bitter attributes make them untenable to market sale, could be vastly improved in taste quality without sugar. For example, exotic wines such as kiwi wine, with the organoleptic characteristics of a white grape wine, require sugar to mask the bitterness of quinic acid (Kume et al., 1993). New vegetables with healthful benefits are constantly being recommended for consumption despite their frequent unpleasant taste.

The success of a taste-specific bitterness inhibitor may be more certain if it finds use in the trusted approval of a home condiment (e.g., for table-top use, akin to salt and pepper). Furthermore, in a table-top format, consumers would have a more conscious choice and the ability to assess their need for the condiment on a particular food based on their personal taste preferences or taste sensitivity.

We must also be mindful of the increasing population of elderly whose taste acuity may be failing. Naturally bitter substances such as caffeine and inosine 5′-monophosphate are useful for potentiating the tastes of salts, amino acids, carbohydrates and nonnutritive sweeteners when ingested by elderly people (Schiffman and Warwick, 1989) for whom a bitterness inhibitor's action upon the taste receptor might interfere with such desirable potentiation of taste. On the other hand, competitive inhibition of the bitterness receptor by a universal bitterness inhibitor might itself potentiate taste in the elderly. Masking gives new insight into the role of aging in taste: older (66–90 yrs.) subjects' thresholds, regardless of masking concentration (of sugar, acids, salts, bitters admixed in pairs), always measured a constant factor higher than younger (18–29 yrs.) subjects' thresholds (about 2–7 times higher, depending on the pair target tastant). Age affects perception relatively mildly in sucrose and citric acid, moderately in sodium chloride and strongly in quinine HCl (Stevens, 1996).

Labels have become more and more obscure with new ingredients despite more scrupulous labeling requirements. Can producers accept another new ingredient? The inconsistency brought about by the addition of another ingredient may not be welcome. The likelihood of a new and synthetic substance gaining acceptance as a food additive is unlikely unless substantiated governmentally as GRAS. The U.S. Nutrition Labeling and Education Act challenges processors to provide an easy way to understand ingredient label information. A synthetic, universal bitterness inhibitor will take years of costly safety studies before U.S. Food and Drug Administration approval. Naturally derived substances will provide more expedient approval and present the consumers with a welcome addition to the ingredients list. Therefore, the universal application of a bitterness inhibitor should have the use of natural substances to permit their expedient use in improving the palatability of normally bitter food products. Lead compounds from a research-oriented pharmaceutical or food company by (SAR) molecular design; or new naturally derived or processed compounds found to be desirable as a universal bitterness inhibitor or bitterness inhibition cocktail will require a full Food and Drug Administration Food Additive Petition and take $50 to 100 MM of investment. Less costly approaches are new naturally derived or processed compounds that are desirable as a universal bitterness inhibitor or bitterness inhibition cocktail.

More systematic knowledge of bitter-masking potential is constantly surfacing in the pharmaceutical literature. New information for substance-universal inhibition with bitter-specific inhibitors is spawned by The Roche Institute of Molecular Biology (Nutley, NJ). Dr. Margolskee's research (now terminated) has published biochemical information to indicate a more thorough understanding of bitter taste transduction. This could inevitably result in the discovery of what a universal bitter inhibitor should possess chemically to be effective. Screening of final products for palatability may teach new structure-activity relationships for classes of compounds. We may yet learn more of a populational difference in sensitivity to bitterness. Indirectly, sensory evaluation of the oral forms of bitter drugs will be improved. When possible with humans, these evaluations should then be tested with new inhibitors in a more formalized environment and include a "compound taste perception" threshold (Mogensen and Adler-Nissen, 1988). This approach would expedite the commercialization of a successful bitter substance-universal and taste-specific inhibitor. Hokkaido University and Kao Corporation must be commended on their diligence in research and discovery of phosphatidic acid and β-lactoglobulin as potential commercially viable substance-universal bitterness inhibitors for foods, beverages and oral pharmaceuticals. Any research for a new bitter inhibitor should focus on using natural, well-known substances to permit commercial use in improving the palatability of foods, beverages and oral pharmaceuticals.

As is the case with ingesting all bitter substances encountered by civilization since man walked the earth, bitterness reduction and inhibition is risky. When formulation chemists begin to practice new bitterness reduction and inhibition with new plant sources, we increase the chance of bitter substance abuse. This is a result of the less bitter product becoming more palatable and raising a significant concern of physiology in utilization of the more palatable ingestions.

Several of the bitter substances found in foods are toxic and some consideration will have to be given to the potential indirect health effects of the widespread use of bitterness inhibitors in cases where that bitter substance may have a historical and detrimental effect on human anatomy and physiology. This profound thought was brought to light by Dr. Michael S. Kellogg of Monsanto Corporation, St. Louis, Missouri. As new food or beverage products with bitter taste are introduced to the marketplace, they will need certain toxicological studies anyway to avert an abuse potential that should not be overlooked. Thus, the approach of specific bitter component removal versus universal ingredient bitter inhibition will have a different cost/benefit for formulation chemists.

REFERENCES

Kume, H., Tsutsumi, S., Matsuo and K. Tsuro, Y. 1993. Kiwi wine preparation from kiwi fruit, in JP 05,236,931 to Tachibana Wain Kk; *Chem. Abstr.*, 120: 29835h.

Mogenson, L. and Adler-Nissen, J. 1988. Bitterness intensity of protein hydrolysates—chemical and organoleptic characterization, *Proc. 5th Int. Flavor Conf.*, July 1–3, 1987. *Frontiers of Flavor,* G. Charalambous, ed., Elsevier, Amsterdam, p. 63.

Sanyo. 1985. Masking bitterness, irritating taste, and odor of mono-glycerides by adding enzymic essence obtained from glutathione rich enzymes to monoglyceride emulsifying agent, Sanyo Kokusaku Pulp, Japanes Patent JP 198648.

Schiffman, S. S. and Warwick, Z. S. 1989. Use of flavor-amplified foods to improve nutritional status in elderly patients, *Ann. N.Y. Acad. Sci., 56,* 267–276.

Stevens, J. C. 1996. Detection of tastes in mixture with other tastes: Issues of masking and aging, *Chemical Senses,* 21:211–221.

Index

Editor's Biography

As the son of a German-bred, chemist-turned-perfumer for a time, Glenn Roy began training in the science of flavors and fragrances in 1971 with Tombarel Freres in Grasse, France. As a field laborer from before sunrise to noon, he harvested botanic raw materials and processed them to produce essential oils and absolutes. Perfumery studies continued as part-time employment with Givaudan Corp. and Roure, Bertrand, Dupont, Inc., both in New Jersey.

In 1975 he graduated with a BA in Chemistry from St. Michael's College in Winooski, Vermont. There, Mr. Roy researched and published, with the late Dr. Gilbert Grady, a new method for the purification of eucalyptol 1,8-cineole from its olfactorally undesirable 1,4-isomer. Subsequently, he graduated with MS (1977) and Ph.D. (1980) degrees in Organic Chemistry from study and research at The Ohio State University, Columbus, Ohio, under Dr. Philip D. Magnus, Royal Chemistry Society Fellow. A group discovery of new organosilicon reagent methodology was illustrated in the syntheses of Frontalin, a beetle pheromone and Latia Lucifern, a firefly luminescence principle. The organosilicon reagents are now commercially available from Petrarch Co.

In 1980 Dr. Roy entered a career in the food industry. Six years of research were conducted in flavor generation and synthesis of high potency sweeteners and taste modifiers at (then) General Foods Corp. (Tarrytown, NY). In 1986 an opportunity arose to pursue research at The NutraSweet Co. (Mt. Prospect, IL) in the area of nonfermentable noncaloric sugars and fat macronutrient substitutes. Additional research in taste modification resulted in the discovery of bitterness inhibitors. The impending breakup of The NutraSweet Co. R&D with patent expiration prompted a move in

1990 to a managerial position in Food Processing Technical Services with Calgon Carbon Corp. in Pittsburgh, Pennsylvania.

There, Dr. Roy authored a book on activated carbon processing and applications for the food and pharmaceutical industries which was published by Technomic Publishing Co., Inc. and was granted a U.S. patent for a continuous food-frying oil treatment to reduce color and polymer formation. The process is to be commercialized when regulations for oil contaminants emerge.

In November 1994, the annual corporate downsizing of Calgon Carbon Corp. resulted in unemployment. In April 1995, Glenn Roy joined The Pepsi-Cola Company (Valhalla, NY) as a Principal Research Scientist in Product/Ingredient Technology of Nutritional and Botanical Beverages. Dr. Roy continues to maintain a position as Field Editor of Food Science & Technology with Technomic Publishing Co., Inc.